GIANT TELESCOPES

W. PATRICK McCRAY

GIANT TELESCOPES

Astronomical Ambition and the Promise of Technology

HARVARD UNIVERSITY PRESS
Cambridge, Massachusetts, and London, England

Copyright © 2004 by the President and Fellows of Harvard College
All rights reserved
Printed in the United States of America

Library of Congress Cataloging-in-Publication Data
McCray, Patrick (W. Patrick)
Giant telescopes : astronomical ambition and the
promise of technology / W. Patrick McCray.
p.   cm.
Includes bibliographical references and index.
ISBN 0-674-01147-3 (alk. paper)
1. Large astronomical telescopes.   2. Large astronomical
telescopes—Technological innovations.   I. Title.
QB90.M33 2003
522'.29—dc22   2003059141

Contents

	Introduction: Beautiful and Cantankerous Instruments	1
1	Leo and Jesse's Changing World	13
2	Tradition and Balance	50
3	Visions of Grandeur	84
4	Paper Telescopes	114
5	Growing Pains	143
6	Astropolitics	172
7	Smoke and Mirrors	204
8	Joining the 8-Meter Club	237
9	Point-and-Click Astronomy	265
	Conclusion: Telescopes, Postwar Science, and the Next Big Machine	290
	Giant Telescopes	309
	Sources	311
	Abbreviations	317
	Notes	319
	Acknowledgments	357
	Index	359

GIANT TELESCOPES

All past events are more remote from our senses than the stars of the remotest galaxies, whose own light at least still reaches the telescopes . . . Knowing the past is as astonishing a performance as knowing the stars.

George Kubler, *The Shape of Time: Remarks on the History of Things*

Know thyself. Know your telescope . . .

Sign in the hall of Cornell University's Astronomy Department, 2002

Introduction: Beautiful and Cantankerous Instruments

Aden B. Meinel was six years old in 1928, the year that George Ellery Hale convinced the Rockefeller Foundation to finance the construction of the world's largest telescope. As a young man growing up in Pasadena, California, Meinel wanted to become an aeronautical engineer. One day he wandered into the Mount Wilson Optical Shop where he met opticians who were polishing the telescope's glass mirror blank, 200 inches in diameter, before it was shipped to the top of Palomar Mountain.[1] Meinel walked up to the polishing machine and gently touched the mirror with his fingers, an experience that sparked his lifelong interest in optics and telescope design.

A quarter of a century later, Meinel was an astronomer and optics expert at Yerkes Observatory in Wisconsin and thinking once again about big telescopes. The 200-inch telescope on Palomar, built with $6 million from the Rockefeller Foundation and named in honor of Hale, had been in operation for five years. Already it was beginning to change people's concept of the universe. In 1952, Walter Baade, an astronomer at the Mount Wilson Observatory, used data collected with the Hale Telescope to show that stars in the Andromeda Galaxy were farther away than scientists had previously believed. Baade's observations "doubled the size" of the known universe, a feat that fired the minds of scientists and caught the public's attention. The mystique of the 200-inch telescope was enhanced by its exclusivity. Only a select cadre of scientists from the California Institute of Technology and the Carnegie Institution of Washington had access to it.

Meinel was walking home from his office at Yerkes one evening in 1953 when he had an idea. There was a driveway in front of the observatory that encircled a large swath of grass. Meinel could see the domes that protected the observatory's instruments. Astronomers there were readying for the night's observing run. He imagined the light from distant stars and galaxies traveling for eons, only to be lost as it fell not on the telescopes' mirrors but on the grass. "Wouldn't it be nice," thought Meinel, "to get a big bite of those photons?"[2]

Designing and building the telescope Meinel wanted—one with a mirror some 500 inches in diameter—demanded the right combination of community interest, technological prowess, and support from willing patrons with abundant resources. These conditions never coalesced to enable the construction of what Meinel and others came to call the "X-inch" telescope. In fact, it would be four more decades before American astronomers had access to a telescope with a bigger light-collecting area than the 200-inch telescope on Palomar.

During these decades, astronomers and science managers thrashed out plans for a new, larger national observing facility, eventually named the Gemini 8-Meter Telescopes Project (hereafter Gemini). Due to political and financial forces, the United States built Gemini in partnership with six other countries—the United Kingdom, Canada, Australia, Chile, Brazil, and Argentina. In June 1999, on the barren, windswept summit of Mauna Kea in Hawaii, the first Gemini telescope, one of the largest in the world, was dedicated. In 2002, with the completion of its twin companion telescope in Chile, astronomers and dignitaries celebrated the completion of the entire Gemini Observatory.

Between the dedication of the Hale Telescope in 1948 and the completion of the Gemini Observatory, astronomers' most iconic symbol, the telescope, was transformed as a research tool. The importance of changes in telescope design and technology results from a deceptively simple relationship: in astronomy, technological innovations often lead to new discoveries. Galileo's spyglass once revealed the surface of the moon in startling detail and the mysterious movement of Jupiter's moons. Today, a new generation of eyes—flesh and blood, glass and steel—has turned to the sky with revitalized powers to observe the universe's diverse phenomena.

Since the 1970s, scientists have brought about a rebirth of the traditional optical telescope with observatories like Gemini. For observational astronomers, few things could be more important. Far from being relegated to third-class status by space observatories like the Hubble Space Telescope and postwar innovations such as the radio telescope, ground-based optical telescopes—the workhorses of astronomy for over four hundred years—are as essential to the astronomer as they have ever been. New telescopes and technologies have radically reshaped how we see the cosmos, literally.

The Gemini telescopes cost close to $200 million to build. Yet, they are only two of a dozen new, very large telescopes completed after 1990. Others will see their first starlight in the next few years. Never before has so much glass and metal been pointed at the night sky. At the dedication ceremony for the first Gemini telescope, Rita Colwell, the National Science Foundation's director, remarked, "This observatory represents the journey by scientists, engineers, and administrators to a symbolic summit."[3] This book explores the long journey to which Colwell referred, the quest of astronomers and engineers to build bigger and more powerful telescopes. More than simply a success story about how scientists prevailed in their crusade for new telescopes in the face of technical, bureaucratic, and financial hurdles, this book highlights the disappointments and triumphs of planning, designing, and building a modern facility for cutting-edge science. It describes the challenges faced by engineers, the scientific goals of the astronomers, the interests of universities and observatories, and the machinations of prominent scientists, funding patrons, and politicians as these giant, new tools for astronomy were built.

There are several different lenses through which you could view the period of postwar astronomy this book addresses. You could focus, for example, on the history of a particular observatory, a famous astronomer, or the development of a revolutionary scientific idea. While aspects of all these elements are discussed in varying degrees, I have centered this book on the telescope, the astronomer's most visible and readily identifiable tool. More specifically, the story focuses on the history of astronomers' plans to build the pair of telescopes that together constitute the Gemini Observatory.

You might rightly ask, "Why Gemini?" After all, over a dozen new telescopes worldwide, each with a collecting area greater than the 200-inch telescope on Palomar, are now collecting data every night. The answer is simple: Gemini's history, more than that of any other large telescope project, provides an exceptional opportunity to examine the broader changes that define postwar astronomy and science in general.

Astronomers' plans for Gemini's precursors in the United States were marked by conflict over aspects ranging from the details of telescope design to debate over whether the national observatory could meet the challenge of building such a complex and expensive research facility. The resolution of these disagreements profoundly shaped the landscape of postwar American astronomy and tested the science community's determination and cohesiveness. The history of the Gemini telescopes and their place in the larger context of contemporary astronomy is an excellent vehicle for examining the relationship between science and technology as reflected in telescope design, technology development, and astronomical research agendas. Gemini's history, in other words, is the history of recent astronomy in a microcosm.

Scientists' conflicts over resources—between those, for example, who have ready access to large telescopes and those who don't—played a critical role in the planning and building of new observatories after the Second World War. The introduction of federal money into postwar astronomy provided research grants and new national telescope facilities for the entire science community. It also introduced increasing competition and divisiveness among astronomers about how such resources should be used and distributed. Not all astronomers saw federal support for a national observatory, for example, as a positive development. The availability and allocation of research funding and telescope time are part of the larger economy of astronomy. Scientists, naturally, maintain strong feelings about how resources should be allocated. These issues, which the book considers carefully, are central to understanding the history of large telescope projects and postwar astronomy in general.

Gemini itself emerged as a major international science endeavor at a time when astronomers voiced growing concerns about what has become known as "Big Science"—large-scale research marked by an emphasis on big machines, generous government funding, and team-based research. Astronomy has always had some traces of Big Science and "big," of course, is a relative term. Tycho Brahe's sixteenth-century observatory on the island of Hven was large-scale science for its time. Aspects of Gemini's history offer insight into the tension postwar astronomers perceived between a more traditional style of building instruments—one that valued the hands-on, entrepreneurial skills of the individual scientist—and a more corporate model that placed a premium on efficiency and effective management.

More broadly, the history of the Gemini Observatory can inform us about the organization of contemporary science and the development of the technology needed to practice it. Politics, funding, and technology development at the national and international level were increasingly interwoven into the fabric of contemporary astronomy. Looking at Gemini and the development of other large telescopes in the last half-century offers a path to understanding the bigger picture of how social, fiscal, and institutional forces influenced astronomy and other sciences and, in some cases, how these disciplines reflect larger cultural patterns.

Many scientists claim that contemporary astronomy is in a golden age, thanks to the relatively sudden availability of so many new, large, and powerful telescopes. There is a general belief among researchers and science managers that the trend toward Big Science represents the normal and natural evolution of astronomy. At the same time, contemporary astronomers express considerable unease and apprehension about how these technological changes have altered, in ways subtle and profound, the nature of astronomical observing and what it means to be an astronomer. This book addresses the issues and questions associated with the important and far-reaching changes in astronomical practice that have occurred in the last half-century.

The paradigmatic example of postwar Big Science is, of course, high-energy physics. In the 1950s and 1960s, under the guidance of physicists at places such as Berkeley, Brookhaven, and CERN, particle accelerators and bubble chambers grew enormously in size, complexity, and cost. As this happened, the relationship between experimenters and their apparatus was altered and physicists came to enjoy progressively less control over their equipment and their experiments. Teams of specialists were needed to build and operate the accelerators and detectors and laboratory work was often routine. Meanwhile, physicists increasingly relied on computers to analyze their data and management and administration became a necessary ingredient of a successful career. While some physicists enjoyed their new power and prestige, others complained about the increasing industrialization of their field.⁴

The history of postwar particle physics has attracted so much attention, in part, because it is such an aberration. In contrast, many areas of postwar science did not have such powerful federal patrons, they did not claim direct links to national security, and they did not rely on big and expensive machines manned by researchers whose publications featured dozens or even hundreds of authors. In short, particle physics, along with space-based experiments and biological research like the Human Genome Project, represents one extreme in the postwar scientific world.

Albert E. Whitford, a respected mid-century American astronomer, remarked that, in his time, using a large telescope demanded "high artistry—doing it yourself. Real mastery of a beautiful and cantankerous instrument."⁵ Such work was challenging. Allan R. Sandage, an influential cosmologist who spent countless nights taking data with large telescopes, noted, "Observing at a telescope, even under the best of conditions, is tedious. Under the worst, it can be cold and miserable."⁶ While being alone with the telescope under the night sky might have been uncomfortable, it nevertheless fostered an intimate bond between scientist and machine.

In the decades following the dedication of Palomar, astronomers' nightly interactions with telescopes irrevocably and profoundly changed. The pace of this change accelerated most noticeably after 1980 as new telescopes and astronomical technology appeared. Electronic instrumentation released astronomers from the cold telescope dome to heated control rooms where images and data were displayed in real time. The image of the solitary astronomer peering through the eyepiece of a large telescope, shown in Figure 1, is now considered romantic, perhaps a sign that it is also rapidly becoming anachronistic.

Many scientists believe that traditional astronomy, with scientists using a ground-based telescope to make images and record spectra, has gravitated at an accelerating rate toward the Big Science pole exemplified by fields such as particle physics. Driven by the need to get as much observing time as possible and the desire to take advantage of the best observing conditions, modern observatories such as Gemini have experimented with new modes of collecting images and spectra. As a result, the telescope has been recast as a factory of scientific data,

Figure 1. An illustration of "traditional" observing made by Russell Porter; an astronomer in suit and tie sits in the prime focus cage of the 200-inch telescope. Courtesy of the Archives, California Institute of Technology.

with scientists as customers whose orders are delivered electronically while they monitor the observing process through Internet links. The Gemini Observatory's twin telescopes are separated by vast stretches of Pacific Ocean, yet they are linked in real time to engineers and astronomers by high-speed data networks, fiber-optic cables, video conferences, and the Web. As a transoceanic system for science research, Gemini is an example *par excellence* of what I refer to toward the end of the book as "hyper-telescopes," almost organic entities with intricate interdependent components whose collective performance astronomers monitor and control.

Telescopes have always been complex research tools, melding mechanical, optical, and electrical components into a single instrument to collect data from the night sky. In the fifty-year period this book covers, many people came to believe that systems engineering—whose proponents use managerial and organizational tools borrowed from aerospace and defense projects—is an indispensable part of building and operating an observatory efficiently. One wonders whether observatories like Gemini could have succeeded as international construction efforts with the telegrams, mimeographs, and couriered letters of Hale's day. Concomitantly, the observatory itself was transformed into a space where daily interactions between groups of specialists—laser experts, software programmers, optical scientists, cryogenic technicians, to name a few—are extensive and even essential. In the late 1990s, there was discussion in the United States about how the country's telescopes were not isolated instruments but part of a larger system of interconnected tools for research.

The act of observing—using a telescope—and the nature of what it means to be an astronomer have also changed dramatically in the last fifty years, as shown in Figure 2. Many of the most far-reaching, rapid transformations have occurred recently. Standing in the midst of all these changes is astronomy's primary technological embodiment, the telescope.

Defining Big Science solely in terms of money, manpower, and machines is a facile exercise. This book goes beyond the traditional metrics of science's scale and explores the effects of new technologies on the astronomer's nightly work. By examining the development of new astronomical tools and the debates scientists have had about how they would be used (even as they were proposed and built), the book sketches a picture of intelligent, committed, and driven people in the midst of tremendous technological change. It includes examples in which astronomy and other fields of research like particle physics share both common and contrasting patterns of change, and then it asks how such changes occurred and whether (as many astronomers seem to believe) they were inevitable. The book also questions the effect on astronomical practice of astronomers' quest for larger, more powerful, and more complex telescopes. If the nature of what it means to be an astronomer has changed, are larger telescopes responsible? If so, what does the current debate concerning even bigger telescopes augur for the future of astronomy in the twenty-first century?

Building a large tool for science has been called, appropriately, an act of faith.[7] Millions of dollars are spent, careers are consumed, and expectations may never be met. A new telescope carries the possibility of tremendous rewards and risks for individuals, institutions, and sometimes even nations. This book is an attempt to understand how

Figure 2. Observing with modern large telescopes. Interior of the control room of Kitt Peak National Observatory's 4-meter telescope, c. 1988. The telescope operator sits to the left and controls the telescope while the astronomer is recording and analyzing data. Photo courtesy of NOAO/AURA/NSF.

such a leap of faith is made and the effects it has on the nighttime work of astronomers. What remains fascinating is why and how men and women make a leap into the dark and how astronomers use the light, collected slowly and silently by a telescope's mirror, to tell us something new about the universe.

Had this book been written in 1950, there would be little confusion about such apparently straightforward words as "telescope" and "astronomy." Scientists and engineers were just beginning to develop radio telescopes as useful research tools, observing with rocket-borne instruments was fraught with uncertainty, and space-based telescopes were just the dreams of a few visionaries.

Times have changed, of course. Given the wide variety of telescopes that contemporary astronomers use—radio, gamma-ray, optical, space-based, and so forth—it is increasingly difficult to call something a telescope without some adjectival modification. In this book, I use "telescope" in the traditional sense unless noted otherwise: ground-based instruments with large mirrors that collect visible and infrared light. In the same manner, when I speak of "astronomers" (or "astronomy") in general, I am talking about the people who use these telescopes to make observations with visible or infrared light.

Electromagnetic radiation from space, which includes visible light, reaches the earth at many wavelengths. Most photons are filtered out by the earth's atmosphere so only a fraction arrives at a telescope's mirror to be collected, recorded, and analyzed. Astronomers who use telescopes on the ground, to be sure, see only a narrow view of the universe. But within the optical and infrared wavelength regimes, almost all phenomena of interest to scientists emit radiation, conveniently offering rich opportunities for scientific discovery.

The optical window extends from about 0.3 microns in the near-ultraviolet (a micron is one millionth of a meter) through the wavelengths of familiar colors such as blue, yellow, and orange, to the red and then the near-infrared regime of about 1 to 5 microns. Beyond this, from about 5 microns to about 30 microns, we are in the mid-infrared region. The astronomers and engineers who designed the twin Gemini telescopes optimized them for making observations in the near and mid-infrared regimes.

Why are scientists increasingly drawn to infrared astronomy? In the near-infrared, hot blue stars seen clearly in visible light fade out and cooler stars come into view. Many interesting parts of the universe are hidden behind clouds of cosmic dust that extinguish visible radiation. Warm interstellar dust starts to shine in the mid-infrared region as does the dust around stars. Sometimes this dust is so thick that the star barely shines through and can be detected only in the infrared. Infrared radiation also reaches back to the very early stages of the universe. Ultraviolet and visible light is shifted toward the red end of the spectrum as it travels from the most distant parts of the universe. As a result, visible light from faraway galaxies may have been emitted so long ago that it can only be seen at longer wavelengths of 10 microns or more.

You may also be curious about the emphasis placed on large telescopes. My interest and attention is indeed focused on what today are the largest cutting-edge tools for astronomical research. This approach, of course, is not meant to demean the work done and discoveries made at more modest facilities. However, since astronomers began to use telescopes, they have desired bigger instruments. A bigger telescope collects more light, enabling an astronomer to observe fainter, more distant, and older celestial objects. A larger telescope also collects light more quickly, meaning astronomers can complete their research in less time. This fact has not been lost on science managers—a large modern observatory costs about a dollar per second to operate and time, literally, is money.

A telescope's primary mirror gathers photons from distant sources —a faraway galaxy or star—and focuses it, using other mirrors and lenses, onto a detector such as an astronomer's eye or a spectrograph (see Figure 3). Analysis of the light and the creation of an image or spectrum are then done by specially designed auxiliary instruments attached to the telescope.

Telescopes produce two basic kinds of information that scientists interpret to better understand the universe. Images are pictures recorded on photographic plates or charge-coupled devices, and spectra are the colored bands produced when light is passed through a spectroscope and recorded. While images from telescopes usually appear in newspaper and magazine articles, often manipulated to display in full color, scientists often derive more useful information from spectra. Spectra of stars and galaxies can provide a wealth of detail, for example, about their chemical composition, speed, temperature, and distance from earth.

Making telescopes bigger, however, introduces a host of difficulties and complications. Much of the debate and effort by astronomers and engineers described in this book centers around how these communities overcame the challenge of building larger yet affordable light-collecting areas. The amount of light a telescope collects is proportional to the area of its main, or primary, mirror. Doubling the mirror's size enables the telescope to collect four times as much light. The theoretical resolution of a telescope—the finest detail the telescope can reveal in observed objects—also should improve with the diameter of the primary mirror. In reality, astronomers have come to rely on complex and expensive high-tech solutions, like adaptive optics, to coax the best possible performance from modern telescopes.

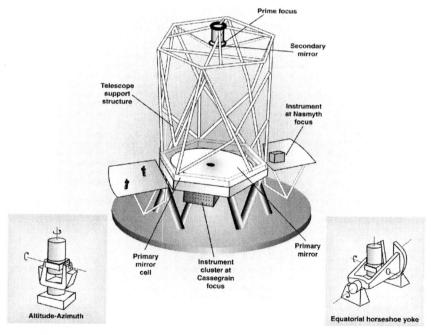

Figure 3. Basic parts of a telescope along with illustrations showing the difference between equatorial and altitude-azimuth mountings. Illustration courtesy of Elliot Plotkin.

CHAPTER 1

Leo and Jesse's Changing World

In August 1936, Jesse L. Greenstein and Leo Goldberg returned to Harvard University to finish their graduate studies in astronomy. In the preceding months, Jesse Owens had punctured the myth of Aryan supremacy at the Berlin Summer Olympics. Bloody battles fought that summer by troops in Spain foreshadowed greater tragedies yet to come. The Dust Bowl and an unemployment rate of 16 percent encouraged Americans to reelect Franklin Roosevelt in a landslide. Some romantics turned to the past in Margaret Mitchell's just-published novel *Gone with the Wind* while other romantics read about the expanding universe in Edwin Hubble's popular book *The Realm of the Nebulae*.

Jesse Greenstein and Leo Goldberg grew up in Brooklyn, New York, but inhabited different social worlds. Born in 1909, Greenstein came from a family of cultured and nonobserving Jews who ran profitable furniture-making and real estate businesses. Greenstein's grandfather gave him a small brass telescope, and as a teenager, he did elementary spectroscopy experiments in the basement and used homemade crystal radio sets at the family's seaside vacation home in New Jersey. Greenstein's early years were pleasant, lacking the "oft-quoted advantage of an impoverished and embittered childhood."[1]

Goldberg's childhood was less comfortable. One colleague described him as the quintessential Horatio Alger character.[2] His parents had emigrated from eastern Poland to work in New York's needle trades before Leo was born in 1913. They were strongly orthodox and Goldberg attended a Jewish parochial school in Brooklyn. When he was eight, his mother and younger brother died in a tenement fire from which he barely escaped. Goldberg's father soon remarried and moved the family to New Bedford, Massachusetts. Decades later, Goldberg still recalled being an outsider—poor and Jewish—when victory in a state spelling bee brought him to a White House dinner where he was left to wonder which items on his plate were kosher.

Both men attended Harvard for their undergraduate education, with Goldberg on a scholarship from the New Bedford Harvard Club. Neither young man planned to make astronomy a career. Greenstein thought he might have science as a hobby after going into the family business while Goldberg imagined a conventional career, perhaps as an engineer. After earning a masters degree from Harvard, Greenstein decided that astronomy was the profession he wanted to pursue. The Depression hit Greenstein's family hard and he was forced to postpone formal training in astronomy. For several years, he worked for his family but kept a hand in science by volunteering for Isidor I. Rabi, a future Nobel laureate, at Columbia University. By 1934, the family finances had recovered, and he returned to Harvard.

Greenstein and Goldberg became friends at Harvard. Their Jewish heritage and New York City upbringing gave them some common ground and Goldberg shed his Orthodox heritage after leaving home. As graduate students, they met the prominent astronomers who came to Harvard to give talks and teach summer seminars. Visitors like Otto Struve and Henry Norris Russell served as important professional contacts for the two young men. The environment at Harvard for astronomers was highly competitive. Not all who finished their degrees went on to have professional careers. While Greenstein could depend on his family income, Goldberg's situation was more precarious. Having given up the security of a conventional career, he needed success as an astronomer.

After graduating from Harvard in 1937, Greenstein received a prestigious fellowship from the National Research Council. Armed with his $2,200 stipend, Greenstein and his wife, Naomi, drove to Williams Bay, Wisconsin, so he could work at the University of Chicago's Yerkes Observatory, one of the world's premiere institutions for astronomy. Astronomers at Yerkes had access to several large telescopes, including the world's largest refractor and an 82-inch reflecting telescope that was nearing completion at McDonald Observatory in Texas. At Yerkes, Greenstein thrived doing both theoretical research and nighttime observing.

Greenstein kept in contact with his classmate and, in contrast to Goldberg's stolid correspondence, sometimes ended his self-deprecating letters with joking reminiscences about Harvard school life. The humor could have a macabre side; Greenstein told Goldberg that he hoped to "be a spectroscopist yet, if Adolph [Hitler] doesn't get me first."[3]

Goldberg was also making progress toward a successful career in astronomy, but his path was harder than Greenstein's. In 1938, as he neared the completion of his doctorate, he worried about the lack of professional opportunities for astronomers. To his surprise, Harvard awarded Goldberg a special postdoctoral fellowship; he later discovered it was a gift from a professor whose son Goldberg had tutored a few years earlier. The fellowship kept his astronomy career alive, and even after his retirement, Goldberg still marveled about how this "100% WASP" generously assisted a "smart and needy Jew."[4]

Goldberg continued his research in solar physics until 1941, when he landed a more secure post at the McMath-Hulbert Observatory, a small facility in Pontiac, Michigan, affiliated with the University of Michigan. It was founded in the 1920s when Robert R. McMath, a young Detroit businessman, began making astronomical observations with a prominent Michigan judge, Henry S. Hulbert. McMath was a wealthy and well-connected scientific dilettante in an era when amateurs could still participate fully in professional astronomy. In 1932, the University of Michigan took over the observatory with McMath as its director. Donald H. Menzel, Goldberg's Harvard advisor and a prominent astrophysicist, convinced McMath that Goldberg would be an excellent addition to the observatory's staff. Compared with Yerkes Observatory's international renown, the McMath-Hulbert Observatory was a relative backwater. Despite this, in July 1941, Goldberg left Harvard to begin his professional career in Michigan, an endeavor soon overshadowed by events on the world stage.

A few days after Pearl Harbor, Greenstein told Goldberg it was "one devil of a time to be worrying about science, but I hope we can stick it out some while yet."[5] Despite such optimism, young scientists from all disciplines soon took on different duties as most civilian research was put on hold. Goldberg and Greenstein did not hesitate to contribute their knowledge of physics and optics to the war effort.

Goldberg originally considered returning to Cambridge to do war-related radar development at MIT's Radiation Laboratory but decided to remain at the McMath-Hulbert Observatory. Robert McMath's Washington contacts helped bring research contracts from Vannevar Bush's National Defense Research Committee to the observatory, which kept Goldberg busy. He spent the war researching anti-aircraft applications and bombsight designs as a consultant to the Navy's Bureau of Ordnance. During the war, he also met and married Charlotte Wyman, a local school teacher. Like Goldberg, Greenstein remained at his observatory during the war. In mid-1942, he and a small group of astronomers studied novel designs for optical systems under the supervision of Yerkes director Otto Struve.[6] One of the more innovative devices Greenstein helped produce was an all-sky camera that could photograph a large swath of the sky at one time. Such activities had longer-term implications. The military establishment came to see scientists such as Goldberg and Greenstein as important resources to be nurtured and exploited. Concomitantly, both astronomers acquired experience in military-related research and consulting, and each made personal contacts that proved important to their postwar scientific careers.

Building Careers in Science

After the war, American astronomy began to change at an accelerating rate. New sources of funding gradually became available to astronomers. Resources from the Office of Naval Research (ONR) and later the National Science Foundation (NSF) opened up research areas that were previously inaccessible. New technologies also reshaped the discipline. Astronomers began to adapt war-surplus electronic, radar, and rocket technology to research, and their efforts slowly yielded scientific payoffs. The demographics of astronomy changed as well. Before the war, the American Astronomical Society had only about 700 members. After the war, this number increased dramatically, thanks to greater resources and the entry of physicists and electrical and aerospace engineers into the field. Finally, the institutional landscape of astronomy changed as federally funded observatories and new agencies such as the NSF and the National Aeronautics and Space Administration (NASA) offered opportunities to astronomers not affiliated with the private observatories.[7]

American scientists adjusted to life in a postwar world that was different politically and technologically. Through their participation in wartime research, many astronomers had been exposed to other fields of study—solid-state physics and electronics, for instance—and had met new colleagues. "From an organizational point of view," recalled Greenstein, "almost every observatory became less insular. People had been elsewhere, knew that there were things being done elsewhere."[8] Moreover, the postwar scientific community was more willing to offer Jewish scientists such as Goldberg and Greenstein opportunities for advancement and recognition.[9] While both men experienced hints of, if not outright, anti-Semitism early in their careers (Menzel once noted his student would easily find a prominent position save for his name), the two helped lead the astronomy community through many of changes.[10]

After the war, Greenstein resumed his research at Yerkes. In addition to more traditional astronomy, he continued to experiment with optical design and the possibilities of making observations with radio telescopes. Goldberg, however, was eager to leave the confines of the McMath-Hulbert Observatory for a university position and was soon presented with an attractive offer. The astronomy department at the University of Michigan asked Goldberg to become its chairman and he began his tenure in Ann Arbor in November 1946.

Goldberg faced several challenges at his new post. He had to rebuild an astronomy department much depleted in staff, students, and equipment. Goldberg was also the chair of an astronomy department without its own large telescope. The best and most modern instruments located nearby were at the McMath-Hulbert Observatory. Goldberg's purchase of new equipment was made possible by a local fund for which Judge Hulbert was secretary. In the late 1940s, before funding from the ONR and the NSF became available, this resource supported the rejuvenation of the Michigan astronomy program by funding a new modest-sized telescope as well as improving instrumentation for solar observing. The new telescope and instruments helped Goldberg attract new students to Michigan's growing astronomy department.[11]

Strengthening his department's staff and instrumental capabilities was part of Goldberg's new responsibilities. He also had to maintain cordial relations with both Robert McMath and Judge Hulbert, a job sometimes complicated by the strained relations between the two men.[12] After the war, McMath became influential in areas of national defense and science policy as a consultant to the Secretary of the Air Force and as a member of the Scientific Advisory Committee to the Secretary of Defense. He also gained further respectability in the astronomy community by serving as president of the American Astronomical Society. While Goldberg did not always appreciate McMath's personality and attempts at micromanagement, the two men regularly discussed scientific and political matters. Given the Detroit businessman's influence in government and business circles, McMath was a valuable ally who could promote Goldberg's career ambitions.

Besides acquiring new students and equipment, Goldberg raised the profile of his department by organizing a series of summer schools in astrophysics. As students at Harvard, Greenstein and Goldberg had attended similar programs organized by director Harlow Shapley that increased their exposure to modern astrophysics. Astronomers and students hailed a one-month symposium, held in 1953, as a "watershed for astrophysics."[13] Walter Baade from the Mount Wilson and Palomar Observatories described his research on stellar populations while George Gamow spoke on "big bang cosmology" and Edwin Salpeter described thermonuclear reactions in stars. These were among the most important and controversial research topics in astronomy, and the meeting Goldberg organized helped the field regain momentum lost during the war.

After the war, astronomers such as Goldberg and Greenstein addressed the broader challenge of responding to the possibilities offered by new technologies and funding sources. Both experimented with captured V-2 rockets as tools to make astronomical observations from high in the atmosphere.[14] In 1946, while still at Yerkes, Greenstein designed a spectrograph that could make measurements using both ultraviolet and visible light. After months of intense frustration with military secrecy and equipment problems, his experiment was launched at White Sands, New Mexico, on the inauspicious date of April 1, 1947. Disappointment soon followed when he developed the film and found it unexposed. Chagrined, Greenstein heeded Otto Struve's suggestion and resumed more traditional research at Yerkes.

Goldberg was also enthusiastic about the scientific possibilities of rocket-borne observing. As he told his former mentor Menzel, doing such research would be worth "shaving my head and working in a cell for the next 10 or 15 years."[15] Goldberg collaborated in rocket-based astronomy with Menzel and Lyman Spitzer, a prominent scientist at Princeton. Like Greenstein, Goldberg found rocket-borne experiments to be a frustrating enterprise. In the late 1940s, most astronomers considered rocket-based astronomy unrealistic and outside the mainstream of scientific research.

Greenstein's and Goldberg's first attempts to make observations with rocket-borne instruments reveal a difference between their approaches to research that would mark the rest of their careers. Goldberg, later motivated by launches of Soviet and American satellites, became a spirited advocate of the scientific possibilities offered by space-based astronomy. Greenstein, meanwhile, remained a staunch champion of astronomy done from the ground. In 1960, when Goldberg returned to Harvard's astronomy department as a full professor, Greenstein congratulated his former classmate but noted his negative feelings toward space research. "I worry often about the future of optical astronomy in this country. I cannot believe the subject is dead," Greenstein said, "I look forward to some real healthy competition [while you are at Harvard] and I hope at least part of this will be in optical astronomy conducted from the ground."[16]

Jesse Greenstein had an excellent incentive to stress the importance of traditional ground-based astronomy. In 1948, Caltech offered him the opportunity to lead their astronomy program. This was like being traded to the New York Yankees just in time for their championship 1949 season. After several years of rural living at Yerkes, Greenstein and his family were ready for a more urbane lifestyle. More importantly, Caltech possessed the undeniable attraction of the world's largest telescope. Greenstein arrived in Pasadena in June 1948, just in time to attend the 200-inch telescope's dedication.

Along with the older but still productive 60-inch and 100-inch telescopes on Mount Wilson, the 200-inch was jointly operated by the Carnegie Institution of Washington and Caltech as the Mount Wilson and Palomar Observatories (hereafter Palomar). A small committee with members from each institution shared the management decisions for what Edwin Hubble called "the astronomical center of the world."[17] Palomar's recently appointed director was Ira S. Bowen, a Caltech professor of physics whose specialty was spectroscopy and optical design. Bowen was not an observational astronomer, a fact that angered Hubble, who had wanted the post himself. Vannevar Bush, Carnegie's president, believed that because "physics and astronomy are so close together," Bowen was the right man to foster a stronger connection between the two disciplines and modernize astrophysics at Palomar. Bowen's first actions after becoming director confirmed Bush's instincts. He announced that traditional observational programs would continue but that the observatories at Mount Wilson and Palomar would get their "teeth into fundamental new problems leading to new concepts" as well. Future advances in astronomy at Palomar, in other words, would come from the integration of physics, astronomy, and modern instrumentation.[18]

Greenstein's talents combined theoretical research and observing experience with big telescopes and, as a result, he was a logical choice to lead Caltech's astronomy program. The telescopes on Mount Wilson and Palomar were the apex of traditional American observational astronomy where "you lived in [a] world of less interpretation, very little theory, largely description at a high level." Greenstein initially found that the staff viewed him as "some kind of wild-haired radical theorist . . . and therefore a little bit dangerous." But he appreciated Pasadena's more sophisticated environs as well as the legacy of Mount Wilson, which he believed had "invented large telescope astronomy." An earlier visit to Pasadena gave him a powerful emotional impression, and he imagined how he could "sit in an office . . . every day quiet, and just have all this wonderful data."[19]

Greenstein had a big task ahead of him as he built up and promoted Caltech's astronomy program. While Carnegie's large staff boasted luminaries such as Hubble and Baade, Caltech had only one astronomer (Fritz Zwicky, a brilliant and sometimes volatile Swiss scientist) when Greenstein arrived in Pasadena. Most of Caltech's early teaching and administrative duties fell to Greenstein, and his program began with only a handful of graduate students who signed up for his year-long courses on stellar evolution and star interiors. According to his former students, Greenstein was an unorganized teacher who did best when he forgot his notes and lectured extemporaneously.[20] To beef up his program, he persuaded astronomers from Carnegie to give courses in their scientific specialties to Caltech students. Caltech's first cohort of students fulfilled Greenstein's hopes by becoming noted and influential members of the international astronomy community—Allan Sandage, for example, became an intellectual leader and sometimes controversial figure in the maturing field of observational cosmology while Helmut A. Abt edited *The Astrophysical Journal* for over 25 years.

Greenstein believed that "a reformed theorist may become the most intelligent observer." To prove his point, he staffed Caltech's astronomy program with men who could combine expertise in theoretical astrophysics with observational programs using the world's biggest telescopes.[21] As he filled his department's roster, Greenstein looked in a very specific direction for new faculty—back to Yerkes. Guido Münch arrived at Caltech from Wisconsin in 1950, followed by Donald E. Osterbrock and Arthur B. Code in 1953 and 1956. Greenstein also attracted famous visitors to Caltech such as Fred Hoyle to do research and teach. The strong bond of physics with astronomy at Caltech meant that new areas of astronomical research were incorporated into the traditional program of optical observing.

As Greenstein and Goldberg met many of the same challenges—recruiting quality staff, generating funding, and building new facilities—they maintained their relationship through correspondence and visits. In late 1955, Goldberg accepted an invitation from Greenstein to give

Figure 4. Jesse Greenstein (left) and Leo Goldberg at the June 1956 meeting of the American Astronomical Society in Berkeley, California. Courtesy of the AIP Emilio Segrè Visual Archives.

a lecture series at Caltech.[22] A photo (Figure 4) taken a few months later shows the two men smiling and standing next to each at a meeting of American Astronomical Society. But while Goldberg, a clever and forceful science manager, labored to modernize his department, forged relations with other prominent scientists, and sat on some of the most influential committees in postwar astronomy, there was one asset of Greenstein's he could never match—the observing power of the 200-inch telescope.

The Palomar Experience

For almost 30 years, the 200-inch was the world's largest telescope. Scientists using it drew upon the considerable financial resources, manpower, and technical know-how of Carnegie and Caltech. Even before the 200-inch was dedicated, the public was fascinated by its size and the science it would do. In February 1948, *Time* magazine featured a cover story about it and the career of Edwin Hubble. While astronomers at Mount Wilson complained about the glorified depiction of Hubble, the public was thrilled. Similar articles in *Collier's* and *Life* presented a romantic image of the astronomer exploring the universe alone at night.

The popular amateur astronomy magazine *Sky & Telescope* presented more accurate but still enthusiastic depictions. For months prior to the telescope's dedication, the magazine featured articles about the 200-inch while its covers displayed Russell W. Porter's fabulously detailed cut-away drawings. In January 1948, an article depicted the journey of the finished 200-inch mirror from Pasadena up to the telescope as a major public spectacle with school children cheering along the way. The 200-inch, another author said, will be "jealously guarded for hunting big game of the universe."[23]

This attention culminated in the dedication of the telescope on June 3, 1948. Several hundred people gathered beneath the telescope's dome, including distinguished scientists, workmen, trustees from the California Institute of Technology, and entertainment celebrities. They sat on folding chairs and benches beneath the recently completed telescope, enjoyed the cool mountain air, and listened as speakers praised the instrument.

Raymond B. Fosdick, the Rockefeller Foundation's president, spoke about the ability of human knowledge to outstrip ethical judgment. With words tempered by concern about the atomic age and the revelation of Nazi atrocities, Fosdick described the 200-inch telescope as a tool to heal an ailing world.[24] Lee A. DuBridge, Caltech's president, announced that the new telescope would be named the Hale Telescope in honor of George Ellery Hale, the late astronomer and statesman of science whose tireless proselytizing brought the project to life two decades earlier.

After the speeches ended, several hundred tons of telescope and dome quietly began to move. Ira Bowen demonstrated how the telescope pointed at objects in the sky as the invited guests watched from the balcony high above the observatory floor. A radio announcer gave a real-time description of the events and many of the dedication's attendees remained after the sunset to view the moons of Saturn with light collected in the massive mirror.

Public interest in the telescope did not subside after the telescope was dedicated. On August 30, 1948, the U.S. Postmaster released a new three-cent stamp showing the telescope's classically styled dome. The telescope became a wonderful tool for promoting astronomy and, as Jesse Greenstein discovered, fund-raising for Caltech astronomy.[25] Public excitement over the telescope was so great, in fact, that Ira Bowen had to develop a plan to accommodate the throngs of visitors flocking to Palomar. The observatory prepared a special publication called "Frontiers in Space," which visitors read while walking among the pines and jays on Palomar. After learning about the 200-inch's engineering marvels and circumnavigating the outside of its imposing enclosure, visitors entered a special gallery to view the telescope.

The most celebrated feature of the Hale telescope was its primary mirror. Corning Glass Company cast the 14-ton mirror blank in 1934 at its New York plant after much difficulty and one aborted attempt. The back of the mirror blank had a ribbed structure like a waffle that reduced some of its weight. The Pyrex glass from which the mirror was made had a lower thermal expansion than conventional glasses. When the mirror cooled at night, the telescope's images were not as distorted by subtle changes in its shape.

Opticians polished the mirror so it had a prime focal ratio much shorter than other instruments of its era. As in cameras, the focal length of a telescope is the distance between the mirror or lens and the point at which these are brought to focus. The focal ratio is this number divided by the diameter of the mirror or lens. A smaller number means the mirror has more steeply polished curves. In the case of the Hale Telescope, the distance from its main mirror to the prime focus point was 660 inches and the diameter of the mirror 200 inches, producing a focal ratio of 3.3. A smaller focal ratio meant that the dome that sheltered the telescope could be made somewhat smaller and would cost less.

Like all optical telescopes of its time, the 200-inch was designed to turn east-west on its polar axis and north-south on the declination axis—a setup known as an equatorial mount. Once the telescope was rotated to point at the right patch of sky, only a slight countermovement about the polar axis was needed to offset the earth's own rotation as astronomers did their observing.

As the telescope moved, the metal superstructure that held the optical system flexed slightly under gravity. When this happened, the primary and secondary mirrors needed to remain precisely aligned. Mark Serrurier, a Caltech graduate student, invented an innovative truss system of metal parallelograms that allowed both ends of the telescope to shift by the same amount, thus keeping the optical system aligned. The solution served as the basis for other large telescope designs. The 500-ton telescope assembly rested between two massive bearings floating on a thin layer of pressurized oil. The entire telescope was enclosed in a towering dome 137 feet high (Figure 5).

The Hale Telescope's design—the massive mirror, the truss system, the huge bearings, the equatorial mount, the semispherical dome—helped establish a design paradigm for large telescope construction, and it set the standard for large telescopes for at least the next quarter-century. It also influenced the thinking of astronomers and engineers about what was possible in a telescope and what was beyond their capabilities. As one astronomer recalled, "I think there was this real feeling of mystery that the 200-inch telescope was something that had been produced by wizards and elves and set down on earth."[26]

Determining who got to use this top-of-the-line telescope was serious business. By 1949, Jesse Greenstein represented Caltech on the joint Carnegie-Caltech Observatory Committee where he helped allocate observing time on all the telescopes on Mount Wilson and Palomar. Besides the 200-inch and the older telescopes on Mount Wilson, these

Figure 5. A 1939 drawing of the 200-inch telescope looking northwest by Russell Porter. Courtesy of the Archives, California Institute of Technology.

included the world's largest Schmidt telescope. Equipped with optics that enabled it to survey and photograph wide swaths of the sky at once and designed to complement the 200-inch, the 48-inch Schmidt telescope began operation in 1948.

Astronomers at Carnegie and Caltech were divided into different research divisions. People such as Milton Humason, Rudolph Minkowski, and Allan Sandage observed "extragalactic nebulae," as galaxies were still called at the time. Greenstein was in the spectroscopy division, an area dominated by older staff astronomers. While boundaries between these divisions diminished over time, some rivalry remained between the scientists. Walter Baade, a member of the extragalactic division, found the spectroscopists' work dreary. "Jesse," he said, "they don't eat, they don't drink, they don't love."[27]

Once a year, Bowen requested short proposals from the Carnegie and Caltech astronomers for telescope time. Through a process of what Greenstein called "friendly throat cutting," the Observatory Committee determined how much time the staff received. "Dark time," when the moon's light doesn't obscure the faintest celestial objects, was the most valuable resource at the 200-inch. This commodity was divided among staff astronomers doing research on faint, extragalactic objects that required the darkest conditions possible. Sandage and Baade regularly received a month or more of dark time annually, a resource unimaginable to many other astronomers. When the moon brightened the sky, astronomers used the telescope for spectroscopic work that didn't require the dark conditions needed for studying faint objects.[28]

By any standard of the day, staff at Carnegie and Caltech had unparalleled opportunities to observe with the world's biggest telescopes. Every year a small number of visiting astronomers could apply to use the telescopes as well. Between 1950 and 1960, about 20 guest astronomers from throughout the United States and abroad visited either Mount Wilson or Palomar annually.[29] Visiting astronomers were typically restricted in the telescopes they could use. The 48-inch Schmidt, for example, was used almost exclusively for the National Geographic Society–Palomar Sky Survey, which wasn't completed until 1956. Most visitors were assigned to the 60-inch or 100-inch telescopes on Mount Wilson. As *Sky & Telescope* predicted years earlier, Greenstein and other Palomar staff had a near-monopoly on the 200-inch for "hunting big game." Bowen explained this policy, saying, "It was a little bit unwise and probably not the best use of telescope time to throw a man directly on the 200-inch."[30] Bowen's comment unconsciously reflects an unfortunate aspect of large telescope use at this time. Equipment at Mount Wilson and Palomar was not available to women until the 1960s, when scientists like Margaret Burbidge, Vera Rubin, and Virginia Trimble broke the gender barrier and were allowed to observe at night.

In the 1950s, many of the observation runs with the 200-inch were carried out at the telescope's prime focus station, located about 75 feet above the primary mirror. The 200-inch was the first telescope with a mirror large enough to permit astronomers to work at the prime focus without cutting off too much light from the main mirror. There researchers sat in the "observing cage" and rode with the telescope all night long while collecting data.

By 1951, astronomers in the cage were using the new prime-focus "nebular spectrograph" designed by Rudolf Minkowski, a German astronomer at Carnegie, and built at the Mount Wilson Optical Shop. Minkowski designed the spectrograph to optimize analyses of light from extremely faint galaxies and stars. The spectrograph's location in the prime focus cage necessitated that it be very compact and lightweight. To help astronomers handle it in the dark, its sturdy design featured smooth edges and controls that cold hands could manipulate. As shown in Figure 1, the spectrograph sat on a pedestal in the observer's cage. Light from the main mirror entered from below and struck a small polished slit at the spectrograph's bottom. Some light was reflected from the slit to an eyepiece that astronomers used to align and guide the instrument. The light passed through the slit and reflected off a concave mirror to a diffraction grating that dispersed the light into a spectrum. Ultimately, light collected by the telescope's massive mirror was recorded on tiny photographic plates less than a square inch in size inside the spectrograph.[31]

Astronomers who used the Palomar telescope, especially in its earliest years of operation, expressed fond, often romantic, feelings about what observing with it was like. Linking much of this pleasure to the experience of being alone with the telescope at night, they spoke of the special privilege of using such a prestigious instrument. Even while proudly boasting of the hardships observing entailed, they rhapsodized about the magically intimate experience of using a large telescope when, as one user reflected, "you sat with your back to the whole universe."[32]

On a typical night of observing, the astronomer walked to the telescope dome after an early dinner. "Each time you go up," Greenstein recalled, "you carry some photographic plates in a tiny little box, and you carry a whole set of dreams of what the object you're going to work on is going to turn out to be."[33] Earlier in the day the astronomer might have prepared his plates by exposing them in a sealed box to nitrogen gas (a process called "baking"), which increased their sensitivity. The astronomer began the night with a list of possible targets, but weather conditions often determined which ones were actually observed. While the telescope operator opened the dome shutters and checked the weather, the astronomer rode the elevator to the prime-focus cage and climbed in. The operator shared the night's experience with the astronomer, operating the telescope from the main control desk on the dome floor where sets of analog dials showed the telescope's position. Throughout the night, the operator pointed the telescope as the astronomer requested through the intercom or simply by shouting.

The observing cage was a cramped space only six feet in diameter where the astronomer sat in an uncomfortable chair while taking photos or spectra. Baade joked that the seat's shape was based on the bear-like Minkowski's rear end. A single exposure might last all night, during which the astronomer looked through an eyepiece almost continually to keep the object centered properly on the photographic plate or spectrograph slit. The work could be quite tiring, especially in the cold winter air, and even inefficient. Fatigue sometimes impaired the astronomer's visual acuity to the extent that he could no longer see the target.[34] Having exceptionally fine eyesight was one of several physical qualities that played a role in determining who became a top-notch observational astronomer. For Greenstein, even having "a tough bladder" was a plus because "if it was a good night, you stayed up [in the cage] from seven o'clock to five." When the night was over, "there is a feeling of utter and complete relief of blood beginning to circulate again."[35]

The personal and physical experience of using a large telescope is revealed in an audio tape (surreptitiously made by a night assistant) of Minkowski talking to himself during a night's observing run:

Where is that thing? I think that's it over there. [sound of slow motion motor] Damn! Wrong button. [motor again] There it is . . . now where's that little double to the left? . . . No, that's not it . . . [to the night assistant] Try a little west . . . Stop! [To himself] Here it comes [slow motion motor]. Yes, I think that's it. Pull it down a little [slow motion motor again] Ah! Too much! [motor sound] Now I've got it! [sound of camera opening; then, to the night assistant] Start the exposure! I'll take three hours on this one. You can rest awhile.

A long sigh followed as Minkowski began to hum Beethoven's "Ode to Joy" before the tape ended.[36]

Using a big telescope was a skill one acquired through experience and perseverance, often while working alone in the dark. Astronomers had their own techniques for collecting data; telescope operators even claimed they could tell who was in the observing cage simply by the types of instructions that were being shouted out to them. Greenstein, according to a former Palomar astronomer, observed like a "bull in a china shop"; he wouldn't always take perfect exposures but he got the data he needed.[37] Caltech's students and young faculty learned big-telescope techniques in a manner reminiscent of craft apprentices centuries earlier. Working first with smaller telescopes in a trial-and-error fashion, they gradually (with tutelage from more skilled astronomers) moved up to help observe on the 200-inch. Between long exposures or on cloudy nights, older astronomers might share their thoughts and experiences with student assistants or new professors over a quick sandwich or cup of coffee. Allan Sandage, for example, spent many nights learning to use the 200-inch telescope with Walter Baade, and Donald Osterbrock had a similar apprenticeship with Minkowski.

In 1950, when the Hale 200-inch was finally debugged and turned over for full-time science operations, photographic plates were the standard tool for recording astronomical observations. But photography was an inefficient technology for recording photons. Only a small percentage of the light hitting the photographic emulsion contributed to the image. For some time, companies such as Eastman Kodak cooperated with astronomers to develop new emulsions suited for fainter sources of light over a broader range of wavelengths. Gradually, astronomers abandoned photographic plates, with their inherent limitations, and turned their attention instead to developing electronic devices that recorded images and spectra more efficiently.

The introduction of electronic light-detection devices and the hiring of people who could marry new technologies to a large telescope were part of what Jesse Greenstein later called the "de-astronomization of astronomy."[38] Observatory directors and astronomy department chairmen realized it was essential to hire not only skilled astronomers but also people with backgrounds in electrical engineering and solid-state physics. In some cases, these people took up research in astronomy as well. As early as 1945, Henry Norris Russell advised Bowen to get a "good man who knows modern electronic mechanisms" and predicted that "in another decade or so, those of us who slaved looking through eyepieces and making measurements will be pitied by the folks who use the new devices."[39] By 1960, traditional optical astronomers could see that their research would inevitably be intertwined with technological innovations developed by people they began to call "gadgeteers."

A few astronomers had experimented with comparatively primitive electronic light-detection systems before the war. Later, aided by the availability of new equipment, these astronomers quickly exploited their wartime experience with electronic devices to do research more efficiently. For instance, while photographic plates could require three or four hundred photons to darken an emulsion grain, new photomultipliers were much more effective at recording the light captured by the primary mirror. Photomultipliers were basically evacuated electronic tubes that converted light into measurable electric current. Light from the telescope entered a glass tube and struck a photocathode. Electrons were ejected that traveled toward an anode, striking intermediate photocathodes along the way, a process that caused a cascade of thousands of electrons. Photomultipliers enjoyed a relatively linear relationship between the energy of incident light and the electric current it produced. These new tools enabled astronomers to do photometry, the measurement of a star's energy output, in a more straightforward fashion.

Astronomers were interested in using photomultipliers to explore phenomena whose energy output changed over time. By accurately measuring how the light intensity of Cepheid variable stars changed, for example, astronomers hoped to obtain better measurements of the distances to remote galaxies and to map their distribution in space.[40] By the early 1950s, William A. Baum led a vigorous program in photoelectric photometry at Palomar. Originally trained as a physicist, Baum developed a photon-counting system that gave improved measurements of star magnitudes photographed with the 200-inch. Baum and Palomar astronomers soon found that the recorded magnitudes of certain types of faint variable stars were brighter than previously thought, a result that contributed to Walter Baade's revision of the actual distance from earth to the Andromeda galaxy.

Photomultipliers, however, only provide a measurement of the total amount of light that is striking them and do not record an image in the same way that a photographic plate does. Astronomers were keen to join the light-amplifying capability of the photomultiplier with a device such as a photographic plate that could record an image. They began to devote considerable resources to technologies that could record two-dimensional images electronically. Along with scientists at institutions like Lick Observatory and Yerkes, astronomers and engineers from Caltech and Palomar researched and built many new electronic devices for astronomy in the 1950s.

These new electronic tools helped redefine the astronomer's interaction with the telescope. J. Beverly Oke, a Canadian astronomer, first worked at Palomar in the late 1950s. He recalled how the old manual telescope guiders used to direct the telescope while tracking targets limited the astronomer. "Because the new instruments could go fainter much faster than the guider," Oke said, "you were getting into the domain where you were making observations on objects you literally couldn't see."[41] New auto-guiding systems relieved the astronomer of having to keep the celestial object of interest centered on the slit of a spectrograph or a photographic plate manually. In 1937, Albert Whitford and Gerald E. Kron built an early version of such a system at the University of Wisconsin. Engineers using the latest photoelectric and amplifying devices installed an improved version on the 200-inch in the mid-1950s.[42] These new tools, according to Oke, made observations easier and more efficient. But the telescope's complexity increased because now "astronomers had to electrify all the things they used to do by hand."

The introduction of electronic computers was another change in the toolkit of the observational astronomer. Astronomers previously used computers to do modeling of systems and sheer number crunching, but they soon began reducing the data they collected at the telescope with electronic computers. In 1959, for example, one of the first graduates from Greenstein's astronomy program, Halton C. Arp, demonstrated how a central computer on the Caltech campus could reduce observational data from the 200-inch in a fraction of the time it took an astronomer using a calculator.[43]

The appearance of computers in the laboratories of the 1950s was certainly not unique to astronomy. Physicists first began using computers on the Manhattan Project and continued to rely on them after the war. Automated data-reduction machines were essential to postwar high-energy physicists. Bigger and more powerful particle accelerators produced floods of data that only computers could efficiently sift through. Physicists using the new 72-inch bubble chamber at the Lawrence Berkeley Laboratory, for instance, relied on machines that could automatically measure particle tracks many times faster than a human operator. Experts in data-processing told physicists that they could envision the "elimination of the humans, function by function" through computerization.[44] Berkeley physicist and future Nobel winner Luis W. Alvarez defended these techniques as a necessary and pragmatic part of modern physics, in which high efficiency comes from "production line organization."[45]

While astronomers' initial experiences with electronic devices may not have been as disorienting as those of particle physicists, new technologies such as photocells and auto-guiders subtly altered their working relationship with the telescope. Jesse Greenstein reflected that "the kind of lone wolf effort which was popular [in the 1950s] disappeared when the technologies became difficult."[46] By 1960, the adoption of increasingly sophisticated electronic devices had introduced a new layer of technology between researchers and the phenomena they observed. Traditional observing practices had to be altered accordingly. For example, an astronomer using the new electronic photometry system at the 200-inch alternated between measuring light from an object of interest and a blank patch of sky. By doing this in intervals of a few minutes, the astronomer was able to subtract the signal from the night air glow and moonlight from that of a star or galaxy.[47] For researchers, electronic technologies increased the level of technological sophistication associated with efficiently doing cutting-edge astronomy while, for telescope engineers, these technologies represented another system that had to be considered when designing or upgrading an observing facility.

Jesse Greenstein was especially interested in using the 200-inch for spectroscopic observations. In 1950, another new spectrograph was installed at the telescope's coudé focus that Greenstein used frequently. The coudé focus lay along a telescope's polar axis; light from the primary mirror was directed to a separate room beneath the telescope where the astronomer collected spectra in the dark. The coudé spectrograph complemented Minkowski's prime focus instrument as it could highly disperse the light from bright sources that astronomers could study in greater detail. One of Greenstein's major research projects was to use the 200-inch telescope and its instruments as a tool to probe the chemical composition of stars. During the 1950s, astronomers were keenly interested in understanding how nuclear reactions inside stars formed elements heavier than helium. Greenstein's research on the relative abundance of certain elements and isotopes was aided by close associations with physicists at Caltech's Kellogg Radiation Laboratory such as William A. Fowler (who later won a Nobel Prize for his studies of how chemical elements formed in the universe) who were interested in the stellar creation of chemical elements.

Greenstein also engaged in a long-term research program to study white dwarfs. These form when nuclear burning at a star's interior stops and its core begins to contract under its own gravity. This compression continues until the star becomes extremely dense, with all of the star's material, called degenerate matter, squeezed into such a small volume that its electrons cannot be packed any closer together. Besides being astronomically interesting, white dwarfs provide "laboratories" for measuring the behavior of matter under extreme conditions that cannot be duplicated elsewhere. Greenstein was drawn to the study of white dwarfs for several reasons, not least because he believed they were extraordinary celestial objects with "all the romance astronomy should have."[48] By the mid-1950s, Greenstein had discovered dozens of new white dwarfs.

Throughout his career, Greenstein maintained a pattern of research shaped by a "low threshold of boredom" and an "inability to resist the use of newly available equipment" on astronomical problems. He frequently plunged into new areas of investigation, published several papers, and moved to another area that caught his attention. Greenstein's ample observing time on the 200-inch telescope enabled him and other scientists to skim the cream from a wide range of research interests. "We had a big telescope, we could work on faint objects," he said, "You can't resist the fact that you've got an unbeatable gadget that nobody else has."[49]

Courting New Patrons

In May 1957, Leo Goldberg wrote his long-time patron Henry Hulbert about new developments in Michigan's astronomy department. Goldberg apologized for not being a more diligent correspondent but, as he explained, "I have also been rather busy during the last few months in helping to organize the National Observatory in Arizona . . . I expect that some day the Arizona observatory will be the most important one in the world, and unlike other observatories, it will be open to all astronomers on the basis of scientific merit only."[50] Goldberg was speaking, of course, about the establishment of AURA, a consortium of universities that would soon begin operating Kitt Peak National Observatory (Kitt Peak hereafter), the national center for optical astronomy. Goldberg's activism, as his letter noted, was a prime catalyst for the establishment of Kitt Peak.[51]

The founding of Kitt Peak was a major milestone in the history of American astronomy. The new national observatory offered any astronomer, regardless of institutional affiliation, an opportunity to receive telescope time and compete with scientists from the major private observatories. In establishing Kitt Peak, the federal government continued the Cold War trend of supporting science by investing in large-scale national facilities. Not all astronomers approved of so much federal money going to national facilities and space-astronomy programs instead of to individual researchers. The priorities of Goldberg and Greenstein often represented opposite sides in an increasingly polarized debate about the proper balance between private and federally funded observatories.

In the 1950s, astronomy metamorphosed from a science largely funded by state and private money to one in which federal dollars had an essential role. Greenstein and Goldberg were active participants in this transition. Both men received some of the first grants from the

ONR and the NSF. This money helped them build up the institutional capabilities of their departments and fund their own research. More importantly, the two men also helped determine how federal money for all of astronomy would be allocated.

During the 1950s, as astronomers started to take advantage of federal funding, Goldberg and Greenstein began to espouse differing opinions about how astronomy should be done—by individuals or teams, from space or the earth, and, most importantly, at private or national facilities. These differences can be partially traced to their institutional affiliations and attendant goals. Greenstein felt passionately about his stewardship of Caltech's astronomy program while Goldberg was intent on fostering national astronomy centers that could complement and compete with private institutions.

Before World War II, American astronomy was a relatively coherent discipline. Some cracks in the façade existed before 1940, such as longstanding mistrust between the major East Coast and West Coast observatories.[52] However, the community was small and united by its shared heritage of using traditional telescopes. Before the advent of the national observatory system, the nation's largest telescopes were controlled by a few private institutions, including Yerkes-McDonald, Lick, and Palomar. Astronomy's patronage system still relied on philanthropy and state funds. The directors of these observatories—men such as Henry Norris Russell, Walter S. Adams, Otto Struve, and Harlow Shapley—wielded considerable power. They could influence the careers of other scientists, shape research agendas, and control the distribution of critical resources like observing time and funding. At night, observations at the telescope were made primarily at visible wavelengths and team-based research was rare.

After the war, the astronomy community fragmented at an accelerating rate. Traditional optical astronomers soon found themselves collaborating and competing with scientists using radio telescopes and rocket-borne and space-based instruments. New technologies and more specialized equipment diluted the importance, in some scientists' eyes, of the large, all-purpose optical telescope.[53] National advisory panels during the 1950s and early 1960s also placed more emphasis on stellar and galactic astronomy than on solar system studies, thereby alienating still other astronomers.[54]

It might appear that American astronomy after 1945 simply witnessed the same changes experienced in high-energy physics or oceanography with the influx and influence of government funding and the construction of new and elaborate facilities. But the extensive private and state investments that defined and shaped astronomy before the war did not die out after generous federal funding appeared. Instead, American ground-based optical astronomers had two considerable and competing sources of support. This was (and still remains) a unique situation in the context of international astronomy. Astronomers in other countries were limited to government and university support. America's dual patronage system for astronomy had (and continues to have) a tremendous impact on debates concerning new, large telescopes.

American astronomers, especially optical astronomers, were slow to take advantage of federal patronage after World War II, responding neither as quickly nor as enthusiastically as the American physics community did. Astronomers' initially tepid response to federal funding can be attributed to "an elite infrastructure" more interested in "preserving authority and control" than adding new members or embracing new technologies made possible by federal money.[55] Harlow Shapley and other members of the community had ethical misgivings about accepting military funding, while others worried that astronomers would become dangerously reliant on outside resources. Observatory directors who had shepherded their institutions through war and economic depression feared becoming dependent on federal patronage and losing control of their research programs and resources. Gradually, the reticent attitude of the astronomy community changed, partly due to the efforts of younger astronomers like Leo Goldberg and Jesse Greenstein.

In early 1946, the ONR began an aggressive program to support basic science research. Within months, three out of every four dollars for basic research came from the ONR.[56] Besides providing funding for research in areas with potential military value, such as solar physics and radio astronomy, the ONR also gave astronomers about $80,000 a year in small grants.[57] This was a pittance, of course, compared to the millions of dollars the physics community was receiving from the government. Physicists had successfully promoted their discipline as having a value beyond scientific knowledge that touched the core of America's national security.

The National Research Council administered the ONR funds for astronomy. This arrangement maintained an established pattern of control dating back to World War I and the committee that advised the ONR was almost entirely composed of directors of elite observatories.[58] These men felt that federal support should not replace traditional sources and were content to keep government contributions at modest levels.

Consequently, the first members of the NRC's advisory panel held conservative views of the postwar astronomy community's needs. In 1950, C. Donald Shane, chair of the NRC committee and director of Lick Observatory, prepared a report that suggested astronomers' annual needs could be met with about $540,000 from the nascent National Science Foundation. Shane's conservative estimate was based on the belief that universities, the traditional supporters of astronomy, should not use federal funds "to escape the expenditures for astronomy they would otherwise make."[59] Moreover, Shane believed advisor-astronomers, not Washington bureaucrats, should control the resources offered by the federal government. Only a few astronomers, including Leo Goldberg and his mentor Donald Menzel, successfully lobbied the Navy directly for tens of thousands of dollars to build equipment and fund long-term research programs. The majority of astronomers merely requested small ONR grants to fund short-term projects with clear goals.

Soon after President Truman signed the bill that established the NSF in May 1950, the new agency asked Jesse Greenstein to lead its Advisory Panel for Astronomy. Goldberg, meanwhile, directly advised Paul Klopsteg, head of the NSF's Mathematical and Physical Sciences Division. A major question was whether the older NRC committee would continue to allocate the NSF's astronomy funding or whether the NSF would exercise autonomy over its own budget. Greenstein preferred the ONR-style of favoring small grants to individuals while Goldberg believed that the NSF needed to establish its own protocols and sphere of influence. Goldberg was also aware of the NSF's plans to establish large-scale national facilities, something not easily done within the ONR model.

In 1951, the NRC's advisory committee had a changing of the guard. Leo Goldberg became its chair and the committee included younger astronomers like Lyman Spitzer and Fred L. Whipple who advocated aggressive plans for expanding astronomy. One of the issues they discussed was how and to what extent the NSF should support astronomy. There was strong disagreement among the committee members about Shane's earlier appraisal for funding from the NSF. Greenstein, who also served on Goldberg's committee, believed Shane's estimate was too large while scientists like Goldberg and Whipple argued that astronomers should ask for at least twice that amount. While Goldberg did not believe federal sources should provide all funding for astronomy, he was much more enterprising than many of his colleagues in pursuing government support for research. He had already seen the benefits of this approach; his astronomy department at the University of Michigan had half of its faculty salaries paid by federal funding, the highest proportion of any program.[60]

Goldberg's and Greenstein's differing viewpoints made sense given their home institutions. Greenstein ran a program that already had access to the world's biggest and best telescopes. Moreover, the Carnegie Institution of Washington barred its observatories from accepting federal funds. On the other hand, Goldberg's Michigan program relied heavily on federal funding while Goldberg himself favored the establishment of large-scale national observatories. Goldberg argued to Greenstein that the NSF's support could improve the overall infrastructure of science, which included building big facilities, and ensure that basic research remained free of military control.[61]

Greenstein remained concerned about how the NSF was approaching the question of funding astronomy and advocated that astronomers themselves keep control rather than ceding it to the agency. Who would be responsible, Greenstein asked, if the NSF made poor choices in funding proposals? Frustrated that the NSF Program Director for Physics and Astronomy, Raymond J. Seeger, was not a practicing astronomer, Greenstein also complained that his advisory panel was not sufficiently informed about the NSF's plans to build a modest-sized national telescope for photoelectric research.[62] In addition, Seeger had appointed another committee that diluted the advisory role of Greenstein's committee. This ad hoc Panel on Astronomical Facilities was chaired by Robert McMath with Struve, Bowen, and Whitford as members. The NSF, acting on a suggestion from McMath's ad hoc panel, organized a conference in August 1953 to discuss plans for the new national facility.

Leo Goldberg attended the August meeting with thirty-five other astronomers at Lowell Observatory in Flagstaff, Arizona, under what he later called "false colors." His acceptance letter noted "that my thinking is much more ambitious than the proponents of the photoelectric telescope."[63] Goldberg believed that astronomy could and should benefit from the NSF's growing financial clout, especially after the original $15 million cap on the agency's budget was removed. In the closing session of the Flagstaff meeting, Goldberg told his colleagues that the NSF should think beyond just building a small telescope for specialized measurements. "What this country needs is a truly National Observatory to which every astronomer with ability and a first-class problem can come on leave from his university."[64] Significantly, Goldberg proposed that this new national, publicly funded telescope rival the size of the large private telescopes in use or under construction.

Goldberg made his recommendation for a national optical astronomy facility at a propitious time. In the mid-1950s, the NSF was eager to develop what it called "large-scale facilities." Besides astronomy, the agency was interested in major undertakings in nuclear science, biological field stations, and university computing centers.[65] Greenstein's advisory panel endorsed plans for what was initially called the Inter-University Astronomical Observatory in January 1954. Soon afterward, Seeger at the NSF assembled yet another committee, the Advisory Panel for a National Astronomical Observatory.

This new committee was again chaired by Robert McMath. McMath was completing his term as president of the American Astronomical Society and, more importantly, had helped negotiate with the Eisenhower Administration for a substantial increase in the NSF's funding. Aided by McMath's intervention and Cold War pressures, the NSF's funding soared during the 1950s, from $8 million in 1954 to $134 million five years later. The NSF allocated more than 12 percent of its $40 million budget to astronomy in 1956–1957 alone.[66] These significant and growing resources made it possible for the NSF to consider investing in large-scale science facilities.

McMath's panel met for the first time in November 1954 to continue negotiations for the national optical observatory. Goldberg helped McMath select committee members who represented the major U.S. observatories and he attended their meetings in Ann Arbor. One of the first tasks of McMath's group was to determine how the

NSF could legally operate a national optical observatory. The legislation that brought the agency into existence did not allow it to manage its own facilities directly. McMath, Goldberg, and the other members of the panel recommended the NSF base its plans on the model used by Associated Universities, Inc., a consortium of universities that operated Brookhaven National Laboratory for the Atomic Energy Commission. At the same time, astronomers (including Goldberg and Greenstein), science managers, and the NSF were hotly debating the merits of a similar approach for what would become the National Radio Astronomy Observatory in Green Bank, West Virginia.

McMath's committee also had to determine where a national observatory for optical astronomy would be located. In November 1954, they recommended that Aden Meinel lead the search for a suitable site. Meinel had also attended the 1953 Flagstaff conference. The Yerkes astronomer and optics expert recalled having whispered discussions with Goldberg while listening to his colleagues debate—too timidly, Meinel thought—the NSF's plans for a new national observatory.[67]

With funding from the NSF, Meinel began to explore sites in the Southwest using survey photos taken from rockets launched from New Mexico. In the early summer of 1955, he and Helmut Abt conducted a more thorough survey over New Mexico and Arizona using a small plane and jeep.[68] Abt thought Kitt Peak looked most promising and, in 1955 and 1956, he and Meinel returned to test the site more thoroughly.[69] While Meinel and Abt were lugging equipment to the tops of mountain peaks in the Arizona desert, Goldberg and McMath's advisory panel worked out the details of how the observatory would be managed.

In April 1956, Albert Whitford described the panel's five-year plan to the astronomy community. Once a site had been officially chosen, work would begin on a 36-inch telescope followed by an 80-inch telescope. The observatory itself would be managed by "a cooperative association of universities . . . representative of those in the eastern and midwestern sections of the country which would be the most frequent users of the telescope." In short, astronomers envisioned the national observatory largely as a resource for the community's "have-nots." Observers from any other institution, however, could apply for observing time to be awarded on the "basis of merit rather than institutional connection."[70]

In 1957, McMath's panel and representatives from the NSF set criteria for membership in the consortium. A school had to have at least three astronomers, a graduate program, and appropriate institutional support. In March 1957, representatives from seven universities—California, Chicago, Harvard, Indiana, Michigan, Ohio State, and Wisconsin—accepted Goldberg's invitation to meet in Ann Arbor. Caltech was invited to participate, as was Princeton. Both schools declined; Caltech cited fears that it might unfairly dominate the fledgling organization and Princeton decided to put its resources into balloon and space-based astronomy.[71] In October 1957, the Association of Universities for Research in Astronomy, Inc. was created, with board members selected from the seven participating universities. The next year, AURA's board approved building the new national observatory on Kitt Peak and selected Aden Meinel as its first director.

Years later, Goldberg gave a more colorful account of Kitt Peak's creation. He recalled attending a Washington dinner party in January 1962. One of the guests was Senator Warren Magnusson, chair of the committee with budgetary authority over the NSF. Magnusson noted that a few years earlier his committee was examining the NSF's budget request when Representative Albert Thomas, Magnusson's House counterpart, proposed eliminating the "Mount Kitt Observatory." Goldberg said Magnusson then nodded toward another dinner guest present that night, Senator Carl Hayden of Arizona. Magnusson described how he had reminded Thomas that he couldn't delete the budget line because "that's Carl's observatory." Thomas, on hearing this, said, "Well, why didn't you say so? I'll put it back in!"[72]

While Goldberg's story may be apocryphal, it illustrates the extensive negotiating and political wrangling that he perceived as necessary to establish national observing facilities. The move by the NSF to develop and manage large-scale science facilities was part of the larger postwar trend to invest federal money into a broad infrastructure for science. Edged out of the accelerator-building business by the Atomic Energy Commission, the NSF turned to other fields such as astronomy, geophysics, and biology to build national research centers and help establish itself as a major supporter of basic research.[73] Goldberg, in later years, noted that that the NSF wasn't explicitly interested in supporting astronomy during the 1950s. Instead, he believed, as probably some at the agency did, that building large facilities for astronomy would enable the agency to move into other areas of research.[74] The NSF's ambition in this arena would reach its apogee with Project Mohole, an infamous and expensive decade-long attempt to drill through the earth's crust that, in 1966, became the first basic science project terminated by Congress.[75]

Astronomers and the NSF established the national observatories for optical and radio astronomy simultaneously. Goldberg and Greenstein consulted and served on committees for both national centers, and Goldberg was later offered the directorship of NRAO but chose instead to remain at Michigan. The path that led to the first optical telescopes on Kitt Peak was peaceful compared to the stormy debates over how NRAO would be managed.[76] Unlike optical astronomy, which retained its tradition of private and state patronage into the postwar era, scientists established radio astronomy almost entirely *ex nihilo* from federal sources.

Ironically, scientists' differing reactions to federal support for radio and optical astronomy in the 1950s would reverse itself over time. After a contentious beginning, American radio astronomers put their faith almost entirely in a federally funded system of facilities while remaining relatively united about their priorities for new telescopes. In stark contrast, optical astronomers, burdened by their rich, prewar heritage of privately funded telescopes, became increasingly divided. They struggled to allocate federal funds among individual investigators, private observatories, and the national optical observatory. The continuing private and state support of astronomy caused many to question the role of the national centers. Such disagreements directly affected debates decades later when the American astronomy community was considering how to build the next generation of giant telescopes.

Deciding Astronomy's Priorities

In 1960, Leo Goldberg accepted an offer from Harvard University's Astronomy Department and returned to Cambridge. Six years later, he became director of the Harvard College Observatory. Both moves further increased his stature and influence in the science community. Goldberg and Greenstein remained in sporadic contact with each other, mostly on professional matters, but playfulness occasionally surfaces in their letters. In the late 1960s, when Greenstein was on Harvard's Board of Overseers and Goldberg was, as he described, just "lowly faculty," Goldberg asked his former classmate for extra tickets to Harvard's football games. Goldberg, acknowledging that Greenstein was one of the "least likely people to use the tickets," promised not to swap them in Harvard Yard for "a case of marijuana." The two astronomers also discussed major events in American politics such as John F. Kennedy's 1960 election and Senator Eugene McCarthy's 1968 presidential campaign, candidates Goldberg supported whereas Greenstein tended toward middle-of-the-road, anti-war, Republican candidates.[77]

During the 1960s, astronomers became increasingly divided over how federal funding should be allocated and which facilities and areas of research should be supported. Consequently, Goldberg and Greenstein came to have ever-differing views about the management of astronomy. Greenstein felt that the best science came from funding elite astronomers and observatories and worked to ensure that private institutions remained powerful. Goldberg, a scientist at a university without access to large telescopes, supported healthy, federally funded national observing centers that would be available to all qualified astronomers.

Their differing opinions were a microcosm of a larger schism that came to bedevil the American astronomy community. They also reflected long-standing differences among American scientists and politicians about the best way for the government to support research. In the 1940s, Vannevar Bush favored an elitist approach through which federal funds would support "autonomous, high-quality research, self-governing units" that represented "small science."[78] He was swayed by the belief that federal agencies were incapable of top-notch science. Countering Bush's elitist views, which were laid out most famously in his 1945 report *Science—The Endless Frontier*, were people like West Virginia's Senator Harley M. Kilgore. He wanted to support research for the general welfare, favored government labs, and opposed supporting research at only a few prominent institutions. Like Bush and Kilgore, in the 1960s, Greenstein and Goldberg became divided over issues such as support for private versus national laboratories and the amount of *laissez-faire* management desirable in science.

The Soviet launch of *Sputnik I* in 1957, the increased funding of the NSF, and the creation of NASA in 1958 brought more government money to astronomy. With it came a need to distribute resources equitably. By 1960, astronomers began to devote increasing attention to how this money should be spent and what their community's priorities were. Initially, panels such as Jesse Greenstein's Advisory Panel for Astronomy determined the NSF's allocations. As the NSF's program officers took a more prominent role, astronomers gradually lost this avenue of control. In time, many influential scientists began to publicly advocate for some degree of long-term planning for astronomy. Furthermore, rapid growth of the astronomy community, tripled in size during the 1960s, strengthened the need for centralized planning.[79]

Not all astronomers were enthusiastic about this strategy. Some equated planning with regimentation and believed that any strategic plan would squelch the ability of the individual researcher to determine a research agenda. Gerard P. Kuiper, a prominent planetary astronomer at Yerkes Observatory, argued for researchers' personal autonomy, saying, "The most important problems of astronomy are those which the best astronomers are now trying to solve."[80]

The astronomers' anxiety was understandable given developments in related disciplines such as physics. During the 1950s, it was increasingly common for teams of physicists and engineers to work for months on a single experiment at one of the AEC-run national labs. Papers authored by large teams became an accepted practice as younger physicists were seen as "organization men" who found satisfaction not in pure research but in the size of their equipment, their access to resources, and the glamour associated with their field (circumstances one historian has called the "suburbanization of postwar physics").[81] By the late 1950s, astronomers recognized the need to set their own priorities while retaining their autonomy.

As early as 1954, Goldberg discussed the possibility for a "survey of the requirements of astronomy" with Otto Struve. Goldberg supported the idea but felt any study had to be carried out by people actively engaged in research (in other words, not the NSF). Goldberg, who carbon-copied his letter to Greenstein, was bullish about what the United States could afford in terms of national facilities and equated American leadership in astronomy with "the expenditure of very large sums of money for equipment."[82] Struve soon published a widely read article urging that astronomers take control over their own destinies.[83]

In the wake of the Korean War and *Sputnik*, the federal government began to take an interest in reforming the management of the basic research it sponsored. Much of the initial attention fell on high-energy physics and space science; only later did its supervisory eye shift to astronomers.[84] In 1962, the National Academy of Sciences (NAS) established a Panel on Astronomical Facilities. As the NAS assembled its advisors, Goldberg suggested that the Academy look at whether federal funding had caused imbalances between radio, space, and optical observing. He also strongly encouraged Albert Whitford, who recently had become director of Lick Observatory (which, coincidentally, had just completed the world's second-largest telescope with a 120-inch mirror), to lead the effort.[85]

By late 1962, the NAS panel was established with Whitford as its chair. It produced the first of what would become known as astronomy's "decadal surveys." These reports describe the field's health, summarize important scientific advances of the past decade, and set research goals for the next decade. More importantly, through a process of debate and negotiation closed to the public and the general astronomy community, the decadal survey committees presented a prioritized and influential list of instruments and facilities that should receive federal funding.

The committee's conservative mandate was based on what Whitford later called a "provincial view of what astronomy was in the early 1960s." The NAS confined Whitford's study to ground-based astronomy and charged it with examining the need for major new facilities in the next five or ten years.[86] Whitford assembled a small group of seven astronomers to carry out the study. They represented all of the major regions of the United States and included people like Bruce Rule from Caltech, W. W. Morgan from Yerkes, and Allan Sandage of Carnegie. While neither Goldberg nor Greenstein were members of Whitford's committee, both offered it substantial input.

Greenstein's biggest concern was what he called the "enormous disproportion" between federal support of astronomy at private institutions such as Caltech and that going to the national centers. The "need for balance" became, in fact, something of a mantra for Greenstein in his correspondence and formal statements throughout the 1960s. In response to a request from Whitford, in March 1963 the Caltech astronomer sent a letter to the NAS. Greenstein expressed "little doubt that a substantial fraction of . . . astronomical research" came from large state and private observatories such as Palomar.

While traditional techniques might have lacked the "space-age cachet" of rocket-borne observing platforms, Greenstein argued that large, ground-based telescopes still had a crucial role in astronomy. If the NSF wanted to build more publicly accessible telescopes, Greenstein saw no reason these could not be managed by a single university (such as Caltech) as a "regional facility supported by government funds." Central to his message was the warning that, unless checked, the national centers would continue to grow at the expense of private facilities.[87]

That same month, Goldberg and Greenstein personally crossed swords over the NAS decadal survey. Greenstein told Goldberg that funding pressures had recently forced the observatories at Mount Wilson and Palomar to discontinue their guest-observer programs. New ground-based telescopes, Greenstein insisted, were of pressing importance while their cost would be "less than . . . a single orbiting astronomical observatory." Greenstein tactfully explained that he "meant no invidious comparison" to Goldberg's interest in space astronomy but simply wanted a balance between "large observatories devoted to ground-based astronomy and the national facilities."[88] Goldberg replied that great observatories such as Palomar could certainly not afford to stand still. But, he concluded, Kitt Peak was founded to meet the growing national need for observing facilities for astronomers without regular access to the private telescopes on the West Coast.

Goldberg also engaged members of Whitford's committee directly about the future of astronomy. For example, he had a pointed exchange with Allan Sandage about putting resources into space instead of ground-based astronomy. He questioned Sandage's assertion that the bulk of future astronomical research would be done from the ground and asked that Whitford's committee take "cognizance of the whole observational field of astronomy rather than of ground-based astronomy alone." Sandage sharply replied that East Coast astronomers were "biased toward selling the space effort as the major hope" for astronomy. "Ground-based astronomy cannot be allowed to stagnate for lack of equipment," Sandage wrote, while "space propaganda" went unchallenged.[89] Years later, an associate of Goldberg's recalled that the Harvard astronomer once visited Palomar, where Sandage gave him a tour. When Sandage asked what he thought of the world's biggest telescope, Goldberg supposedly quipped that he was "mentally weighing it" for launch into space.[90]

Other astronomers shared Sandage's view. In January 1963, Lawrence Aller, a former member of Goldberg's Michigan department, publicly complained how space and radio astronomy were lavishly supported while traditional ground-based optical telescopes languished. "In no other scientific field," Aller charged, were researchers expected to "bring home the bacon with capital equipment dating from 1908, 1888, or earlier." While space and radio techniques were impressive, the opening of new electromagnetic regions did not mean traditional tools should be allowed to "blacken with the dust of obsolescence." Many of astronomy's most pressing research problems, Aller argued, could be attacked only with "instruments of large, light-gathering power."[91] Goldberg did not challenge cries that ground-based astronomy was suffering as resources were directed toward space astronomy. Rather, he took a more expansive tone, suggesting that all of astronomy would be afflicted if either space or ground-based observing was promoted at the expense of the other.[92]

In April 1963, Geoffrey R. Burbidge, a scientist at the University of California in San Diego, gave a provocative presentation to Whitford's committee. In 1957, Burbidge, along with Margaret Burbidge, Willy Fowler, and Fred Hoyle, had published a seminal paper on stellar nucleosynthesis that brought the English physicist-turned-astronomer great renown. Not known for shyness, Burbidge criticized astronomers' inherent conservatism in failing to embrace more efficient ways of doing research. Team-based research was yielding dramatic payoffs for physicists, while astronomers working alone at telescopes were being left behind. Changes were needed in the infrastructure, organization, and practice of astronomy, including a nationally available telescope with a size rivaling the 200-inch and managed like the national accelerator labs. He urged that astronomers shed their cautious nature and adopt more of the traits of high-energy physics "where people are literally eating each other up in order to do something of significance."[93]

Greenstein received a copy of Burbidge's polemic, read it with "interest and some pleasure," but disagreed with most of it. Team research, he said, had some value, but it was usually not of the same quality as that done by individual astronomers. Greenstein argued that instead of putting more large telescopes, optical or radio, under control of the national observatory, "the only proper method is to see whether Lick or Mount Wilson-Palomar deserve another big telescope and give it to them." Instead of nationally funded facilities, Greenstein pushed for "the benevolent dictatorship of the elite. If you don't think they move fast enough, give the elite more backing."[94]

In August 1964, the Whitford panel presented its final report to the NAS. So far as the construction of new large optical telescopes was concerned, Whitford's major (and most expensive) recommendation was that three new large telescopes be built at a cost of about $20 million apiece. The NSF had already funded one of these, a telescope with a 4-meter mirror (recall that the Hale Telescope on Palomar had a 5-meter mirror) for the national observatory at Kitt Peak.

Who was to operate all these new telescopes? The report didn't identify specific institutions, instead calling for free and open competition. As Whitford later recalled, it was in the community's best interest to have some rivalry between individual institutions for new instruments.[95] Both Caltech and Lick, for example, had set their sights on a large telescope in the southern hemisphere.

Perhaps recognizing the delicate question of whether new telescopes should go to private institutions or national centers, Whitford's report was intentionally noncommittal. Each future telescope presented a unique situation. Federal or inter-university operation might work in some cases, while private management might be better suited in others. It concluded that the quality of the people who built and used the new telescopes was more important. "More than one arrangement," the report noted, "has been made to work well by the nuclear physicists" in running big accelerators.[96]

While, on its face, the report was neutral as to the balance between national and private telescopes, a more careful reading shows an implicit bias toward elite ownership of large telescopes. Many astronomers at private observatories believed that they were doing the best, cutting-edge research. Even Leo Goldberg conceded that Kitt Peak's early scientific staff was mediocre as it focused initially on building new telescopes.[97] Even though "the greatest good for the greatest numbers doctrine" was compelling, some astronomers at private observatories told Whitford that what astronomy needed most was "theoretically well-grounded, astronomically-oriented physicists, not third-rate observatory-trained . . . technicians."[98] So the Whitford report's statement that the "principle of equality of opportunity must be subordinated to the . . . requirement of excellence" could be interpreted as evidence that private observatories should reap the rewards of new telescopes.

As the 1960s marched on, Greenstein and Goldberg continued to present disparate views of how astronomy should be funded and done. In 1966, AURA's board invited Goldberg to join it after he completed his term as President of the American Astronomical Society. In this position, he advocated further improvement of the national observatory to enable its equipment and staffing to rival the private observatories. Goldberg also continued to push for a more central role for space-based observatories. The first attempts were heart-breaking failures; but by 1969, he and his students were getting good data from space. Goldberg also chaired NASA's Astronomy Missions Board, which helped determine space astronomy priorities from 1967 to 1970. Panel members discussed a myriad of plans for federally funded telescopes in orbit and on the moon but clashed with more traditional astronomers who favored greater NASA support of ground-based projects.[99]

Greenstein maintained his belief that cutting-edge astronomy was best done at a few private or university-run observatories. He continued to argue that the balance between national and university facilities was uneven and worked to see a new policy put into practice. Greenstein also served for a period of time on the Astronomy Missions Board but resigned in 1969 when he realized that NASA was not going to provide the support he thought necessary for "the future of modernized, ground-based optical and radio astronomy." Greenstein's resignation letter made it clear that he "did not believe that all astronomy will be done from above the atmosphere" and that NASA's rewards from space astronomy would be limited if "very large optical telescopes" were not built on the ground as well.[100]

He would not have to wait long before having an unparalleled opportunity to put his vision of astronomy's future into practice. A few months after stepping down from Goldberg's NASA panel, the National Academy of Sciences picked Greenstein to chair its next decadal survey of astronomy, with a mandate to chart the community's priorities for the 1970s. This set the stage for a contentious battle over how new telescopes should be funded and built and who would have access to them.

CHAPTER 2

Tradition and Balance

Blueprints and designs for innovative ways to make bigger and better instruments litter the history of science. One prophet of larger telescopes was George W. Ritchey. An American telescope maker and astronomical photographer, Ritchey labored for years polishing and testing the mirrors for Mount Wilson's telescopes. His dour temperament and perfectionism brought him into conflict with George Ellery Hale and Walter Adams, a prominent scientist at Mount Wilson. In 1919, after the 100-inch was completed, Ritchey was fired amid charges of disloyalty, and he moved to France to pursue his dream of building ever larger and more powerful telescopes.

Ritchey, like all astronomers, knew that the heart of any telescope is its primary mirror. Besides determining a telescope's performance, the mirror is the most difficult and expensive part of a telescope to make. Typically made of glass, the mirror blank is exquisitely polished for months and eventually coated with only a few micrometers of metal. In the world's biggest telescopes, thousands of pounds of glass costing millions of dollars serve to support a few ounces of reflective aluminum.

Ritchey explored ways to make large but lighter weight mirrors as part of a more ambitious goal to overturn traditional ideas about telescope design. In time, Ritchey predicted, astronomers would see "how inefficient, how primitive it was to work with thick solid mirrors, obsolete mirror-curves, equatorial telescope-mountings . . . requiring enormous domes and buildings, and similar anomalies in a progressive age."[1]

One avenue Ritchey explored was what he called "cellular mirrors." Instead of a monolithic mirror blank of solid glass, Ritchey handcrafted and joined many thin glass ribs to separate top and bottom glass face sheets as part of a lightweight, rigid, honeycomb-like structure. He predicted that mirrors built this way would lead to the "great telescopes of the future," and he believed nations should unite to build several "super-telescopes" at different latitudes at sites believed to offer superior observing conditions (such as the rim of the Grand Canyon). He envisioned these "mighty guns of Peace," boasting 8-meter mirrors, as the basis for an international observatory system operated under "democratic management" for astronomers by "specially-trained technicians."[2]

Unfortunately Ritchey, an outsider in the astronomy community, could not mobilize the support needed for his ambitious and idealistic concepts. He also had the misfortune of suggesting that telescope designers adopt his ideas when current approaches toward building and designing large telescopes were still quite useful.

Like Ritchey, Aden Meinel also had innovative visions for larger telescopes. Meinel was born in 1922 in Pasadena, California. In 1940, while attending Pasadena Junior College, he met Marjorie Pettit and switched his major to match hers in astronomy. Marjorie's father, Edison Pettit, was an astronomer at Mount Wilson. While they dated, Meinel assisted in the Caltech physics department and the Mount Wilson Optical Shop. Here he learned firsthand about optical design from John Anderson, who was leading the effort to prepare the 200-inch's mirror.[3]

After America entered the war, Meinel and Pettit participated in Caltech's rocket weaponry program. Working together became a pattern that they retained for the rest of their careers. In 1944, after marrying Marjorie, Meinel was assigned to Patton's Third Army and helped inspect German V-2 facilities. After finishing his Ph.D. at Berkeley in 1949, Meinel and his family took up residence at Yerkes, where he designed new instruments and researched auroral phenomena for the Air Force.

At Yerkes, Meinel began to explore alternative designs for larger telescopes. He had just helped design what would become the Arecibo radio telescope, a 1000-foot wide instrument nestled in a bowl-shaped valley in Puerto Rico. He contemplated how to build its counterpart for optical astronomy. Imagining a telescope with a mirror as large as

500 inches across, Meinel made a model, shown in Figure 6, to help convey his idea. He based his design on a massive light-collecting surface composed of several hundred smaller mirrors. These would direct light to a secondary mirror assembly and instruments located above the bowl.[4] His goal was not so much a telescope with the best optical precision, but rather a giant "light bucket" to collect photons.

In the spring of 1953, Meinel visited observatories in California to pitch his idea. Telescope engineers such as Bruce Rule at Palomar and William Baustian at Lick supported Meinel's concept for a 500-inch telescope and thought it technically feasible, but he found astronomers unreceptive and even hostile. In Pasadena, Ira Bowen told Meinel there was no pressing need at that time for telescopes larger than the 200-inch. The director of Palomar questioned Meinel's cost estimates, which varied from $10 million to as high as $50 million, as both too high and too vague. Meinel maintained the "psychological

Figure 6. Home photo of Meinel's 1953 design for the "X-inch" telescope. Picture courtesy of Aden and Marjorie Meinel.

factors" associated with building such a grand facility would help generate support. Bowen countered that the astronomy community would be better off to wait a decade and develop a stronger case for the scientific problems a giant "X-inch" telescope might tackle. Disappointed, Meinel left California thinking that the astronomers' belief that "bigger telescopes are scientifically undesirable is as dangerous as rushing blindly into a larger telescope project."[5]

After returning to Yerkes, Meinel began to address more seriously the concerns that the California astronomers had raised. He assembled a notebook of sketches and calculations of the "X-inch" telescope and discussed them with Gerard Kuiper, an influential Dutch astronomer at Yerkes. Kuiper, in turn, prepared a confidential memorandum that supported the technical feasibility of Meinel's design. Kuiper noted that a cautious approach to building bigger telescopes might be the "most economical way of doing science," but it lacked boldness and missed the opportunities that came with "pushing the technical facilities to the limit." Nuclear physicists certainly had not taken the same cautious path, Kuiper said, as "no effort is spared to do the best this generation [of physicists] is capable of doing." Building the "X-inch" telescope would probably mean that "principles of institutional and international cooperation [would] be incorporated into the project from the outset," a fairly novel idea for the time. It was reasonable, Kuiper concluded, that astronomers not forego the "advantages of a really powerful telescope, provided that a very good case can be made for its construction."[6]

In late 1955, further plans for the "X-inch" telescope were put on hold when Robert McMath's Advisory Panel asked Meinel to lead the site survey for what would become Kitt Peak National Observatory. In many ways, the young optics expert was the ideal person to carry out this task. The energetic Meinel threw himself headlong into ambitious projects. He was also skilled at finding novel solutions to technical problems, such as devising a simple yet elegant technique for automatically measuring the quality of possible telescope sites as he and Abt traveled throughout the Southwest.[7]

Meinel's relationship with McMath and his panel was not always harmonious during the site survey for Kitt Peak, partly due to Meinel's occasionally unfocused enthusiasm for the new project. Acting without authorization or funding, Meinel initiated contacts with powerful persons in Arizona, including the president of the University of Arizona and Senator Barry Goldwater. The more senior McMath and Leo Goldberg listened with growing impatience as Meinel told them how the national observatory should be bolder and more forward thinking in its plans.[8] Incidents such as these angered McMath, who believed Meinel was acting rashly and unilaterally. Relations between the two men were not improved when McMath learned that Meinel's father-in-law was Edison Pettit. The two men had had a falling-out years earlier when McMath had removed Pettit's name from a paper they had co-authored.[9]

Despite these difficulties, in December 1957, AURA's board chose Meinel to be Kitt Peak's first director. At that time, Kitt Peak was an observatory without any major telescopes. Construction soon began on a 36-inch telescope and the NSF approved funding for another with an 80-inch mirror. Along with his new directorial duties, Meinel advocated an innovative design for the 80-inch telescope. Building on techniques that opticians had developed during the war to polish mirrors with steeper curves, Meinel designed a lightweight telescope with a very fast focal ratio. Fond of clever acronyms, he named his design MI-AMI, short for Minimum Inertia and Mass Instrument.[10] When Meinel presented a model and watercolor sketch for this telescope to McMath and the other AURA officers, they praised its innovative features but decided the newly formed (and politically vulnerable) national observatory could not risk building a telescope with an unproven design.

Meinel still managed to incorporate some novel design features into the Kitt Peak 80-inch telescope. More traditional telescopes, like the Hale Telescope, use what is called a Cassegrain optical design. A Cassegrain telescope has a primary mirror polished so its profile is a parabola. A convex secondary mirror, as shown in Figure 3, is at the telescope's prime focus above the primary mirror. This reflects photons back down through a small hole in the primary mirror where they are detected. Instead of a traditional Cassegrain telescope, Meinel based the Kitt Peak 80-inch on a modified optical design first developed by George Ritchey and Henri Chrétien, a young French astronomer. Using mirrors with hyperbolic rather than parabolic curves, the Ritchey-Chrétien design gave a wider field of view free from the optical errors present in the superficially similar Cassegrain design. Meinel's efforts made the 80-inch telescope more suitable for astro-photography and inaugurated a period in which telescope engineers preferred the Ritchey-Chrétien design.

In September 1964, when astronomers first began using the new telescope, built with a slightly larger 84-inch mirror, Aden Meinel was no longer Kitt Peak's director. In 1960, the AURA Board learned that Meinel had exceeded his budget by a considerable amount. This was not a complete surprise because the NSF and AURA had complained about his management approach and fiscal control earlier. According to Abt, "Meinel was an idea person. This was good for science, but difficult for planning budgets."[11] AURA could not afford to give potential opponents of the national observatory any reasons to criticize the fledgling institution, and it asked for his resignation two weeks after the national observatory was dedicated in March 1960. Later, Meinel admitted that he was "a little too aggressive," noting that when Kitt Peak began operating as an observatory with national visibility "it called for a different type of director."[12]

AURA soon picked Nicholas U. Mayall, a well-respected stellar astronomer from Lick Observatory with no administrative experience, to be Kitt Peak's new director. Meinel was named an Associate Director in Kitt Peak's Stellar and Space Divisions, where he designed a precursor to the Hubble Space Telescope. Chafing after his rebuke from AURA, however, he resigned in August 1961. A few months later, the University of Arizona, located across the street from Kitt Peak's offices, asked Meinel to chair its astronomy department.

Arizona's university administrators saw that Kitt Peak's presence might benefit their school and they decided to give the astronomy department greater support. They approved plans to expand its presence on campus, hired new staff, and discussed building a new telescope much larger than their old 36-inch reflector. This was the perfect environment for the enterprising Meinel. As Arizona's astronomy department began expanding rapidly, he again postponed plans for the giant "X-inch" telescope.

In 1964, when the first decadal survey was presented to the National Academy of Sciences, it noted that a giant "X-inch" telescope would probably cost at least $100 million and take 15 years or more to build. The report concluded that the astronomy community should instead give priority to "proven designs" between 150 and 200 inches in size. Delaying the decision to build anything bigger, the panel emphasized, would not present astronomers with any "suddenly breached threshold analogous to . . . a particle accelerator since the cutoff of any large telescope is not a sudden one."[13]

Albert Whitford's panel did recommend that, once three new 150- to 200-inch telescopes were underway, AURA assemble a group of astronomers and engineers to do design studies for "a telescope of the largest feasible size." Whitford's report suggested that a number of important technical questions needed answers before astronomers could consider an "X-inch" telescope. For example, how could glass companies make the primary mirror? Once built, how would the mirror be transported? Could large optical telescopes be based on the design of radio telescopes?[14] These were all crucial questions, and the panel suggested that federal agencies spend at least $1 million before 1968 to answer them. In spite of this recommendation, little real progress toward making bigger telescopes happened in this period. Instead, astronomers and engineers continued to plan new telescopes, with some modifications, on the venerable design of the 200-inch.

Consolidation and Conservatism

After the 200-inch was dedicated in 1948, astronomers built no larger telescopes for more than 25 years. For modern science, this was unusual. High-energy physicists perpetually advocated bigger and more powerful accelerators, for example, while electron microscopes continued to improve in versatility and resolving power. Why were the 1960s largely a time of consolidation when astronomers and telescope engineers opted for only slight incremental modifications to the 200-inch's design?

The successful completion of a large telescope project requires a happy confluence of community interest, technological capability, institutional support, and financial resources. The right combination of these factors did not exist in the 1960s. Most of the older and influential members of the astronomy community were, according to Meinel, conditioned by memories of the Great Depression and cautious in accepting bolder designs.[15] As one astronomer-turned-telescope builder explained, "There is a mystique to building telescopes. You had to have a lot of balls to come along and say, 'I can make one better than the 200-inch.'"[16]

Funding realities also tempered the ambitions of astronomers: a telescope twice as large as the 200-inch could cost $100 million or more. Physicists could secure that level of funding—in 1961, Congress approved $114 million for Stanford's new linear accelerator—by appealing to fears about national security and the desire for international prestige. The NSF and astronomers simply did not have comparable resources. The Whitford panel's report recommended spending only $224 million for a decade's worth of research and facilities. Astronomers found the possibility of allocating half of this funding to a bigger telescope, especially one with a novel design, difficult to imagine.

The rapidly improving performance and expanding possibilities of electronic detectors and other auxiliary tools shaped astronomers' considerations as well. By the mid-1960s, observatories were reporting gains of a factor of ten over traditional photographic methods. This was equivalent to increasing the size of the primary mirror dramatically and at a fraction of the cost of a bigger telescope.[17] Astronomers recognized that further gains in performance were possible and that building better detectors rather than bigger mirrors was more cost effective. Even astronomers using well-designed but smaller telescopes, such as Kitt Peak's 84-inch, could compete with the big mirrors at Palomar and Lick by using better electronic tools.

Change was also difficult because any new telescopes that astronomers designed would probably not be completed for at least fifteen years. Once built, an instrument would have to serve the science community for several decades. Consequently, astronomers believed they should wait until they had a clearer sense of the improvements made possible by changes in instrument design and electronic detectors. To do otherwise might, as one observatory director warned, burden astronomers with telescopes whose designs were "out of date before they are completed."[18]

Finally, astronomers during the 1950s and 1960s could point to irrefutable evidence that the data that had been collected with existing large telescopes had resulted in many seminal papers. In 1956, Milton Humason, Nicholas Mayall, and Allan Sandage published the results of a multiyear study.[19] The authors pooled data collected at Lick and Palomar to present the first in a series of publications that explored the large-scale nature of the universe. Three years later, Jesse Greenstein co-authored an influential paper on stellar abundances.[20] This research was enabled by generous Air Force funding and relied heavily on data collected by the 200-inch's coudé spectrograph. To gather that data, Greenstein and his colleagues sometimes sat in the dark and exposed a photographic plate for twelve hours on a single object.

Perhaps the most shocking result came in 1963 when Greenstein and Maarten Schmidt, another Caltech astronomer, reported their observations of two quasi-stellar radio sources (now called quasars), innocuously named 3C48 and 3C273.[21] While radio and optical astronomers had previously observed the two objects, Greenstein and Schmidt were the first to ascertain their cosmological significance. After reinterpreting the overexposed spectra Schmidt took at the 200-inch's prime focus, the two astronomers realized that the position of the spectral lines implied that 3C48 was "redshifted" by 37 percent. This placed it more than five billion light years from earth, making it one of the most distant objects ever observed. The discovery put Maarten Schmidt on the cover of *Time* magazine and revealed a universe far stranger and more violent than astronomers had suspected. It was clear that the limits of existing telescopes' light-gathering ability did not yet pose a barrier to new scientific discoveries.

When he retired as director of Palomar, the American Astronomical Society honored Ira Bowen by asking him to give its Henry Norris Russell Lecture, the society's award for lifetime achievement. Bowen's 1964 address—simply called "Telescopes"—exemplified the cautious approach astronomers took toward larger telescopes and new designs.[22] Bowen's talk discussed the optimal design for large telescopes and reflected an increased interest among astronomers in quantifying the performance of their telescopes and instruments.[23] The design criteria Bowen advocated became the basis of a whole generation of telescopes planned during the 1960s and early 1970s.[24]

Speaking to hundreds of astronomers gathered in Flagstaff, Bowen noted that there essentially was no limit to how faint an object could be recorded by a telescope because this could be extended indefinitely by observing for longer times. But the interest in space exploration, the growing number of astronomers and physicists, and the overall "postwar upsurgence of all science," Bowen said, had seriously strained observatories' ability to provide telescope time. Bowen's lecture challenged "the concept that the best answer to all of our problems is the construction of bigger and bigger instruments" and suggested that astronomers instead pursue other alternatives for the immediate future.[25] In the face of uncertainty about the future performance of electronic detectors and the time required to improve them, the real question, according to Bowen, was whether the community should build "one huge telescope . . . or a number of identical smaller telescopes" with the equivalent light-collecting power.

Astronomers adopted this more conservative path in their plans for new telescopes. For example, in the mid-1960s, the Carnegie Institution of Washington considered building what was basically a copy of the Hale Telescope, then almost two decades old in the southern hemisphere. The proposal for the telescope rhetorically asked why "a 200-inch telescope . . . rather than a large instrument of about 400-inch aperture?" The answer was that, while most astronomers believed that a larger facility should be built, "the time for it has not yet come."[26] In the mid-1960s, about half a dozen groups were seriously pursuing plans to build telescopes with mirrors ranging from 3 to 4 meters. All of the designs were heavily based on the Palomar paradigm—traditional dome enclosures, equatorial mountings, relatively conservative focal ratios, and mirrors made from single massive pieces of glass.

One notable exception was a design Soviet astronomers and engineers adopted for a 6-meter telescope, a project begun in 1960 and finished in 1976. Instead of a conventional equatorial mounting, Soviet engineers designed their telescope with an altitude-azimuth (alt-az) mount. As a result, the telescope's engineers did not have to design around the tilted polar axis of the traditional equatorially mounted telescope. Instead, one axis was horizontal and the other was vertical, each driven simultaneously like the guns in an anti-aircraft mount. Improved computer systems controlled the telescope's movement with the necessary accuracy. The Soviet 6-meter project was the first to break away from a standard telescope mounting. Large mirror and innovative mount notwithstanding, in all other aspects, the overall telescope design was still relatively conservative.[27]

In 1965, less than a year after Bowen's talk, engineers and scientists met to discuss telescope design at a symposium sponsored by the International Astronomical Union.[28] Engineers and astronomers heard talks at the week-long meeting and visited Kitt Peak, Lick, and Palomar accompanied by representatives from firms such as Westinghouse

Electric and Corning Glass. The number of engineers and astronomers actively involved in designing and building telescopes was fairly small; only about 70 names appear on this meeting's roster. Bruce H. Rule was probably the most prominent telescope engineer present. After graduating from Caltech in 1932, Rule became the chief engineer of the 200-inch project. In the early 1960s, Rule served on Whitford's astronomy survey committee, an unusual appointment for a nonscientist. Throughout the 1960s and 1970s, Rule remained at Caltech and worked on several major telescope projects while helping train an entire generation of telescope designers.[29]

Most members of the telescope design community migrated from finished projects to new ones underway, making the community somewhat insular. Members of the telescope-building community were not explicitly trained in telescope design as no university offered degrees in this esoteric subject. Instead, observatories hired mechanical or electrical engineers who worked with older, more experienced staff to learn the craft skills and quirks involved in building telescopes. For example, after working with Bruce Rule, William Baustian was hired by Lick Observatory in 1946 as the designer and chief engineer for their 120-inch telescope project. Baustian worked on the telescope until 1957 when Aden Meinel lured him to Kitt Peak. There, Baustian trained Larry K. Randall, a recently graduated mechanical engineer from the University of Arizona. Randall went on to master the technical details of telescope engineering on Kitt Peak's 4-meter telescope project and would later apply his skills to the Gemini telescopes.

The 1965 meeting offered scientists and engineers their first formal chance to evaluate the state of telescope building since the completion of the large telescopes at Palomar and Lick. Bowen's opening address explained that the conference's goals were to evaluate conventional wisdom about large telescope design and start dialogues among astronomers, engineers, and industry about where to go next. Reiterating his belief that it was too costly to build larger telescopes, Bowen emphasized improved efficiency and performance through better instrumentation. Almost all of the meeting's talks described how projects already underway were dealing with design and construction challenges, making the meeting more of a consolidation of existing knowledge rather than an exploration of new ideas. When it came to thinking about the telescope itself, most of the meeting's participants took a reductionist view. There were separate sessions discussing mirrors, domes, drive systems, and so forth, and only hints of the integrated systems engineering and total project management approach that astronomers and engineers would adopt for building giant telescopes in the 1980s and 1990s.

Breaking the Rules

By 1963, Aden Meinel was in a propitious position as both the chair of the University of Arizona's astronomy department and the director of its Steward Observatory. A few years earlier the university had lured Gerard Kuiper to Tucson. Kuiper and his associates formed the nucleus of what became Arizona's Lunar and Planetary Laboratory. Meinel and Kuiper coordinated their efforts and the lab soon had over $1 million in funding and a new building, courtesy of NASA. Kuiper's lab parlayed these resources into several modest-sized, new telescopes in mountains north of Tucson in preparation for NASA's upcoming lunar missions.

Meinel's astronomy department expanded under his direction. In 1965, the NSF gave Steward Observatory $1.4 million for a new 2.3-meter telescope to be built next to the national telescopes on Kitt Peak. In February 1964, Meinel met with a representative from the Advanced Research Projects Agency, the Defense Department's research division. Noting the rapid growth of Kitt Peak National Observatory, ARPA asked why the university wasn't more involved in the field of optics and hinted at possible funding for such a program. By April 1964, Meinel had the first installment of funding for an exploratory three-year program in optical sciences.[30]

Meinel decided to devote his energy full-time to building what became the university's Optical Sciences Center. In 1967, this center received additional money from the Air Force as part of an effort to fund nonclassified research at universities. Additional state and corporate sponsorship contributed to its growth, and it soon had over two dozen faculty associated with it.[31]

By 1970, the University of Arizona possessed a vigorous research program spanning several related fields: astronomy, planetary science, and optical science and engineering. Federal money flowed freely, which enabled the construction of new campus facilities. The location of Kitt Peak's headquarters in Tucson and the construction of the national 4-meter telescope nearby was another strong magnet. The synergy between these different institutions and areas of research attracted several bright and ambitious young scientists.

One of the people drawn to Tucson in the 1960s was Frank J. Low. As a graduate student at what was then called Rice Institute in Houston, Low did research on low-temperature physics and became proficient at building complex experimental apparatus.[32] After graduating, Low went to work for Texas Instruments where he researched the behavior of semi-conducting materials at low temperatures. Low began to investigate whether germanium kept at cryogenic temperatures could be used as a bolometer, an exquisitely sensitive energy-measuring device. When light struck Low's bolometer, the electrical conductivity of the germanium changed in a predictable fashion that he could measure and correlate to the incident radiation's energy. He also applied his instrument-building skills to the construction of dewars suited to the rigors of nightly use at a telescope. These were insulated flasks, much like thermos bottles, which contained the liquefied helium that kept Low's bolometers sufficiently cold. In November 1961, Low described his invention in a brief paper and remarked, almost as an aside, that it was now possible to build detectors comparable in sensitivity to photomultiplier tubes and radio receivers that could work in all regions of the infrared.[33]

This was exciting news for astronomers looking for better ways to measure the infrared energy emitted by stars and galaxies. For years, astronomers had used lead-sulphide cells as infrared detectors. Often obtained as military surplus, these worked only over a limited range of wavelengths. Some astronomers began using Low's germanium bolometers for infrared measurements at increasingly longer wavelengths. The results were exciting. They found, for example, that interstellar dust played an important role in the extinction and reddening of stellar light. Astronomers were also electrified by news that Caltech astronomers had observed many stars that emitted far more energy in the infrared than they did in the visible. That sources of energy could exist that were so bright in the infrared, yet were invisible or almost invisible to the traditional optical astronomer, suggested many new discoveries to come.

Because of findings such as these, infrared astronomy became increasingly important. Gradually, the sharp distinctions between the optical and infrared regions disappeared, and more optical astronomers began to pay attention to the neighboring infrared wavelengths. By the late 1960s, infrared astronomy was a burgeoning field driven by the appearance of specially designed telescopes, new detectors, military interest, and growing competition between researchers at places such as the University of Arizona, Cornell, and Caltech.

Low joined the migration of physicists into astronomy. In 1965, after spending a few years at the national radio observatory in West Virginia, he took a faculty position at the University of Arizona where he began to make infrared observations with moderate-sized telescopes. For infrared astronomers, conventional telescopes like the 200-inch emitted excessive background thermal radiation. As a result, the telescope itself produced enough thermal noise to swamp the signal from the object of interest. The night sky itself was also a large source of undesirable background noise. To overcome these extraneous sources, astronomers such as Low designed infrared telescopes that produced less background noise. They also developed new ways of using telescopes. Rapidly wobbling a small secondary mirror back and forth, for example, offered astronomers a way to switch between observing a star and a patch of blank sky. Astronomers then electronically subtracted the sky's signal to get the data they wanted.

In 1967, after noticing that smaller telescopes frequently outperformed larger conventionally designed telescopes for infrared work, Low began to consider the possibility of using several small telescopes, each designed for infrared observing, and combining their light at a common focus. When he presented this idea to Aden Meinel, he found a receptive listener.

During the 1960s, Meinel had also considered this idea when he advised the military and the CIA in the design of their classified satellite reconnaissance systems. He recalled a meeting at the Pentagon where people were "sitting around, blue-skying about what we could do with mirrors," and he suggested it might be possible to make a big telescope by linking several smaller ones.[34] In 1967, Meinel attended a summer seminar at Woods Hole on synthetic aperture optics convened at the request of the Air Force. Representatives from industry, academia, and the military presented both classified and unclassified material at the month-long workshop.[35] Afterwards, Meinel built a small model of what he called Project COLT because its arrangement of six mirrors looked like the pistol's cylinder.

For years, radio astronomers had employed aperture synthesis, whereby several smaller antennae simulated a single large radio dish, to get better resolution. The optics community working on classified projects was interested in extending this technique to optical wavelengths—a considerable challenge because the wavelengths of visible light are shorter by a factor of a million than those of radio waves. Meinel's idea attracted the attention of the Air Force. He soon obtained several surplus mirrors from a secret spy satellite program, each 1.8 meters in diameter and with identical focal lengths. They featured an internal cross-ribbed structure that separated and supported fused silica face and back sheets (much like Ritchey's earlier plans for cellular mirrors). The final products weighed only one-third as much as solid mirrors, a desirable feature because their original purpose was to be launched into orbit.

Low discussed the possibility of combining several smaller telescopes into a single facility with prominent astronomers and funding representatives. His recent research showing that a number of galaxies emitted far more energy in the infrared than expected helped secure their attention. Meanwhile, Meinel investigated how to design a telescope that could combine light collected by several separate mirrors. He was not interested simply in building what astronomers pejoratively called a "light bucket," an instrument that just collected lots of photons. Instead, he envisioned a telescope that would superimpose the images from six mirrors to replicate the light-collecting ability and resolution of a much larger single mirror. To do this, light from each mirror would have to be carefully combined at a common focus so all six images could be accurately stacked on top of one another. In the spring of 1970, Meinel wrote a paper describing how this might be done.[36]

At the same time, scientists and engineers at the Smithsonian Astrophysical Observatory (SAO) in Cambridge were also preoccupied with finding an affordable way to build a bigger and better telescope. Fred L. Whipple, SAO's director, was not pleased that his institution lacked a "real observatory."[37] As a remedy, SAO built a few small telescopes on a ridge near Mount Hopkins, an 8,500-foot peak south of Tucson within sight of Kitt Peak. Whipple and the SAO staff reserved Mount

Hopkins' summit for a future large telescope, although they had no design in mind or funding in hand. SAO first considered building a telescope with a segmented primary mirror similar to the one Meinel proposed for the "X-inch" project in 1953, a concept that SAO staff explored with few tangible results.[38]

Through conversations with Gerard Kuiper, Whipple learned of the designs Meinel and Low had proposed. In February 1970, as Meinel, Kuiper, and Low were wondering how to get funding, Meinel received a telephone call from Whipple. "I said, 'Fred, I've got the mirrors here,'" Meinel recalled, "and he said, 'Well, I think I can get the money.' It just flowed together beautifully."[39] Meinel and Low discussed the idea in more detail with Whipple and other SAO staff over the summer and, in November 1970, SAO presented Congress with a proposal wisely titled "A Large Astronomical Telescope at Low Cost."[40]

In the 1960s, astronomers estimated that the cost of a new telescope was proportional to at least the size of its primary mirror squared or even cubed. This meant that doubling the primary mirror's size would increase costs up to eightfold. In the early 1970s, astronomers estimated that a large telescope built using the 200-inch's design, the standard metric of the time, would cost about $25 million.[41] Meinel and Whipple were proposing to construct one of comparable light-collecting power for a fraction of the price.

Whipple and top-level Smithsonian administrators gave presentations on Capitol Hill concerning the new telescope. Their statements to Congress argued the multiple-mirror approach would "pave the way for the scientific community to build even larger, more powerful telescopes" more cheaply than conventional telescopes. Failure to provide support would sacrifice the opportunity to "convert an experiment in technology into a powerful operational scientific instrument."[42] As Congress considered SAO's bold request, Ray J. Weymann, the new director of the Steward Observatory at the University of Arizona, successfully lobbied Arizona politicians to endorse the project.[43] By July 1971, Congress approved the first installment of SAO's request for $1.5 million, half of the project's estimated cost. Five months later, SAO signed an agreement with the University of Arizona to build the Multiple Mirror Telescope (MMT).

As scientists and engineers explored optical and mechanical designs, they were guided by two goals. According to Nathaniel Carleton,

SAO's project scientist for the MMT, one was to produce a high-quality telescope at a reasonable cost. Carleton and his colleagues also wanted to "shake up the community by showing them that there were other ways to do things."[44] Their final design became both a working research tool and a test bed for new telescope technologies.[45]

The mirror blanks that Meinel received as Air Force surplus were the key components of the MMT. Six 1.8 meter mirrors, arranged together in a hexagonal pattern on a common mount, had the light-collecting power of a 4.5-meter telescope. Each primary mirror had its own secondary mirror, creating, in effect, six separate telescopes. Some astronomers joked the project should really be named the Multiple-Telescope Telescope. Photons, after striking the primary and secondary mirrors, were reflected to a beam-combiner that brought light from all the mirrors to a common point.

Building on Meinel's extensive experience with advanced optical systems, project engineers designed a complex system of lasers, electro-mechanical servos, and optical sensors coupled with closed-loop computer control to align the MMT's images. The optics system that combined the images from the six primary mirrors depended on automatic computer sensing and control. Three different electronic systems, linked together, controlled the movement of the telescope, the co-alignment of the mirrors, and its scientific instruments. Computers were essential to the most basic functions of the MMT, and engineers went so far as to describe the telescope itself as a "large and unique peripheral device" for a computer system.[46]

From the outset, Carleton and other project staff paid close attention to the integration of control and data collection systems of the telescope. Unlike some conventional telescopes, the MMT had no observer cage where the astronomer could ride and guide the telescope while collecting data. The MMT's designers envisioned that astronomers would operate the telescope remotely from a control room and move the telescope using a television system while monitoring the output of data on a computer display.[47]

The telescope's location atop the summit of Mount Hopkins forced the telescope's designers to "think small." A narrow rising isthmus of land connects the small and exposed summit with the lower ridge and the sides of the mountain fall precipitously away from the telescope site. Partly as a result of space limitations, Arizona and SAO staff opted for a more compact alt-az mount for the telescope's six mirrors instead of the traditional equatorial arrangement. While this design required a more sophisticated system to track objects in the night sky, computer systems were available by that time to make it feasible. According to the project's chief engineer, Thomas Hoffman, big telescopes were like "large precise machine tools" that could be accurately controlled by a variety of techniques. As Hoffman and other project members conceptualized the MMT, they were inspired by large radio telescopes whose designs successfully incorporated the advantages of the alt-az mount.[48]

Perhaps the most striking difference in the MMT's appearance was its enclosure. Rather than adopting the classic lines of a telescope dome, Carleton and Hoffman advocated an unglamorous but inexpensive rectangular building that looked more like a barn. The novelty continued. When faced with the difficulty of enclosing a moving telescope on a small plot of land while still providing space for a control room and laboratories, project engineers solved the problem by rotating the telescope simultaneously with the closely fitting building. Therefore, while the telescope tracked objects at night, the entire 450-ton building moved with it. Years later, this feature caused an unusual insurance claim when the moving building struck a parked car.

Any one of the MMT's features—multiple, lightweight mirrors, the alt-az mount, the box-like building, the extensive incorporation of systems engineering and environmental control—broke from the traditional Palomar paradigm of telescope building. Combined, they made the MMT, shown in Figure 7, a singularly bold departure in telescope design. This break in tradition brought criticism from some in the astronomy community. Some astronomers were nervous about taking so many new design directions at once. Many questioned the wisdom of enclosing the telescope in what looked like an oversized shed and wondered whether light from several mirrors could be accurately combined.

Funding the MMT's novel design was more difficult than solving its technical challenges. By 1975, the project's cost had risen to almost $7 million, far above the $3 million that Whipple and Meinel first thought adequate. Meinel's hopes for military support never materialized and the NSF was reluctant to provide funding. Astronomers also questioned whether the MMT's unconventional design would affect its

Figure 7. The Multiple Mirror Telescope on Mount Hopkins in southern Arizona in its original configuration. Note its six 1.8-meter mirrors and barn-like enclosure on a rotating platform. Image by H. Lester, MMT Observatory. Used with permission.

research potential.[49] Originally, scientists like Frank Low saw the MMT as a tool to collect infrared radiation for photometry or spectroscopy of individual faint objects such as quasars.[50] Astronomers didn't design the MMT with direct imaging or photography in mind. In fact, its novel optical system gave it a small field of view compared to general-purpose telescopes such as the 200-inch or the Kitt Peak 4-meter. As its design and construction progressed, however, project staff expanded their vision of the telescope beyond its primary goal of doing infrared astronomy.[51]

Although some in the astronomy community continued to harbor doubts about the multiple-mirror concept, the MMT integrated several innovative features into a facility that was part research instrument and part technological experiment. Astronomers began to imagine that the MMT concept might lead the way to even larger telescopes.[52] For astronomers contemplating building large telescopes, the MMT showed that the traditional cost-size relationship for telescopes could be broken. In 1979, when the MMT was dedicated, SAO and Arizona astronomers triumphantly claimed they had built it for less than half the cost of a conventional telescope and compared its charmed life to a small child who wanders into a dangerous neighborhood and "later walks out of it, miraculously unharmed."[53]

Technology Supplants Artistry

While scientists at the SAO and University of Arizona were engrossed in building the MMT, other members of the astronomy community grappled with uncertainty about their discipline's future. Their unease was partly associated with the rapidly changing technology of astronomy and the resulting changes in practice. Not all of the technological evolution took place at large observatories such as Palomar, Lick, or Kitt Peak. An electronic detector 25 times more efficient than a photographic plate could make an ordinary 40-inch telescope equivalent, in some ways, to a 200-inch telescope, encouraging astronomers at almost all institutions to improve their auxiliary equipment. As engineers and astronomers introduced new electronic tools into the observatory, astronomers had increasingly vocal discussions about how these innovations might alter the nighttime traditions of astronomy.

Planning a long-term policy for astronomy also raised critical issues about the future of the telescope as a research tool. As detectors continued to improve, scientists wondered whether there was any reason to advocate more large telescopes. Astronomers and engineers pondering these questions were caught in the tension between past technological experiences, current and future technological expectations, future research needs, and political and funding realities. In 1971, Lick astronomer George Herbig remarked that, when considering the next generation of telescopes, "one should try to profit from past experiences." But, Herbig continued, "let us not hold too conservative a philosophy and be accused . . . of always planning to wage the next war with the weapons that won the last."[54]

The design of the MMT was at the forefront of telescope innovation, but most astronomers were not affected directly by it. Instead, they remained busy observing with conventional telescopes and developing new electronic gadgets to improve their performance and efficiency.

In the process, astronomers' interactions with their telescopes continued to evolve.

Like the soul-searching debates astronomers had in the 1950s about the role of federal funding, discussions about new ways to use telescopes were largely about control. Some scientists wondered whether the craft of traditional observing would vanish, claiming that "technology [was] supplanting artistry."[55] Astronomers faced questions that went beyond technical issues as they experimented with a bewildering variety of electronic and computer systems: How would astronomers use telescopes in the future? Would individual talent and skill at the telescope still be relevant? Where would the astronomer fit into the observatory of the future?

One cause of astronomers' apprehension was the rapid appearance of computers in the observatory. In the 1950s, astronomers began using computers for routine tasks such as data reduction. These large and costly machines were centrally located on university campuses. With the introduction of minicomputers in the mid-1960s, modern observatories could install powerful, yet relatively, inexpensive computers that astronomers could link to other electronic equipment or even to the telescope itself. As this trend continued, observatories were obliged to hire scientists and engineers with the skills to make the new devices work.

Edwin W. Dennison was one such person. After getting his Ph.D. in astronomy from Leo Goldberg's department at Michigan, Dennison did solar physics research for several years in New Mexico. In 1963, he was hired as a researcher at Caltech. He soon turned his attention toward the development of electronic systems for telescopes. In 1966, Dennison directed what came to be called the Astro-Electronics Laboratory at Palomar. By 1969, when the Mount Wilson and Palomar Observatories officially changed its name to the Hale Observatories, Dennison's laboratory employed a dozen people. One of the lab's main projects was adapting minicomputers for direct use at the telescope.[56] Dennison's devices found a welcome reception at Palomar, and by 1970, astronomers at the 200-inch used a digital data-recording system on one out of every three observing runs.[57]

The degree to which technological enthusiasts believed that modern electronic instruments could change astronomers' nightly work is reflected in a 1971 article Dennison published in *Science*. It outlined two "philosophical principles" that guided his lab's efforts: to enable the astronomer to make observations that would otherwise be difficult or impossible and to improve the "operating efficiency of the instruments."[58]

Besides describing the computer and electronic systems at the Hale Observatories, the article offered an overall philosophy that guided the development of electronic instrumentation. Dennison, while recognizing that no machine could have "the flexibility or ingenuity of the experienced observer," expressed concern that a fatigued and uncomfortable astronomer could accidentally cause mistakes during the night, wasting increasingly valuable observing time. As a result, all telescope and electronic controls should be as simple as possible. Only those the observer would need during the night should be available. The observer, according to Dennison, should be free to monitor the data as it was collected in order to "eliminate the feeling that he is being manipulated by the dictates of the equipment." But, to avoid systematic errors, the observer "must never be a link in the data-collection chain."

Besides computing systems, a new and much more complex instrument for the 200-inch was also undergoing development at Palomar in the late 1960s. A multichannel spectrometer, designed and built by Bev Oke and funded by a grant from ARPA, used 32 photomultiplier tubes to cover the visible spectrum from the blue all the way to the near-infrared. This device, as well as other new instruments, was designed for use at the telescope's Cassegrain focus, located under the primary mirror. Unlike the prime focus cage located high above the primary mirror, the Cassegrain focus was easily accessible to astronomers and technicians. This was an important advantage for engineers when they were installing and debugging more complex instruments, which might weigh several hundred pounds. Oke's new spectrograph automatically subtracted background signals from the night sky and, when linked with a minicomputer system developed by Dennison's lab, transmitted information collected by the 200-inch directly to a data room. Meanwhile, the astronomer sat at the Cassegrain focus in a specially designed observing platform that resembled a barber's chair and monitored the process.[59]

When astronomers used the multichannel spectrometer, they still had to ride with the telescope for extended periods of time while keeping their eyes adapted to the dark.[60] Astronomers and electrical engineers had systems in mind, though, that would completely remove the astronomer from the telescope. Television-based guiding systems brought astronomers in from cold observing cages to warm, well-lit data rooms while enabling them to track faint objects they could not normally see. The electrical output of these sensitive, yet bulky, systems was displayed in a separate control room and could also be used to control the telescope automatically.

As they built new electronic systems, Dennison and other scientists were guided by their desire to improve the efficiency of telescope use and to ensure a more productive and comfortable data collection process. Not all astronomers were at ease with the new technology. At a conference on telescopes, Dennison gave a paper in which he showed a block diagram of the Hale Observatories computing system. On one side of the drawing, set apart from boxes labeled "teletype" and "16K core memory," was an icon identified simply as "telescope." After Dennison's talk, one astronomer remarked that he was afraid scientists "would end up with a telescope that is a big computer with a large optical analog-input at its periphery."[61]

As computer equipment became more common in astronomy, astronomers recognized its potential to affect other areas of science practice besides data collection. For example, in 1966, astronomers at Northwestern University published a paper that described the use of a computerized information retrieval system that could provide answers to technical inquiries.[62] By processing up to eight questions a minute, the system analyzed queries posed in English about stellar astronomy, searched a star catalog, and provided numerical answers.

In 1971, the European Southern Observatory (ESO) and CERN, the European Organization for Nuclear Research that operated large particle accelerators, jointly organized a meeting in Geneva about large telescopes. This was a logical partnership; ESO's Telescope Project Division was based at CERN from 1969 to 1975 and hoped to benefit from CERN's experience in building large-scale tools for science. At the meeting's opening, the directors of both organizations predicted that optical astronomy would profit from the knowledge CERN had with complex instruments, electronics, and automation.

Astronomers listened as speakers explained how computers could automatically control telescopes. The scientists were aware of experiments in which telescopes were remotely operated from miles away. Three years earlier, scientists using punched paper tape and a computer in Tucson had remotely collected data with a small telescope on Kitt Peak for several nights. Not all astronomers were sanguine about such technological feats. Some experienced observers still preferred "manual controls, manual guiding, and fairly extensive human interaction with the instruments."[63]

Looking beyond the use of the telescope to its design, engineers described how computers were assisting with the elementary engineering analyses for several new 4-meter telescope projects. Previously, computer-aided design and analysis programs had not been widely used, and telescope design had been as much art as science. One telescope engineer recalled, "As a mechanical engineer, you did the best you could. You analyzed the stresses and deflections and when you were done, multiplied by three. All the telescopes built up through about 1975 were, generally speaking, engineered by eyeball."[64] According to another engineer, "People drew pictures of the whole telescope and then started drawing parts. You couldn't quantify the design and the best thing was never to step too far away from what had been done before."[65] By separating the telescope into different "analytical segments," computer-assisted modeling enabled efficient modeling of the mechanical behavior of telescopes before construction. This prompted one person at the 1971 conference to ask, "Now we have got rid of the observer, do we get rid of the engineer too by putting everything into a computer?"[66]

Talks presented at the 1971 ESO-CERN meeting suggested that the astronomer's traditional techniques for collecting data alone in the prime focus cage were changing. David Crawford from Kitt Peak, for example, described AURA's two 4-meter telescopes under construction in Arizona and Chile at a cost of $10 million each. Both instruments "adopted many of the design features of the 200-inch and 120-inch" telescopes. But, when it came to how astronomers anticipated using the telescopes, Crawford emphasized the Cassegrain rather than the prime focus station.[67]

George Herbig, an astronomer at the 1971 meeting, went even further and asked, "Why is it necessary for the observer to be at the prime focus or at the Cassegrain focus at all?"[68] While this location might be traditional, was this still "an acceptable way for a scientist to perform his science"? Referring to innovative new instruments and electronics developed at Lick and elsewhere, Herbig made a case for complete remote control and TV monitoring of telescope operations. Acknowledging that these systems were not cheap, Herbig reasoned that well-trained astronomers were also an "expensive instrumental investment," making any system that promised greater comfort and efficiency worth considering.

Three years later, ESO and CERN organized a follow-up conference to explore how scientists should use large telescopes. An entire morning session was devoted to the philosophy of telescope use. Speakers such as Jesse Greenstein assured the large international gathering that the telescope was still their "central research tool." Nonetheless, Greenstein told colleagues that, while he was impressed and comforted by the extraordinary gains in data-gathering ability afforded by new electronic devices, they often caused him to "return from observing in a state of personal rage."[69]

Greenstein and some of the other astronomers were not consoled when they heard Geoffrey Burbidge's typically provocative talk on the last day of the conference. Burbidge began by referring to the traditional "star system" of astronomy whereby most of the important discoveries were made by talented individuals working alone. While recognizing the talents of people such as Edwin Hubble and Walter Baade as first-class observational astronomers, Burbidge said that much of astronomy's progress had resulted from access to the best and biggest telescopes—as he put it, "the telescope frequently made the man."[70]

Burbidge argued that "there was no political justification for a policy of exclusiveness" in an era when many new telescope projects were financed to some degree by the public. Astronomy, Burbidge claimed, was rapidly moving to an era of team research, and for many projects, "individual research with a large telescope is going out very fast."[71] At Lick Observatory, Burbidge noted, collecting data with a new and sophisticated image-tube scanner required the coordinated effort of a half-dozen or more astronomers, electronic engineers, and technicians. Besides advocating more competitive team-based research in the style of experimental physics and radio astronomy, Burbidge opined how future telescopes should be managed. Many optical astronomers, Burbidge conceded, would prefer a "peaceful life with their own observing time assured" rather than competition and teamwork. Astronomy's future, however, required that major national and international facilities be used in some sort of democratic fashion with "talent, imagination, and creative ability being the ultimate deciding factor when it comes to observing time."[72]

After Burbidge's talk, Adriaan Blaauw, ESO's director, chaired a lively panel discussion that reflected the state of unease and transition in the astronomy community. Broadly speaking, the astronomers who spoke that spring morning in Geneva reflected two different visions for the future of large telescope use. Harry van der Laan was one of the scientists at the meeting familiar with both radio and optical telescopes. The Dutch astronomer, later ESO's director, commented on the new skills astronomers might need. Eventually, he noted, only a few astronomers would be "familiar with the whole complex system" as telescopes increasingly relied on computers and other electronic devices. Van der Laan suggested that it might be better, as in radio astronomy, if the astronomer "doesn't go near the [observatory] . . . he communicates by telephone and telex and gets his data back." Of course there would be exceptional cases in which interaction between the astronomer and the observatory staff might be necessary, but "big modern optical telescopes are evolving . . . where it should be mostly a hands-off policy as far as the using astronomer is concerned."[73]

Another ESO astronomer, Jaap Tinbergen, made even more pointed remarks, saying that optical astronomers were "just too conservative, and too lazy in some cases, to try it any other way." He then proceeded to tell a story about a neophyte astronomer who wasted a night's worth of observing because he had not properly aligned his target on the spectrograph's slit. "Now that wouldn't have happened," Tinbergen concluded, "if he had allowed his spectra to be taken by an assistant."[74]

These comments elicited sharp responses from many at the meeting. Peter A. Strittmatter, from the University of Arizona, argued that having "an expert on the machine is going to be absolutely essential" as astronomers bring new instruments into service. In some cases, having an observer collect data for the scientist was acceptable, but "most astronomers want to be around when their observations are being taken." Bev Oke and Maarten Schmidt, both Caltech astronomers, supported the view that the best science is done when astronomers collect their own data at the telescope. Schmidt refuted the suggestion that most observations could be taken by remote control or by a service staff whose duty was to execute astronomers' instructions. "There are so many variables in optical observing," he said, "I want the observer to be at the site and to make the instant decisions that are necessary from hour to hour."[75]

Greenstein reacted angrily to the debate between his American and European colleagues about how telescopes would be used in the future. He warned the European and Australian astronomers present, many of whom were planning new telescope projects, not to make the mistake of believing that "they will be able to send messages" over the phone to the observatory and get reliable data. "I wouldn't trust anybody else," he contended, "Either you believe that the 'Establishment' has been a bunch of idiots for all these years and that there are nasty kids that like to be up on a freezing mountain, or you must believe that there was some point in it."[76]

For scientists like Greenstein, the idea that the astronomer might not directly participate in data collection was anathema. For others who lacked the unparalleled access to big telescopes that Greenstein and his colleagues enjoyed, times had changed and there was not nearly enough telescope time to satisfy the needs of the community. Furthermore, time allocation committees were granting ever-smaller blocks of observing time to individual astronomers. As a result of these pressures, the possibilities of using telescopes in new and supposedly more efficient ways intrigued many astronomers and induced apprehension in others.

Greenstein's Dilemmas

To Greenstein, the 1960s had not been a good time for the health of beloved elite science institutions. While astronomers at observatories such as Palomar and Lick continued to work at the frontier of research, the majority of the NSF's support for astronomy went to the national centers. The fraction of astronomy funding that the NSF spent on the national observatories increased from about 50 percent in 1964 to 73 percent five years later. Annual funding for national optical facilities was over $10 million in 1969, far in excess of the NSF's entire budget for astronomy at universities.[77]

Greenstein had also learned that plans to build a copy of the 200-inch telescope in Chile had fallen apart. In 1963, Horace W. Babcock, soon to become the new director of the Mount Wilson and Palomar Observatories, started discussions with members of AURA's board and Kitt Peak staff concerning what was called the Carnegie Southern Observatory (CARSO) in Chile.[78] AURA, at the time, was already building the Cerro Tololo Inter-American Observatory in Chile. The following year, the Carnegie Institution of Washington approached the Ford Foundation with a plan to build a 200-inch southern telescope. Between 1964 and 1966, Carnegie continued to meet with AURA representatives and discuss how they might cooperate. These negotiations were secretive because of delicate issues. A major one was how astronomers not affiliated with Carnegie would get access to the telescopes. Meanwhile, Carnegie representatives continued to discuss the possibility of receiving $19 million from the Ford Foundation for CARSO and believed negotiations were on track.

On March 1, 1966, McGeorge Bundy, a former Harvard dean and influential advisor to Presidents Kennedy and Johnson, became the Ford Foundation's new president. A few days after taking office, Bundy called Leo Goldberg at his Harvard office to discuss the CARSO proposal. On March 14, after talking with Goldberg and other prominent astronomers, Bundy and the Ford Foundation trustees decided to postpone Carnegie's proposal. They cited the project's cost and, perhaps more importantly, the fact that "the proposed provision of access for astronomers other than those of the Mount Wilson-Palomar staff was considered inadequate."[79]

Babcock, distressed at the news, requested letters of support from prominent astronomers, including Goldberg. Goldberg had only learned of the CARSO negotiations after he joined AURA's board in early 1966. Like many of his colleagues, the news surprised him. Greenstein recalled how Goldberg and Martin Schwarzschild, a Princeton astronomer, "gave me hell" when they first talked about the plan. In their view, any plot to build the largest telescope in the southern hemisphere without getting support from the entire astronomy community was "crazy."[80]

Goldberg's major reservation about the CARSO project was its provision to grant access to scientists not affiliated with Carnegie. In April 1966, he wrote Babcock that it was "absolutely essential that a 200-inch telescope in the southern hemisphere be operated as a national or international facility."[81] Goldberg carbon-copied this letter to Bundy; a month later, Goldberg sent Bundy a second letter. While he agreed that building no large telescopes in Chile would be a disaster, he expressed concern about the proposed arrangement between Carnegie and AURA. Such an arrangement, he said, might inhibit AURA from building its own telescope in Chile. Goldberg suggested instead that the Ford Foundation and the NSF pool their money and fund an AURA-run telescope in Chile.[82]

Bundy soon told Julius Stratton, the president of MIT and chair of the Ford Foundation's board, "that it would be a grave mistake to place the second 200-inch telescope in the world under the same management as the first . . . this proposition would be so simple as to be self-evident in the world of high-energy physics, and it says a great deal about the current state of astronomy that people are so hesitant to reach this obvious conclusion." Bundy also noted that statements from the Mount Wilson and Palomar astronomers "reflected once again [their] unconscious self-centeredness."[83]

When it became clear that the Ford Foundation was not going to support the CARSO 200-inch, Lee DuBridge, Caltech's president, told Goldberg that some in Pasadena believed that AURA (and, more specifically, Goldberg) had ruined the deal. Goldberg objected to the implication that either he or AURA had been the spoiler. The entire Carnegie-AURA arrangement, he said, was a "shotgun marriage" from the beginning, and there was not the "slightest evidence that AURA ever adopted a 'dog in the manger' attitude toward the CARSO proposal." DuBridge's diplomatic reply assured Goldberg that any competition between Carnegie and AURA was settled and that if Caltech and Carnegie ever managed to build a large, private telescope in Chile, all would be able to use it.[84]

Bundy, concerned about any single institution having a monopoly on large telescopes, supported the idea of making a grant directly to AURA for a new telescope that would be open to a wider community of scientists. Throughout the summer and fall of 1966, Ford Foundation representatives met with the NSF (unbeknownst to AURA) to discuss this. Julius Stratton, also a member of the National Science Board which oversees the NSF, helped broker the deal. At their December 1966 meeting, the trustees of the Ford Foundation agreed to give $5 million toward an AURA-operated, 4-meter telescope in Chile provided that the NSF could secure the remaining funds from Congress.

In January 1967, the Ford Foundation invited AURA to submit its proposal. Eleven months later, President Johnson authorized the project to proceed.

Greenstein learned how plans for CARSO's proposed 200-inch telescope fell apart through discussions with colleagues at Caltech and Carnegie. He believed that the intervention of a few senior astronomers connected with AURA, including Goldberg, was "decisive in the personal feeling that Bundy arrived at."[85] Many of Greenstein's colleagues in Pasadena shared this view; and, despite the lack of supporting documentary evidence, it became accepted folklore in the Caltech and Carnegie communities that AURA had derailed the deal with the Ford Foundation. Years later, Babcock still recalled, "A last minute plot by the competition resulted in the diversion of a $5 million grant from the Ford Foundation; instead of coming to the Carnegie Institution, which had done all the ground work . . . the grant went to [AURA]."[86]

To Greenstein, the loss of the Ford Foundation's grant cost American astronomy much more than just a new telescope. "The real shame," he said, "was the loss of an enormous, private fund for an observatory which I think could have fulfilled the public responsibility." Since the release of the Whitford Report in 1964, no new, large, privately run telescope projects had started. To Greenstein, the construction of AURA's 4-meter national telescope in Chile was another example of declining support for elite, university-based astronomy and the work of the "brilliant individualist" that he favored.[87]

Carnegie's failed bid for a second 200-inch occurred as national centers for science and debates about their dominance were becoming more prevalent. Greenstein observed federal funding and facilities squeezing out university leadership in one field of science after another. Radio astronomy, which began with the work of a few pioneering individuals and flourished at schools like Caltech, had gravitated toward more national facilities. More notably, the most powerful accelerators for high-energy physics were run as national or international facilities. In the early 1960s, plans to build the Stanford Linear Accelerator Center paralyzed that institution as debates ensued over the merits of science done at large centers versus smaller-scale department-based research. While the CARSO plan was foundering, the physics community was readying plans to build the National Accelerator Laboratory—later named Fermilab—as a national facility. Unlike

Brookhaven or Livermore, outsiders would have equal access to accelerator time, abolishing the "hegemony of the elite."[88] Caltech's own electron-synchrotron facility closed in 1969, just after ground was broken for Fermilab, as Caltech physicists traveled in increasing numbers to the big national accelerators. All of these changes were, as Greenstein saw it, blows to university research that left him depressed.[89]

Greenstein became increasingly vocal about the need to redress astronomy's priorities. After the release of the Whitford report, the national optical astronomy centers at Kitt Peak and Chile had each received several million dollars for new telescopes. In April 1967, in a short handwritten note to Nick Mayall, Kitt Peak's director, he confessed, "I do not believe that further telescope construction at Kitt Peak is in the national astronomical best interest." When Mayall demanded an explanation, Greenstein circulated a letter to AURA's entire board stating that the "steady drain for continued operation for National facilities is in direct competition with university research applications. I do not feel that the balance has been maintained and I am not enthusiastic about any major expansion in these centers until a new policy has evolved." Federal money, were he in control of it, would go first to Lick and then Caltech.[90] Moreover, he saw no reason why individual universities should not manage large, nationally funded science facilities.[91]

In July 1969, the National Academy of Sciences asked Greenstein to take responsibility for astronomy's next decadal survey. The new survey differed from the earlier effort chaired by Whitford in two important respects. First, it considered astronomy practiced from the ground, funded mainly by the NSF, along with space-based research supported by NASA, which the Whitford report had not addressed. Second, Greenstein and his committee had a strong mandate from the Academy as well as the Office of Management and Budget to recommend which new facilities should be funded first. "It was no longer possible to send a shopping list to the government." Greenstein recalled, "They wanted priorities."[92]

Greenstein recruited panel members, including Goldberg, and solicited advice from other eminent members of the astronomy community. Greenstein, as he told Goldberg and other senior astronomers, believed that one of the more pressing issues they had to consider was the "balance between . . . government laboratory and university research, or between a big federally supported facility and an individually operated laboratory," describing the situation as "very delicate."[93]

For two years, he and a group of twenty-two internationally recognized scientists met and argued over astronomers' priorities for new research facilities and how the funding agencies that supported astronomy should spend their money. Helmut Abt, a Kitt Peak astronomer and Greenstein's former student, chaired the optical astronomy subpanel. Meeting several times between late 1969 and June 1971, Abt and the other panel members soon recognized the science community's strong sentiment that the first recommendation of Greenstein's report should be a major new facility for radio astronomy. Greenstein, of course, remained sympathetic to the optical astronomers. He told Abt, "You know I like radio telescopes but I love optical telescopes and I think they have a fine future," but he predicted that future large optical facilities would probably have to wait.[94] By early 1971, Abt's panel had decided their first recommendation for optical astronomy was not new and larger telescopes but funding for better instruments and electronic equipment. On other issues, optical astronomers expressed little agreement as to priorities.

In June 1971, Greenstein received the reports from the various subpanels and his main committee met for two weeks at the Institute for Theoretical Physics in Aspen. After days of heated discussion, Greenstein held two secret ballot votes. He personally voted for more large optical telescopes, but the majority's (including Goldberg) first priority was the Very Large Array, a national facility for radio astronomy in New Mexico estimated to cost at least $60 million. In fact, almost all of the committee's major recommendations were proposed as national efforts. Greenstein recalled, "It ended with all of my claims for a balanced program . . . disappearing down the maw of big science, the death of university astronomy."[95]

As Greenstein counted the votes, he got "mad as hell . . . It just was so much against my own life experience and my own intuition of what's good for science."[96] Disturbed and dismayed, Greenstein told his committee he would step down as chair and refuse to sign their final report, an unprecedented act of protest. For Greenstein, it was the "death of everything I held dear."[97] He later wrote, "The major recommendations so clearly required management by NASA Centers or by National Observatories . . . [and were] personally antithetical to my style of research and management."[98] Defeated, Greenstein resigned. That evening, a delegation of astronomers came to his room and argued that his withdrawal would only hurt astronomy. After an emotional discussion, he agreed to resume his leadership role.

When the National Academy of Sciences released Greenstein's final report in April 1972, new tools for optical astronomy appeared second after the VLA on the prioritized list of recommendations. The survey recommended continuing to "vastly increase the efficiency of existing [optical] telescopes by use of modern electronic auxiliaries and at the same time create the new large telescopes" of the future.[99] Greenstein personally added a final chapter that offered his view of how the recommendations should be implemented. In the face of expanding national astronomy centers, he argued for the continued importance of university-based astronomy. There was no size limit for telescopes, Greenstein argued, beyond which a facility had to be nationally managed. Not all the committee members agreed with his approach. Abt and Mayall, representing Kitt Peak, charged that his wording unfairly argued for greater support of university research by denigrating the national centers.[100] Kitt Peak and Cerro Tololo, they said, were extensions of university-based astronomy, not competitors.

One crucial duty of the astronomy survey's chair was to take the committee's message to the science community. Over the next few years, Greenstein gave talks that outlined how the report was prepared and his interpretation of its recommendations. In June 1973, he presented the report to the National Science Board. In the margins of his handwritten speech, Greenstein scrawled "Tears!" before describing the major policy dilemmas facing astronomy. Whether he was reminding himself to appear sorrowful or merely reflecting the personal difficulties he had with this issue, Greenstein's message was clear: the university, not a national center, was where the best scientists did frontier research. Without adequate and balanced support, "the universities for which the national centers were built," he said, "will die on the vine at the greatest period in astronomy history."[101]

While it rarely surfaced in ways as dramatic as Greenstein's resignation, the strain between the national and private observatory systems had been increasing since the establishment of AURA and Kitt Peak. This tension placed scientists such as Greenstein and Goldberg at loggerheads while it directly affected astronomers' plans to build bigger future telescope facilities. By the early 1970s, cracks within the optical astronomy community had become fissures. Scientists found themselves having widely diverging opinions, their positions often a reflection of their institutional affiliations. Should new large telescopes be managed by national centers or by private or state institutions? Was the relationship between different observatories competitive or complementary? Was it better to build several smaller telescopes or one large instrument? What was the best way to use new facilities?

Within the astronomy community, there were also polarized opinions about the national observatory system as a whole. Some astronomers from smaller universities gladly depended on it and advocated its expansion, while scientists from prestigious institutions (who sometimes used the facilities at Kitt Peak or Cerro Tololo) resented the amount of support the national centers received. Some, like Jesse Greenstein, even feared that, left unchecked, national centers might dominate astronomy as they did other fields such as high-energy physics and radio astronomy. In the early 1970s, astronomers' strongly emotional, and sometimes conflicting, opinions about what was best for their field became an essential consideration when Leo Goldberg proposed that Kitt Peak build a giant, new telescope that would dwarf any other in the world.

CHAPTER 3

Visions of Grandeur

Leo Goldberg returned to Harvard in 1960 as a professor of astronomy. True to his interests, he devoted his research efforts and generous NASA funding to building a series of orbiting solar observatories. In 1966, Goldberg turned down an offer to direct NASA's Goddard Space Flight Center in Maryland and instead took the directorship of Harvard College Observatory and chair of Harvard's astronomy department. The Smithsonian Astrophysical Observatory had moved to Cambridge a decade earlier; Goldberg's former professor, Fred Whipple, sat on the council that oversaw Harvard's observatory. Whipple had a strong voice in its management and Goldberg's position was politically delicate. In time, he became increasingly unhappy with the complex administrative arrangement. When AURA offered him the directorship of Kitt Peak in January 1971, Goldberg readily accepted. As he began his new duties, he believed he had a mandate from AURA and the NSF "to transform the observatory from a sort of manufacturing facility [for new telescopes] into a working scientific laboratory."[1]

After Leo Goldberg became Kitt Peak's new director, Jesse Greenstein congratulated his former classmate and expressed his wish that Goldberg's new position would bring Kitt Peak into "direct competition with us dinosaurs in Pasadena."[2] What was offered as a gesture of friendly rivalry ultimately became prophetic. When Goldberg arrived at Kitt Peak in September 1971, the national observatory was in a crucial period of transition. For almost 15 years, Kitt Peak had pursued an aggressive program of telescope construction. By 1970, Kitt Peak operated several telescopes of small to moderate size, and its 4-meter instrument was nearing completion. In Chile, engineers and astronomers were building another 4-meter telescope for AURA at the Cerro Tololo Inter-American Observatory. As these new telescopes entered service, many expected the national observatories to switch from building facilities to operating them efficiently, serving the user community and, some insisted, doing frontier research rivaling the work done at places such as Palomar and Lick.

The American astronomy community lacked a unified vision for the national observatories. For astronomers from small universities or those located at places where observing conditions were poor—six of AURA's original seven members were Midwestern or East Coast schools—Kitt Peak offered telescopes and instruments essential to their careers. Many astronomers from these schools came to Kitt Peak to use the smaller telescopes and were less interested in state-of-the-art instrumentation. The primary concern of the "have nots" was for Kitt Peak to build and maintain simple and reliable facilities where they could collect a few nights' worth of data and to employ a staff to assist them during observing runs.

Those astronomers with access to their own large telescopes viewed Kitt Peak as a competitor for resources and prestige. By 1970, in an attempt to strengthen the national infrastructure for science, for every $1 the NSF gave to astronomers as individual research grants, Kitt Peak and Cerro Tololo received $3. This deeply angered some astronomers. Their resentment was fueled by the strong sentiment that neither the research done at Kitt Peak nor the instruments on its telescopes were cutting edge. Astronomers at Caltech remarked in confidence that they could think of "no innovative night-time instrument which has been developed at Kitt Peak." Even scientists in Goldberg's department at Harvard, one of the AURA's charter universities, complained that, for the amount of money the NSF was spending at Kitt Peak, "it is disturbing that there is not more first class optical astronomy coming out."[3]

There was disagreement about the national observatory's mission even among the Kitt Peak staff. Some placed emphasis on serving visiting scientists while others believed Kitt Peak was in danger of simply becoming another large and mediocre government laboratory. A solution, some insisted, was to produce scientific research of such quality that "we cannot be looked down upon and . . . become completely subservient to direction from scientists outside Kitt Peak. We will never be liked by our well-endowed, but jealous, critics but we should be able to command their respect . . ."[4] What is remarkable about the views held by scientists at Kitt Peak and those at institutions with private observatories is that each group believed the other was somehow better off. This dissonance in perception between the advocates and the critics of Kitt Peak hindered attempts to develop a coherent strategy that satisfied both groups.

The astronomy community's inconsistent and divided perception of Kitt Peak's mission affected the choice of its new director. When Nicholas Mayall announced he would retire in 1971, AURA asked prominent astronomers to suggest possible successors. Greenstein, comparing Kitt Peak with "a national trust, like the national parks," believed it should be administered for the overall good of astronomy. Because private observatories like Palomar could not match the financial support the NSF gave Kitt Peak, he believed only research of the highest quality should be done there. This could mean rejecting a large fraction of astronomers who requested observing time. Such a move would be "quite an anti-democratic change" but one that the substantial national investment in national observatories demanded. Greenstein suggested that, unless astronomers wanted to have "black days," Kitt Peak hire "a statesman of science, a strong protagonist for all of astronomy . . . the type of salesman for science that Robert A. Millikan [one of Caltech's founders] was 40 years ago."[5]

Even before he began as director, scientists approached Goldberg and expressed their views of what Kitt Peak's mission was. One especially outspoken person was Stephen E. Strom, an ambitious scientist and recent graduate from Harvard's astronomy program. Strom wrote Goldberg that "Kitt Peak should have scientific staff of the highest quality; it does not, at the present." The awarding of observing time in ever-smaller amounts also concerned Strom, who thought it "appalling to see astronomer X from a small southern university having his three or four nights on the 150-inch by reason of pure political convenience."[6]

As he began his directorship, Goldberg contended with these competing visions for the national observatory. Goldberg personally believed Kitt Peak's scientific staff should be organized much more like an academic department, with eligible researchers receiving tenure and so forth. To accomplish this, he advocated a three-fold approach. Hiring exceptionally talented scientists would ensure that first-rate science would result. The new talent would then develop state-of-the-art instruments that they had a personal interest in using. Visiting observers would reap the benefits from the availability of first-class instruments and high-caliber staff with broad research interests.[7] In addition to following his intuition about how a first-class laboratory should be operated, Goldberg also responded to recommendations made by a special Visiting Committee that AURA created with Goldberg's encouragement. In December 1971, its first report noted that if the observatory wished to provide "facilities for frontier optical research," it needed a "staff of first-class scientists themselves engaged in front line scientific research" and personally committed to building "frontier instrumentation."[8]

Goldberg soon reorganized Kitt Peak with a single scientific staff and a prioritized roster of engineering projects. Heading his list was the completion of advanced science instruments for the new 4-meter telescope as soon as possible. He attracted a number of promising young astronomers, including Steve Strom. Goldberg also insisted that Kitt Peak develop a strong infrared program of research and instrumentation. For this effort, he hired Donald N. B. Hall, another Harvard student, and Frederick C. Gillett from the University of Minnesota. Each of these scientists would play a role in the coming years as debates about a new national telescope and the role of the national observatory intensified.

Goldberg declined to expand into other research areas, insisting that the main mission of Kitt Peak should remain observational optical astronomy.[9] Goldberg also altered the research focus of Kitt Peak. Believing that the observatory's staff was too focused on planetary astronomy, he increased staff hires and research programs oriented toward galactic and extra-galactic objects; by 1974, 40 percent of Kitt Peak staff worked in what was called the "stellar" program.[10]

Goldberg and the observatory's other scientists believed that the major tool for research would be Kitt Peak's new 4-meter telescope. The new national telescope was dedicated in June 1973 and Jesse Greenstein gave an invited talk called "The Exploration of a Strange and Beautiful World" at the ceremony. The following year, a second 4-

meter telescope, located at Cerro Tololo, took its first images. AURA now operated a pair of the world's second-largest telescopes.

The fact remained that the national observatories operated telescopes smaller than the 200-inch run by Caltech and Carnegie. This was the opposite of what scientists observed in other fields such as high-energy physics and radio astronomy, in which national facilities were typically the biggest or most powerful. For many astronomers, this was an odd, if not suspicious, situation that affected the national observatory's mission and status.

A Telescope for the Next Generation

Even before AURA had officially formed, Goldberg had big plans for the organization. In 1957, he boasted to Lee DuBridge that he looked forward to the day when the national observatory might operate very large telescopes "perhaps 200 inches in diameter or greater."[11] By 1974, his reorganization of Kitt Peak was complete, and scientists began collecting data with the new 4-meter telescope. Building the telescope had required a large number of engineers and support staff; but when the telescope was finally debugged and handed over to the astronomers, the core team of engineers began to disband.

In the summer of 1974, Goldberg encouraged his staff to consider possible ways that Kitt Peak might build a new telescope. Their goal was a light-collecting area of 25 meters (about 1,000 inches). A telescope this big would have 25 times the light-collecting area of the 200-inch telescope. The stunning size was "bigger than anything yet produced and it was bigger than anything we felt we could easily make by just extrapolating an existing design. It was an exercise in forced innovation."[12] The 25-meter telescope program was not created *ex nihilo*. For years, a nebulous item called the "X-inch telescope" that harked back to Aden Meinel's investigations had existed in AURA's long-range plans.

In supporting plans for a 25-meter telescope, Goldberg accomplished two things. First, he was able to keep the talented engineering staff who had coalesced around the 4-meter telescope engaged on a new project at the technological frontier instead of losing them to other institutions. Lawrence ("Larry") Barr was one of the people Goldberg hoped to keep. Barr was a mechanical engineer who came to Kitt Peak in 1968 after spending several years in industry. While working with Larry Randall, Bill Baustian, and others on Kitt Peak's 4-meter project, Barr learned the craft of telescope engineering. Barr and some of his colleagues wanted to build new telescopes. "Once the 4-meter was built," he recalled, "there were too many damn engineers around here!" Some, including Barr, were especially attracted to the 25-meter program—which, Barr admitted, was dubbed by skeptics at the observatory as a "Works Progress Administration for engineers." Some criticized Goldberg for looking to build telescopes simply to keep his staff busy.[13]

The 25-meter project also represented an opportunity for the future expansion of the national observatory and U.S. astronomy in general. To Goldberg, failure to pursue future projects was tantamount to stagnation and a sure recipe for poor morale and eventual decline. Goldberg argued that, as other countries were beginning to build copies of Kitt Peak's 4-meter telescope, a failure to think and build for the future would eventually make "the U.S. position in optical astronomy . . . competitive rather than preeminent."[14]

Exploring new telescope concepts was not a major priority for most of Kitt Peak's staff. Larry Barr, whom Goldberg appointed as the 25-meter telescope's chief engineer, recalled working on it during the summer months in 1975 and 1976 when Arizona's afternoon rainstorms helped make time available. Don Hall took time away from his regular duties of nighttime observing and designing infrared detectors to serve as the project's chief scientist. Together with input from other Kitt Peak scientists, Hall developed a scientific justification for a 25-meter telescope. Still, Hall and Barr remembered that the staff's involvement on the hypothetical 25-meter telescope was largely detached from their regular duties at Kitt Peak. Few people clamored to join an exploratory project that would not be completed for many years, if ever.

By September 1975, Goldberg could tell the Kitt Peak staff that engineers and astronomers had arrived at a tentative concept for a 25-meter telescope.[15] The design's name was PALANTIR, a loosely fitting acronym derived from Program for a Large Aperture Novel Thousand Inch Telescope. *Palantiri* were the magical seeing stones in J. R. R. Tolkein's *Lord of the Rings* that allowed their users to see through time and space. PALANTIR's design, shown in the upper left of Figure 8, had more in common with large-dish radio telescopes like the Arecibo Observatory in Puerto Rico than with other optical telescopes.[16]

As with all large telescopes, the major challenge was deciding how to make the primary mirror. PALANTIR's gargantuan mirror surface would be 75 meters long and 25 meters wide. Instead of a single massive glass mirror, the design of PALANTIR's primary mirror used hundreds of polished aluminum segments that, when combined, created a massive light-collecting surface.

While the idea of a segmented primary mirror was something astronomers had considered for well over a hundred years, the engineers' final vision of PALANTIR was a telescope that was massive by any measure. Barr and other Kitt Peak staff designed PALANTIR so that light striking the metal mirror segments would be reflected to a secondary mirror 3 meters in diameter (not an inconsiderable size for a primary mirror in the mid-1970s, let alone a secondary mirror). Placed at the end of long movable arm, the secondary mirror "saw"

Figure 8. Four concepts developed for Kitt Peak's Next Generation Telescope project. Courtesy of Rick Showalter/NOAO/AURA/NSF.

only a portion, about 25 meters worth, of the giant primary mirror at any given time. Additional optics would relay light to spectrographs and other instruments. The entire primary mirror would be supported on a mammoth, "shoe-like" structure—the design's unofficial name was "the Rotating Shoe"—and covered with a dome over 80 meters in diameter.

An advantage of PALANTIR that Kitt Peak staff touted was that the primary mirror segments remained immobile with respect to gravity and could not be distorted as the telescope moved. Despite its gargantuan size, Kitt Peak engineers insisted that PALANTIR did not require significant technological innovations. PALANTIR would be big, but as Kitt Peak's January 1977 proposal to the NSF stressed, it was a "technologically conservative approach." Moreover, the Kitt Peak staff advocated PALANTIR as a "practicable, cost-effective solution to [building] a telescope of a size unattainable by classical design."[17]

Kitt Peak's proposal estimated that building PALANTIR would take a decade and cost about $160 million.[18] By 1977, Kitt Peak's two new 4-meter telescopes had just been completed, and the NSF lacked the resources and political will to undertake another major project for the optical astronomy community. The agency was already funding the Very Large Array of radio telescopes, a project prioritized by Greenstein's 1971 decadal survey and a major commitment in its own right.

There were also technical objections to the PALANTIR design. Despite its massive size, PALANTIR's field of view—the amount of sky astronomers can view at one time—would have been quite small, far less than Kitt Peak's 4-meter, for example. Moreover, the optical resolution of PALANTIR's primary mirror would have been only about one arcsecond, which was poor compared to smaller telescopes already in operation.[19] Finally, critics also noted that the PALANTIR's tremendous size would have required a flat mountaintop site some one hundred meters in diameter. Given these design limitations, few people were surprised when the NSF reviewers cited the project's cost as problematic and refused to fund it.

Goldberg, Barr, and Hall were not surprised by the proposal's rejection. According to Barr, their main goal was to generate interest and funding for a very large telescope program at Kitt Peak. Working on PALANTIR gave them an opportunity to experiment with new design tools, including finite element analysis and computer-aided design and modeling for large structures. In spite of PALANTIR's design shortcomings, the possibility of a 25-meter telescope attracted enough attention that the NSF added $200,000 to the Kitt Peak budget for further study.

Goldberg used these funds to establish the Next Generation Telescope (NGT) program. He also formed an NGT Scientific Advisory Committee, chaired by Geoffrey Burbidge, to evaluate the research for which a 25-meter telescope might be suited. Goldberg was cautious, however, about how this program might be perceived outside of Kitt Peak. Even before the PALANTIR proposal was sent to the NSF, he warned staff that any press releases about the 25-meter project should stress the modest level of work being undertaken and staff were asked not to discuss future projects with the press.[20]

Once the NGT program was underway, the engineers and scientists involved with the project literally threw the doors open and considered a cornucopia of designs. Throughout 1977 and 1978, they were guided by the belief that any future giant telescope would probably not have a single, monolithic mirror in the traditional fashion. The overall light-collecting area would somehow have to be divided into smaller parts. The question was how.

One option was an array of smaller telescopes, instead of a single giant machine.[21] The Kitt Peak staff considered a concept called the Singles Array for the NGT.[22] The chief technical advantage, supporters said, was that all the telescopes could be pointed at the same object, and the light sent to a common point if maximum light-collecting power was needed. Otherwise, the different telescopes could be used for separate research programs. This design might appeal to astronomers without access to their own instruments and provide a political advantage. Building many smaller telescopes presented no special technical challenge. Construction could be started immediately and the astronomy community would not have to wait until all the telescopes were built before using some of them.

One proponent of the array concept was Michael J. Disney. He had recently gained notice in the scientific community when, in 1969, he and two other astronomers at the University of Arizona made the first optical observations of a pulsating radio source in the Crab Nebula, a discovery accomplished with an old 36-inch telescope at Kitt Peak. Perhaps buoyed by this success, Disney began to advocate an array of moderate-sized telescopes. Disney believed that a single big telescope provided time only for a "small and fortunate clique of astronomers." Because of the pressure for observing time, he said there often was a "temptation to use the largest telescopes either for bandwagon astronomy, or for measurements which will bring a definite return . . . in other words, to do first class second-rate astronomy."[23] By proposing an array of smaller telescopes, Disney claimed he was trying to avoid the path that particle physicists had taken by building one or two giant machines to address a few key problems.[24]

The designs for a singles array that Kitt Peak staff developed ranged from a half-dozen 10-meter telescopes to scores of 2-meter telescopes. The latter, a section of which is shown in the lower left of Figure 8, earned monikers like the "the pygmy village" and the "mushroom patch" from its detractors. Rude comments notwithstanding, proponents of the array concept realized their idea faced a more daunting hurdle. It was harder to find patrons who would fund several dozen small conventional telescopes than it was to interest them in single enormous instrument.[25]

Barr, Hall, and others at Kitt Peak explored many other NGT concepts. These included a giant version of the MMT and a large "steerable dish" that looked more like a radio telescope than anything else. "We didn't invent a lot of these designs," Barr noted, "We were a focal point for new telescope concepts. People came to us with their ideas, and it wasn't long before we had more than we knew what to do with."[26]

Goldberg did not publicly favor a particular design for the NGT but continued to advocate the broader importance of the 25-meter program both for the continued health of Kitt Peak and the astronomy community. By 1977, Kitt Peak was the leader in pursuing new ideas for bigger telescopes. Astronomers from all over the world received technical reports about the NGT project and Larry Barr and Donald Hall presented papers on their work at conferences on large telescopes. While the 25-meter project received only modest financial support from Kitt Peak's main budget, Goldberg hoped the project would soon be given its own funding line, and "with support in the astronomical community, stand on its own feet."[27] Unfortunately, political and financial pressures appeared that not only threatened the construction of any new telescopes by Kitt Peak but also caused a final confrontation between Goldberg and Jesse Greenstein.

A Very Delicate Question

The showdown between Greenstein and Goldberg was more than just a philosophical disagreement over what was best for science. At its core was a larger question of who would dominate and control astronomical research: elite universities and a relatively few autonomous observers or large national facilities serving a broader community of scientists. In the 1950s, astronomers including Greenstein and Goldberg grappled with the issue of who should control the expanding reservoir of federal money. Two decades later, they faced a crisis of similar magnitude that centered around how American astronomy should be organized and whether AURA should assume a leading, or even dominant, role. The positions taken by the two men toward the mission of the national observatories represented fundamentally different visions for the future of astronomy.

The organizational structure of AURA made conflicts of interest probable, if not inevitable. As a consortium of universities, each member institution chose two people for AURA's board that managed the national observatories. While each board member was obliged to represent the best interests of facilities like Kitt Peak, they were also mindful of their home institution's welfare. It was quite possible that a member of AURA might be called on to recommend whether Kitt Peak should build a new telescope at the same time that the member's school was seeking money from the NSF for its own projects.

As AURA's membership grew, so did the likelihood of such conflicts of interest. In 1972, AURA accepted Caltech, the University of Arizona, and the University of Texas as new members. Along with the University of California (a charter member of AURA), these schools already had bold ambitions for building their own large telescopes. At the same time, AURA's board decided that the organization's growth required a full-time president, and it voted to hire Gilbert L. Lee, a former administrator from the University of Michigan.

That same year, after 24 years of leadership, Greenstein stepped down as head of Caltech's astronomy program. This was a difficult transition for the 63-year-old scientist. The last several years of stressful committee work had left him thinking he had no friends anymore, "only official acquaintances."[28] Greenstein began attending AURA meetings as Caltech's representative. In April, when he visited Tucson for a meeting of AURA's Scientific Committee, he and his wife stayed at Goldberg's house.[29] During the meeting, Greenstein and Goldberg had the chance to talk about the current politics of astronomy.

Like Greenstein, Goldberg was in the twilight of his career and faced considerable pressures. In 1971, he was elected president of the International Astronomical Union. While this was a great honor, it brought a heavy burden of administrative work and travel and he spent less time at Kitt Peak than some of the observatory's staff would have liked. One solution he found was to rely on and confide in Beverly T. Lynds. Lynds was an astronomer from the University of Arizona who had joined Kitt Peak (where her husband Roger was a noted observational astronomer) in 1971 as the assistant to the director. Over time, Goldberg and Lynds developed a close and intimate relationship.

After Goldberg restructured the observatory in the manner he felt appropriate for a prestigious national laboratory, he had to contend with a staff that remained divided over the observatory's mission. Some believed that more attention should be paid to assisting guest observers who came to use the observatory's telescopes. Goldberg did not attach as much weight to this goal as some of his staff might have wished. Doing so, he said, would be "flying in the face of my whole life's experience as an administrator" and it would disappoint the AURA board and the NSF, who "believe that KPNO should become the world's leading center for optical astronomy."[30]

The completion of AURA's two 4-meter telescopes in the early 1970s put a considerable amount of light-collecting power in the hands of a much larger group of astronomers. It also contributed to increased rivalry between the private and national observatories. Astronomers at large private and state observatories often perceived themselves as different from the scientists and staff whose major resource was the national facilities. As the beneficiaries of a "magnificent tradition" of private and philanthropic patronage of science, some of them viewed "with a certain apprehension, the influence and demands of the 'government project' research worker."[31]

The rivalry between the "haves" and "have nots" sometimes emerged as petty charges and complaints. For example, soon after Kitt Peak's 4-meter was dedicated, a Caltech representative took a tour of the new facility. Afterwards, he circulated a memo claiming that the Kitt Peak tour guide had slighted the capability of the 200-inch. "One gets the distinct impression," he said, "that Kitt Peak is trying to gain, by football crowd tactics, the reputation of being 'No. 1.'" Soon after, a vice-provost of Caltech asked Goldberg if there was a policy to promote Kitt Peak at the expense of the 200-inch. Goldberg, embarrassed by this incident, promised that more care would be taken with future tours and press releases. Underneath this seemingly trivial incident were more serious currents of jealousy and fear. "Goldberg knows," the provost wrote Greenstein, "that he is living off Federal funds to the extent of $12 million a year when we receive on the order of or less than one-quarter million."[32]

Goldberg had his own fears. In 1973, the NSF considered requiring that the staff at the national observatories submit requests for telescope time that would be peer-reviewed like those of visitors to the facilities. Goldberg vehemently opposed the suggestion that his scientists might be denied time on their own instruments, a policy that would undercut his plans to transform Kitt Peak into a top-notch science laboratory. The idea, Goldberg said, was political maneuvering proposed by people such as prominent astrophysicist Willy Fowler, a member of the National Science Board, "whose bias and conflict of interest as a Caltech grantee [of NSF research funds for astronomy] are well known."[33] While this proposal quickly faded, Goldberg was bedeviled at the same time by budget concerns. The NSF was not increasing Kitt Peak's funding and inflation was severely eroding the observatory's purchasing power.

In February 1974, AURA's board invited Greenstein to become its new chair. This surprised astronomers who were aware of his concerns over what he perceived as the national observatories' growing power at the expense of university-based science. Greenstein accepted the responsibility to help manage the national observatories in Arizona and Chile. He also had a more personal mission, one "essentially connected with the public relations effort to enlist what support we can from the astronomers of the country for the rightful claims of the Hale Observatories and Caltech for financial support from the Federal

Government."[34] By chairing AURA's board, he hoped, he might also make Caltech's professional relations with other astronomers easier.

One of Greenstein's confidants and patrons was William T. Golden, a childhood friend and lawyer who became a wealthy New York investor. Golden helped establish the office of a formal Presidential science advisor in the 1950s and, as a result, was a well-connected consultant in areas where science, money, and politics met. In the mid-1970s, Golden also served on AURA's Visiting Committee. Soon after his first AURA meeting as chair, Greenstein confidentially told Golden that "there is much deep thinking to be done about AURA, about the relations between AURA and the Director of KPNO. . . . There seems little doubt in the mind of the KPNO director as to who is top dog."[35]

Greenstein's letter to Golden also expressed his vision for astronomy at Kitt Peak. He explained that the private observatories traditionally had the luxury of carrying out valuable long-term research programs that took several years and hundreds of observing hours to complete. Greenstein believed this was the only efficient way to use big telescopes. He even speculated that "privileged astronomers," as he phrased it, from the private observatories might apply for large amounts of time on the national telescopes but noted such moves would be on "dangerous political ground." At the national centers, large-scale programs dominated by a few astronomers could spark a rebellion by shortchanged rank-and-file astronomers from AURA's member schools. "But science," Greenstein wrote Golden, "is not democratic, it is aristocratic, in achievement."[36] As he later expressed it, "the main problem for AURA management was the question of democratic choice versus elitist, snobbish, concentration of effort." How, he asked, does one select a few preeminent astronomers to use the national centers without compromising community support for them?[37]

To many astronomers, debates about the role of the national observatory and how science should best be done, were personified by Kitt Peak's director and AURA's chairman.[38] Each man had different professional responsibilities, many of which were contradictory. The AURA board expected Goldberg to make Kitt Peak a prestigious scientific institution and, at the same time, provide facilities for as many visiting observers as possible. Greenstein was trying to "walk a careful line between excessive partisanship for the university science, and for places like the Hale Observatories, and excessive concentration of effort in one place, such as Kitt Peak."³⁹ He was well aware of how university-based radio astronomy had stagnated as the National Radio Astronomy Observatory grew, a situation he believed was made worse by the NSF's recent decision to build the Very Large Array.

Greenstein was increasingly convinced that AURA was becoming too powerful in the astronomy community and would eventually dominate university researchers. In late 1975, he learned from Goldberg that the Air Force was abandoning support for Sacramento Peak Observatory, a solar physics facility in western New Mexico. The NSF stepped in to provide funding and, as Goldberg predicted, asked AURA to manage the facility as it did Kitt Peak and Cerro Tololo.⁴⁰

Astronomers were also engaged in a rancorous debate over the scientific management of NASA's Space Telescope (later renamed the Hubble Space Telescope). Kitt Peak's scientists and engineers worried that the much-touted space observatory might "signal the end of Kitt Peak's leadership in optical astronomy."⁴¹ Goldberg, in response, suggested that they participate in NASA's billion dollar mega-project by building instruments or processing data. Soon afterward, Goldberg suggested that AURA would be a natural choice to operate the space telescope after it was launched. AURA did not officially win the competition to manage the Hubble Space Telescope until January 1981. Nevertheless, the possibility of it managing the world's first major space telescope led Greenstein to believe even more strongly that AURA was poised to control almost all major research facilities for astronomy.⁴²

By mid-1975, other fissures appeared in the relationship between Goldberg and Greenstein. The Kitt Peak director was displeased that the NSF had not given him any merit or cost-of-living increases. He blamed Greenstein for not advocating this issue more strongly with AURA and the NSF. It was, Goldberg said, the first time he ever had to plead the case for his own salary.⁴³ He encouraged influential colleagues in the science community and Kitt Peak staff to apply pressure to Greenstein, tactics the Caltech astronomer found inappropriate.

Goldberg also complained to Greenstein and the AURA board about what he saw as a lack of involvement on the part of Gil Lee, AURA's president.⁴⁴ Some on the AURA board interpreted his critical assessment of Lee as a sign that he wanted the president's chair for himself. From such a powerful position, Greenstein feared Goldberg (and thus AURA) would possess even greater influence over most major astronomical facilities.[45]

At an AURA meeting in November 1975, Goldberg was surprised to find that, instead of discussing Lee's performance, the AURA board asked him about what Greenstein called "a delicate question"—his impending retirement.[46] Goldberg would turn 65 in the fall of 1978 and faced a mandatory retirement unless the board voted otherwise. Within a few months, the AURA board decided not to continue Goldberg's tenure beyond his 65th birthday. Over two years before AURA rules required Goldberg to step down, it initiated a search for his replacement.

All of these issues—Goldberg's salary woes and forced retirement, the continued debate over the mission of the national observatory, the increasingly powerful role of AURA, and the different visions of how science should be practiced and managed—became festering sources of tension between Goldberg and Greenstein. Their relationship, which had become more fragile since 1971, disintegrated further. Describing the maneuvering taking place as AURA tried to hire a new director as "sheer nonsense," Goldberg wrote Greenstein confidentially, "Just so there will be misunderstanding of my own motivation in this miserable affair, I want you to know that you can dispense with one item on the agenda of the forthcoming Board meeting. . . . I will not accept further appointment of any sort from AURA after I retire as Director."[47]

A lame duck and perhaps feeling he had little to lose, Goldberg became increasingly vocal about AURA's anticipated role in American astronomy. In February 1976, he circulated among senior astronomers a report he penned called "The Future for AURA." Goldberg made the case that AURA should take an even more active role in managing American astronomy. "AURA," Goldberg said, "could be the group that speaks for both university interests and national center interests . . . the entity that communicates freely with universities, centers, federal agencies, and Congress . . . the leader in setting the goals and needs for astronomy . . . The opportunity is there if AURA will take it."[48]

This was exactly the type of centralized control and management of science that Greenstein opposed. As America celebrated its Bicentennial, the two elder statesmen of astronomy clashed repeatedly over the continued expansion of the national observatory. Greenstein began to make his preference for a less powerful national observatory more widely known and argued that AURA should "remain small."[49]

One project he singled out for particular attack was Kitt Peak's 25-meter telescope. Larry Randall, Goldberg's head of engineering at Kitt Peak, recalled how Greenstein told him over lunch that there simply wasn't going to be any 25-meter telescope.[50] Randall, shocked and disillusioned by the announcement, soon left Kitt Peak. Greenstein also told staff scientists at Kitt Peak that, while their current director might be an "activist," Goldberg's replacement would be less inclined in this direction.[51]

Goldberg's call for a more prominent role for AURA prompted a widespread discussion among senior astronomers and science managers. William Golden, for example, asked Goldberg who would be qualified to manage such an expanded astronomical organization. Would not AURA become politically vulnerable if it became more active in Washington? Golden, whose opinions carried much weight in the science community, opposed an expansion of AURA's role as a policymaker and lobbyist. Privately, he told Greenstein that "the situation at AURA and the contention over Leo Goldberg are disturbing to this taxpayer."[52]

In response to Goldberg's report, AURA attempted to define more clearly the mission of the national observatories it managed. When a first draft of this mission statement was available in the fall of 1976, Greenstein circulated it among Caltech and Carnegie astronomers. The paper generated an instantaneous firestorm of protest. One astronomer called it "evil," while another said it "scared the hell" out of him and he hoped that it would never officially leave the AURA board room.[53]

Of particular concern was the claim that the national observatory "should provide uniquely valuable observing facilities which [were] too large, too costly, and too complex" to be developed and managed by universities or private observatories.[54] The mission statement and report did not spell out how AURA (or anyone else) might determine what telescopes were too ambitious for private or university institutions to build. The clear intent of the mission statement, Greenstein's colleagues concluded, was to relegate "all major private observatories

... to a level of mediocrity such that they will not be competitive with Kitt Peak and Cerro Tololo."[55]

The final document, edited in response to comments such as these, was much more modest in scope. The first task of the national observatories, it said, was to provide unique and nationally available facilities that were too large, complex, or expensive to be built or operated by "universities with *small* departments or by *smaller* private observatories."[56] In other words, Kitt Peak and Cerro Tololo were not to dominate the large private or state-funded observatories when it came to building newer and bigger telescopes.

Faced with opposition from Greenstein, members of AURA's board, and influential scientists, Goldberg's move to establish a more powerful presence for AURA and the national observatory fell short. His attempts to transform Kitt Peak into a premiere scientific laboratory that rivaled the best private observatories brought complaints from visitors to Kitt Peak who resented staff privileges and from elite astronomers who opposed a federally funded competitor. Coupled with these professional disappointments was the deteriorating relationship with his wife, and, perhaps, anxiety associated with the growing closeness between himself and Beverly Lynds. In December, he gave notice that he would resign as Kitt Peak's director by September 1977, shortly before his 65th birthday. After he stepped down as director, Goldberg confessed in his personal log that if he had "any inkling of the treatment the Observatory would receive from its sponsor ... I would have been out of my mind to take the job."[57]

The confrontation between Goldberg and Greenstein that culminated in Goldberg's resignation was the last major act in two scientific careers that spanned more than four decades. After resigning as Kitt Peak's director, Goldberg made a long visit to China before returning, uncomfortably, to the national observatory as a research scientist. Greenstein formally retired from Caltech in 1979; but like Goldberg, he remained active in scientific research and offered advice and opinions to the next generation of astronomers planning big telescopes. The events of 1977 effectively ended their relationship. There is no evidence of further correspondence between the two men after Goldberg stepped down. When Caltech held a symposium in 1984 to honor Greenstein, Goldberg was invited but noticeably absent.

New Faces, New Paths

In 1977, oblivious to AURA's painful tribulations, Jerry E. Nelson was a 33-year-old astrophysicist working at the Lawrence Berkeley Laboratory in California. Often clad in Hawaiian shirts and described by a reporter as a cross between Howdy Doody and a political activist, Nelson was a true son of California. He was born and raised in the San Fernando Valley, where his father was a machinist for a local aerospace company. Nelson did his undergraduate studies at Caltech, where he took freshman physics from Richard Feynman. The famous physicist's charm helped persuade Nelson to major in physics. In his sophomore year, he met Gerry Neugebauer and Robert B. Leighton, two Caltech professors interested in infrared observing. Nelson helped them fabricate a novel and inexpensive 62-inch telescope mirror that was used in an important infrared sky survey. Still, he did not plan on being an astronomer. After graduating in 1965, Nelson began graduate school in physics at Berkeley.

Nelson finished his doctoral thesis on elementary particle physics in 1972 and took a position at Lawrence Berkeley Laboratory. The relatively unstructured environment there allowed him to pursue research in a variety of areas. He took part in particle physics experiments but was turned off by large-scale research. He published a few papers in astronomy during the 1970s and spent a fair amount of time making optical observations of pulsars at Lick Observatory and Kitt Peak. While he had some experience in building electronic instrumentation for telescopes, Nelson was not part of the relatively insular community of telescope builders, and he lacked serious credentials or recognition among established astronomers.[58]

In the fall of the 1977, his life changed. The chair of Berkeley's astronomy department asked Nelson to join a committee that was considering the future of astronomy at the University of California. At that time, California astronomers were concerned that increased urban light pollution around Mount Hamilton would doom Lick Observatory.[59] Some University of California astronomers were interested in building a conventional telescope around 100 or 150 inches in size at another site. A number of telescope projects of this size were already underway at other institutions. Another one, Nelson recalled, just "didn't have any sex appeal."[60] The University of California astronomers gradually became convinced that only a really big project, with a mirror 7 or even 10 meters in diameter, would get the attention and funding they needed.

Nelson began to devote all his time to researching the design and construction of large telescopes. He knew a 10-meter telescope could not be built simply by scaling up conventional designs, so he studied blueprints, talked to people in the telescope-building community, and burrowed into Berkeley's library. As he developed his design, Nelson received inspiration from two sources.

Like the engineers and scientists who developed the Multiple Mirror Telescope, Nelson was intrigued with the design of radio telescopes. "The image I had in those naïve days," Nelson said, "was that one would build something that looked like a radio telescope, but the quality of the [mirror] surface would be that of an optical telescope."[61] Nelson knew that the reflecting surfaces of large radio telescopes were not monolithic structures, so he began to consider the optical analog of the segmented radio telescope design. Placed together, Nelson believed, separate lightweight segments of glass controlled by a system of sensors and pistons could form the familiar parabolic reflecting surface of a conventional telescope.

Nelson also began to interact with Kitt Peak engineers who were studying concepts for their 25-meter NGT project. "I realized there was a whole field out there," Nelson said, "which was nice to know."[62] He visited Kitt Peak about once a month to brainstorm and exchange ideas. Like Nelson, Kitt Peak engineers were also inspired by the design of radio telescopes. In the summer of 1977, they were exploring a concept for the NGT called the "Steerable Dish."[63] The design's mirror had over 1,000 glass segments and was a direct descendant of the earlier PALANTIR project.

Nelson's regular visits to Kitt Peak and his examination of the Steerable Dish design helped refine his concept for a 10-meter segmented telescope. Aided by a modest amount of seed money from Lawrence Berkeley Laboratory, he assembled a small team to develop his concept. One recruit was Terry S. Mast, a former Caltech classmate who was now a particle physicist at Lawrence Berkeley Laboratory. They were joined by George Gabor and Jacob Lubliner, engineers with experience in hardware, electronics, and computer analysis of complex systems.

When Nelson first proposed that the University of California's giant new telescope have a segmented primary mirror, he encountered a great deal of resistance. Some astronomers at the University of California already had their own ideas for building a big telescope. One of these scientists was E. Joseph Wampler, an astronomer who had been on the Lick Observatory staff since 1966 and who was 10 years older than Nelson. Well known in the astronomy community, Wampler was the designer of sophisticated instruments and a skilled observer. His idea was to have a company such as Corning cast a large, but very thin, monolithic piece of glass for the telescope's primary mirror. Wampler argued that a thin mirror, as large as 7 meters, would weigh much less than mirrors in conventional telescopes.[64] By using less glass and reducing the weight, the cost and engineering requirements of the entire telescope would be manageable, according to Wampler and his supporters.

In the fall of 1977, Wampler and Nelson paid a visit to David S. Saxon, the president of the University of California. Saxon, a former theoretical physicist, listened to their design pitches. Both the 10-meter "mosaic mirror" and the slightly smaller "meniscus mirror" were clever and innovative alternatives to conventional telescope designs. Saxon found Nelson's idea of a giant, segmented mirror particularly appealing. Encouraged by reports he heard about Nelson from Luis Alvarez and others in the physics community, he allocated $270,000 for an in-depth study of the segmented design.[65] Donald Osterbrock, Lick Observatory's director, appointed a special "graybeards" committee of senior astronomers from the California university system to evaluate the two designs and pick a winner. Nelson now faced an uphill battle to convince his colleagues, some of whom resented the young physicist's boldness and good fortune, that a giant segmented mirror was the technological solution on which they should bet their careers.

Soon after his meeting with Saxon, Nelson attended his first large telescope conference. In December 1977, over two hundred astronomers and engineers convened in Geneva for an ESO-sponsored meeting called "Optical Telescopes of the Future." CERN had just completed a 400-gigavolt particle accelerator and the astronomers were treated to a tour of the new physics facility. They could not help noticing the parallels between existing large accelerators and their own plans for the next generation of giant telescopes.

Nelson described his general concept for the University of California 10-meter telescope to his new colleagues.[66] To save space and money, his design would, like the MMT, use an alt-az mount and as short a focal length as possible. The biggest challenge, of course, was making the primary mirror. Nelson presented a number of possible configurations for the segments, including the one he ultimately chose—an arrangement of hexagonal mirrors around a central hole for the Cassegrain focus. Each of the segments would be actively controlled by actuators to maintain the shape of the light-collecting surface. Nelson's talk does not appear to have attracted much notice. At the meeting, his was only one of several papers given by astronomers with bold ideas for bigger telescopes. Donald Hall, for instance, described Kitt Peak's progress on the NGT project. Several different design concepts were under consideration, and staff astronomers, in conjunction with scientists from other institutions, had begun a parallel effort to articulate the scientific drivers for a 25-meter telescope.

Other issues generated much more controversy among the astronomers gathered in Geneva. Traditional, ground-based astronomers were afraid that the upcoming launch of NASA's Hubble Space Telescope would make their skills and facilities obsolete. This was (and remained throughout the 1980s) a cause of some concern for astronomers, mainly because of the uncertainties associated with new and untested tools. Many at the meeting asked how they could select the best design for new ground-based telescopes when they didn't know what the Space Telescope's capabilities or scientific payoff would be.[67]

Attending his last large-telescope conference, Jesse Greenstein came to the 1977 meeting as the conference's most distinguished and experienced spokesperson. Greenstein attempted to alleviate some of his colleagues' anxieties. On the meeting's last day, Greenstein summarized his 40-year career as an observational astronomer who had used telescopes of every size and had experience with detectors ranging from simple photographic plates to the most sophisticated electronic devices at Palomar.[68] Unlike space telescopes, or even particle accelerators, large telescopes on the ground, Greenstein said, "last forever" and would remain the standard tool of the astronomer. He cautioned that the community was living on "borrowed glory." Astronomers had already skimmed the cream from their exploration of new wavelength regimes—radio, x-ray, infrared—thus making future discoveries that much harder. The fundamental question, as Greenstein posed it, was whether the optical astronomy community could, like physicists and radio astronomers, set aside individual, institutional, and national interests to unite behind a single big project.

The Geneva meeting was the apogee of Kitt Peak's NGT program. At that time, no other group had done as much work investigating possible technologies, exploring designs, and interacting with the community. Over 500 people were on the mailing list for the semiregular NGT reports, and Kitt Peak had taken the lead in the design of future giant telescopes.[69] Soon, Hall, Barr, and other Kitt Peak staff would watch as groups at the University of Texas, the University of Arizona, and ESO (along with the efforts of Nelson and Wampler) closed the gap and moved forward.

Recent management changes at Kitt Peak and AURA did not help the NGT program. After Goldberg stepped down as Kitt Peak's director, the AURA board chose Geoffrey Burbidge to succeed him in November 1978. Burbidge, a theoretical physicist, was seen by some as an odd choice to manage an optical astronomy observatory. While some were turned off by his brash and blustery personality, his wife, Margaret, possessed impeccable credentials as an observational astronomer. Both Geoffrey and Margaret had used the Kitt Peak facilities extensively.

When he arrived at Kitt Peak, Burbidge encountered two major problems. Kitt Peak's budget was declining and the institution still had not clarified its role as the national observatory. Many members of the AURA board (and the Users' Committee it selected) believed that Goldberg had placed too much emphasis on developing Kitt Peak as a first-rate scientific institution at the expense of serving the visiting astronomers who were the primary telescope users. Burbidge, for example, recalled that some of the best instruments at Kitt Peak were only available to staff astronomers. "My charge," he said, "was to make the observatory user-friendly so non-staff scientists didn't feel they were second-class citizens."[70]

Once Burbidge took the reins as Kitt Peak's director, the pendulum began to swing away from leadership in research toward service. This earned Burbidge high marks from AURA's Users' Committee but resulted in the swift departure of scientists whom Goldberg had recruited such as Steve Strom.[71] Burbidge also reduced the role that scientists had in management and setting observatory policy. Kitt Peak's perceptions of its conflicting roles was best illustrated by a cartoon in the *KPNO Newsletter*.[72] The drawing shows a man tugging on the sides of a house as he attempts to install a television antenna. The caption reads: "We are a research institute . . . a service institute . . . a service-research institute." Pulled one way, then another, the house eventually collapses.

While the NGT project fell into disarray, Jerry Nelson and his team continued to explore how to build a reliable and accurate segmented mirror. He had decided that the segments would be hexagons about 1.8 meters on a side; the pieces of glass would be only 7 centimeters thick and consequently lightweight. Nelson and his team chose to make their mirrors from a special glass called Zerodur. Manufactured by Schott Glasswerke, a German company in Mainz, this glass was insensitive to temperature changes; therefore, dropping nighttime temperatures at the telescope would not cause changes in the mirror's shape that would reduce image quality.

Nelson's immediate challenge was to demonstrate that he could fabricate the raw glass segments into the desired shape and control them. The overall shape of the mirror's surface presented a special problem. To focus light from distant sources to a single point, the curvature of a mirror must be a parabola. A mirror with a spherical curvature, like that in the PALANTIR's design, is easier to make but focuses light less precisely due to an effect known as spherical aberration. If a parabolic surface is cut into segments, each piece is asymmetric and has a slightly different curvature from that of its neighbors. Due to this effect, Nelson's fabrication of glass segments would be much more complex.

The solution that Nelson and his team selected was based on a technique developed in 1929 by the Estonian telescope maker Bernhard Schmidt. A Schmidt telescope uses a combination of lens and mirrors to image a very wide field. To fabricate the correcting lens for his telescopes, Schmidt carefully distorted the glass, polished it, and then released the stress on it. The glass then relaxed immediately into the shape Schmidt had predicted.

Nelson decided to use a similar approach to fabricate the 36 subtly different mirror segments needed to create a 10-meter parabolic mirror surface. Throughout 1978 and 1979, Nelson and Jacob Lubliner did complex mathematical analyses to determine what forces they should apply to the edges of the low-expansion glass with which they were experimenting. After months of calculation and experimentation, Nelson's group discovered that, after suspending weights unequally around a thin, 14-inch disk and polishing it into a spherical surface, the glass acquired the desired asymmetric shape and proper curvature when the weights were released.[73]

For Nelson, this was an important milestone. In November 1979, his concept passed an even more important test. The University of California's graybeards committee pitted Nelson's design against Wampler's. After much debate, disagreement, and a close vote, they picked Nelson's segmented-mirror technology. Within a few months, the University of California regents, prodded by David Saxon, awarded Nelson over $1 million to refine his concept and develop a full-size prototype.[74]

While Nelson may remember 1979 as the year his innovative segmented-mirror design was finally taken seriously by his colleagues, it was also a banner year for new telescopes. That year, astronomers and engineers held ceremonies to dedicate four new facilities. Three of these were located on top of a remote mountain called Mauna Kea ("white mountain" in Hawaiian), located in the center of the big island of Hawaii. Rising 13,800 feet above the sparkling ocean 50 miles away, the Mauna Kea Observatory formally opened in June 1970 when the University of Hawaii completed an 88-inch telescope there. Nine years later, several more telescope domes dotted the ridge surrounding an extinct volcano. NASA and the United Kingdom each built 4-meter class telescopes specially designed for infrared observing. A scientific consortium of France, Canada, and the University of Hawaii completed a conventional 3.6-meter telescope as well.

The 1970s, in fact, were a heyday for completion of new telescopes worldwide, most managed by consortia of institutions and internationally based. After the dedication of the 4-meter telescopes at Kitt Peak and Cerro Tololo, another 4-meter class telescope jointly operated by Great Britain and Australia entered service in 1975. A year later, ESO began operating a conventional 3.6-meter telescope in Chile for its European member states.

In 1976, the Soviet Union outgunned all of these when its engineers and astronomers finally completed the Bolshoi Teleskop Azimutal'ny ("large alt-azimuth telescope"). The behemoth featured a massive, 6-meter mirror, making it the world's largest telescope. It was also the first giant optical telescope to break tradition and use an innovative alt-az mount. Most Western scientists, some perhaps motivated by Cold War rivalry, dismissed it as a Soviet megaproject built more as a demonstration of the state's technical prowess rather than as a precision scientific tool. Stories circulated in the astronomy community about the poor quality of the telescope's location and the faults of its massive 42-ton mirror resulted in the telescope's reputation as a curiosity rather than as a first-class scientific tool.[75]

An exception to this burst of relatively conventional telescopes was, of course, the Multiple Mirror Telescope. On May 9, 1979, after a decade of design and construction, staff from the Smithsonian and the University of Arizona gathered on Mount Hopkins to dedicate the facility. Already a buzz of excitement surrounded the new telescope. Earlier that year, astronomers at nearby Kitt Peak had observed two faint blue quasars in the constellation Ursa Major that were separated by only six arc-seconds. Closer examination of their spectra with radio telescopes showed that the quasars were practically identical. Scientists proposed several possible explanations for this phenomenon, including the idea that the objects were the first example of a binary quasar. The MMT would provide the answer.

On the afternoon of April 20, Frederic H. Chaffee, an astronomer at Mount Hopkins, and Ray Weymann from the University of Arizona, loaded an old spectrograph into a pickup truck, bounced up the road to the MMT, and mounted the instrument on the telescope. Engineers and technicians were still debugging the complex computer and mirror-control systems but, for three nights, Chaffee and Weymann collected data with the new telescope. Chaffee, who later became the director of the MMT Observatory, recalled, "Within half an hour of when we had first pointed the MMT, we could see the data coming in and you realized what it was. It was just one of these epiphanies. I remember going out into the dome and just screaming with joy."[76]

The "binary quasar" was actually a single quasar. Between it and the MMT's mirrors was a massive elliptical galaxy. The image of the quasar was split by the intervening galaxy's gravitational field so it appeared to observers as dual objects with identical redshifts and spectra.[77] What had excited Chaffee so much turned out to be the first gravitational lens. The discovery brought the MMT a good deal of attention and some grudging respect from critics as Weymann and Chaffee demonstrated that it could indeed be a productive research tool. Moreover, the news suggested that an even larger version of the MMT might be the best design for future large telescopes.

In January 1980, Geoffrey Burbidge and the Kitt Peak staff hosted a conference at the national observatory in response to the snowballing interest in new telescopes. This was the first major American meeting devoted to telescopes since 1965. Over 200 scientists, engineers, and science managers from all over the world attended and presented papers ranging from the specifics of optical design to how to balance money, science, and politics in astronomy. The meeting took place as the National Academy of Science was carrying out its third decadal review of astronomy, chaired this time by George B. Field of the Harvard-Smithsonian Center for Astrophysics.

The meeting was significant for two reasons. It was the first such meeting that the national observatory organized, thus indicating Kitt Peak's then-prominent role in planning and designing the next generation of telescopes. The conference also demonstrated that astronomers' interest in the design of future instruments had reached a critical mass. "We learned first-hand just how strong the interest was in so many places," one participant said, "It was very clear that lots of groups weren't just talking about big telescopes but that they were beginning to do things. 1980 and onward was when serious development work started."[78]

Several common themes emerged from the presentations, panel discussions, and lunchtime conversations that attendees shared during the six-day meeting. There was a great deal of discussion about the capabilities of new technologies and how these would affect astronomers. One of the new tools was the charge-coupled device (CCD). CCDs resemble photographic plates in that they not only record the amount of light hitting them but also where that light falls. They are made of tiny, light-sensitive capacitors with an array of electrodes sandwiched on the thin surface of a semi-conducting material. Light falling on each pixel of the CCD generates an electric current proportional to the number of photons hitting it. CCDs store this charge and this electronic signal is read out to form an image.

As was the case for early electronic detectors, astronomers found themselves adapting a commercial product originally developed for other purposes for use at the telescope. Researchers at Bell Labs first revealed CCDs to the public in 1970. The military, of course, had already integrated them into its latest photoreconnaissance satellites.[79] Beginning in the early 1970s, astronomers began to experiment with these devices at ground-based telescopes. A few years later, scientists convinced NASA and their colleagues that instruments on the Hubble Space Telescope should use the new solid-state technology.[80]

In the 1970s and 1980s, CCDs were very expensive and, unlike photographic plates, they were small and thus not well suited for recording large areas of the sky. CCDs, however, were much more efficient and sensitive than the photoelectric devices and image tubes that astronomers had developed during the 1950s and 1960s. The electronically generated images could be stored in digital form and later massaged by computer-processing techniques to improve a picture's resolution, adjust the contrast, and remove undesirable noise.[81] By the late 1970s, the number of publications about the astronomical uses of CCDs had exploded. Lauded as the detector of the future, several astronomers at the 1980 Tucson conference insisted that they should follow the space astronomers' lead and incorporate CCD technology into any new telescope designs.

Anxieties and expectations that astronomers had about NASA's plans for space telescopes, then scheduled for launch in the mid-1980s, were powerful. Leo Goldberg emphasized that the Hubble Space Telescope and the NGT (or its equivalent) would and should complement each other, and "erase completely the dividing line between ground-based and space-based astronomers."[82]

Goldberg also stressed that the case would be weak for a national, federally funded telescope featuring anything smaller than a 20-meter mirror. Its support among astronomers and congressional members would be reduced further if it did not have unique capabilities, like being optimized for infrared as well as optical observing. Finally, and perhaps most importantly, Goldberg insisted that unless optical astronomers united behind a single large project (as radio astronomers had done for the Very Large Array 10 years earlier), no NGT, regardless of design, was going to be built.

Goldberg's warning stemmed from a noticeable and, to some, alarming trend seen at the conference. There were now over half a dozen serious efforts underway worldwide to build telescopes larger than the 200-inch and Russia's 6-meter telescope. Who would fund and build all of these telescopes? Speakers lamented the fragmentation of the astronomy community's efforts throughout the meeting and implored their colleagues to get behind a single project and move forward. Neville J. Woolf, from the University of Arizona, took this a step further and made a spirited call for Kitt Peak to shrug off its reputation of having second-class citizenship and take the lead in uniting efforts, even internationally.[83]

Several representatives from the NSF attended the meeting. They were not shy about offering opinions about the challenges facing astronomers. Larry Randall, who took a management position at the NSF after leaving Kitt Peak, told astronomers that they were "very conservative, made up of individualists, and quite parochial."[84] William Howard, a former radio astronomer in charge of the NSF's Astronomy Division, attempted to explain funding and political realities to the scientists who depended on his agency's support. He reminded them that George Field's decadal review could not make precise recommendations if astronomers simply presented it with a "smorgasbord of activity" and failed to support a particular plan. The Office of Management and Budget was even less likely to ignore a lack of unity among optical astronomers; it did not distinguish between space, radio, and optical astronomy. To this important agency, a telescope was a telescope was a telescope.

To help put their ideas and ambitions in perspective, Howard suggested that they take the cost of the Very Large Array (about $106 million in 1980 dollars) and develop a concept around that benchmark. "The instrument ought to be designed in such a way that it is nationally available, to the extent that national funds are used," Howard said, "and it should be sufficiently general purpose to assure broad participation and support by as many different subgroups of optical and infrared astronomy as exist."[85] Congress, he warned, had noticed that "not a significant number of optical facilities have been closed down," and he suggested that optical astronomers be more receptive to closing older telescopes, as the radio community had done, in order to present a more sympathetic case.[86]

Scientists now appeared to have several possible strategies they could follow in order to realize the next generation of giant telescopes.

Astronomers from the University of Texas showed evidence that a thin and lightweight glass meniscus could serve as a light-collecting area if properly supported, while a telescope using several 5-meter mirrors—a scaled-up MMT—on a common mount was an option that astronomers from the University of Arizona liked. Scientists and engineers gave Jerry Nelson a round of applause when he showed data proving that the "bend and polish" technique he championed had yielded a small prototype for a segmented mirror. One engineer was moved to remark, "Given what I see now, plus some money, we engineers will give you guys the moon."[87]

Despite these movements toward consensus, as the Reagan era began, the divide between optical astronomy's "haves" and "have-nots" continued to grow. Astronomers and engineers found themselves wanting not only money but also community consensus and a demonstratively superior technological path to larger telescopes.

CHAPTER 4

Paper Telescopes

In 1980, when Kitt Peak held its "Telescopes for the 1990s" conference, a minority of scientists still contended that a collection of smaller telescopes was superior to a single large facility. "After all," one said, "more astronomers go to work in a Volkswagen than a Rolls-Royce or Ferrari."[1] Most astronomers, however, and engineers agreed that bigger was better, not least because giant telescopes were more likely to generate crucial political and public support. As one advocate of bigger telescopes phrased it, there was simply an "undeniable excitement and allure of very large projects" that was hard for the public and politicians to resist.[2]

Astronomers' desire for bigger telescopes transcended public relations and political strategy. Larger telescopes promised, first and foremost, to be essential research tools. As such, scientists needed to offer compelling scientific justifications for the massive amount of time and money their construction would demand. One of the more outspoken champions of bigger telescopes was Sandra M. Faber. An astronomer at Lick Observatory, Faber also served on George Field's decadal survey committee from 1979 to 1981. She worked hard to develop and impress upon her colleagues the justifications for building bigger telescopes, especially the California 10-meter project.

Large telescopes offered astronomers the opportunity to do more than just collect more light. With increased size came improved angular resolution, the ability to observe fainter objects and to carry out research much more quickly—as much as a factor of 200 times more efficiently at some wavelengths, according to calculations by Faber and others.

Such new capabilities would, Faber argued, "open up virgin territory that would otherwise remain beyond our reach."[3] One area astronomers were especially interested in understanding was the large-scale structure of the universe. Before about 1980, theoretical astronomers generally accepted that, at sufficiently large scales, matter throughout the universe was distributed uniformly as the standard Big Bang model suggested. Photographs taken of galaxies showed them adrift in space, everywhere bunched in small groups. Occasionally, however, astronomers saw thousands of galaxies in superclusters that were millions of light years across and clumped together by gravitational attraction.

About the same time that the Field committee was writing its report, astronomers' observations began to suggest that the universe was a lot lumpier and more confusing place than theorists had imagined. In 1981, for example, a team of astronomers led by Harvard's Robert P. Kirshner found an immense void in the direction of the constellation Bootes. In this region, some 150 million light-years across, the density of bright galaxies was only about 20 percent of their typical value throughout the universe.[4] Initial surveys of galaxy distribution helped astronomers develop three-dimensional maps of galaxies in space that suggested they were not distributed randomly but instead formed great sheets with vast empty regions between them.

Many astronomers became extremely interested in understanding how common such large-scale structures were and how they could have formed. Larger telescopes were seen as one tool to look more deeply into space (and, thus, back in time) to measure the redshifts—the amount by which the wavelength of light from a receding astronomical object is lengthened—of large numbers of distant galaxies and their distribution more rapidly than could be done with smaller instruments.[5] Improved understanding of large-scale structure was, of course, only one of a myriad number of research areas in which astronomers like Faber saw a role for large telescopes. Bigger instruments would also "open the door to qualitatively new science" in the study of quasars, the distribution of so-called dark matter, and the evolution of galaxies. In short, there were few research areas that would not benefit from new and bigger telescopes and astronomers were understandably eager to see them built as quickly as possible.

By 1980, the National Science Foundation had spent almost $900,000 on the Next Generation Telescope program. There was disappointment both at Kitt Peak and in the science community at the slow progress the national observatory was making toward a new national telescope. Some engineers and astronomers at Kitt Peak were just as frustrated by the NSF's apparently contradictory advice that they explore many new telescope technologies and, at the same time, build a working instrument.[6]

In 1979, Harlan J. Smith was the director of McDonald Observatory and chair of the University of Texas's rapidly growing astronomy department in Austin. The 55-year old astronomer was also the chair of Kitt Peak's advisory committee for the Next Generation Telescope. He was concerned about the excessive period of time needed to design, fund, and build a new large telescope. Smith and other astronomers thought that bigger and better space telescopes might soon become cheaper than ground-based observatories. As Smith expressed it, "the several hundred million dollars which the 25-meter NGT would represent might prove to be our equivalent of . . . building battleships rather than aircraft carriers." If true, then these "immensely expensive investments of relatively limited utility" might soon have to be abandoned.[7] Anxious that space telescopes might supplant ones on the ground, Smith and others advised Geoffrey Burbidge that Kitt Peak should start a new ground-based project as soon as possible, even if this meant scaling back its large telescope dreams to the more manageable and affordable size of 15 meters.

Soon after the January 1980 telescope conference, Kitt Peak announced an exploratory development program for a national 15-meter telescope.[8] Larry Barr recalled, "We needed to focus our efforts and scaled down from 25 meters to 15. Not because of any overwhelming justification for the size but because we felt we could probably build a 15 meter telescope where 25 meters was going to be a real stretch."[9] Concomitantly, Kitt Peak soon proposed a different organizational arrangement for building a new national telescope than it had pursued in the 1970s.

At this time, Kitt Peak, along with the state universities of Arizona,

California, and Texas, was among the most serious contenders for building the first of the new generation of giant telescopes. However, none of these institutions had the technological expertise and fully refined designs necessary to win funding or start construction. Burbidge, therefore, initiated a collaboration whereby Kitt Peak and the three universities would jointly propose a Technology Development Plan to the NSF. By late 1981, Kitt Peak announced that the NSF had finally approved the cooperative plan. Kitt Peak now had $500,000 available for a three-year technology development study and its new project had a name—the National New Technology Telescope (NNTT).[10]

This collaborative approach did not provide the prominent leadership role for the national observatory that Leo Goldberg had envisioned when he began the NGT program. Under the new scheme, Kitt Peak would no longer be the sole, or even primary, institution designing the giant telescope. Instead, Kitt Peak would serve as a coordinator, by distributing research and development funding to the three universities and consolidating their knowledge and experience into a final design for the NNTT. While some of the technology development would be done at the national observatory, most of it would become the province of university-based astronomers and engineers.

If the experiences the science community gained from the Next Generation Telescope program and other efforts demonstrated one thing, it was that the main technical obstacle for bigger telescopes was still the primary mirror. Historian Thomas P. Hughes referred to this type of barrier as a "reverse salient,"[11] comparing the continual improvements made to a technological system with an unevenly expanding military front. The development of some system components often lagged behind that of others, thus hindering the overall improvement of the system. After the 200-inch went into service, astronomers focused on electronic detectors. By 1980, their investments of time and energy resulted in older telescopes having greater light-collecting efficiency. To collect more photons, bigger mirrors were the favored solution. Astronomers and engineers advocated three different techniques for making these mirrors. None was clearly superior and competition was fierce, even antagonistic, as the proponents of the different techniques sought to convince the science community and its patrons to adopt their approach.

The first of these approaches was being pursued by the University of Texas. The University, then awash in money from Texas oil fields, wanted to build a new telescope with a 7.6-meter mirror. By the end of 1981, Texas astronomers had raised over $600,000 toward their goal (of an estimated total cost of about $45 million) and were looking for one or two generous donors to make up the rest. The Texans wanted to win the race for the next giant telescope quickly and with a minimum of untried technological innovation. As one of them said, "We would like to design by the process of unabashed piracy. Other people have good ideas. Let's adopt them."[12]

The Texans began designing their telescope around a single, thin piece of glass—a so-called meniscus mirror. In keeping with their design philosophy, they chose an approach that was not entirely novel. In fact, it was similar to what Joseph Wampler had unsuccessfully proposed for the University of California's 10-meter project a few years earlier. The Texans planned to have a commercial firm make the 7.6-meter mirror from zero-expansion glass. Another optics company would then grind and polish the 4-inch-thick piece of glass. The thin profile would minimize the mirror's weight, while the special glass would prevent temperature differences between the glass and air from producing any convection currents at the mirror surface. Such differences might degrade the telescope's performance (think of heat waves rising above a hot road).[13]

Clearly, however, the NNTT project could not take this same technological path. It simply wasn't possible to cast and polish, let alone precisely support, a thin glass disk 15 meters wide. As a result of its prior commitment to its own telescope project, the University of Texas's role in the NNTT project was limited. The Texas astronomers continued to pursue their own project alone with little financial support from Kitt Peak's technology development program. Their plans were derailed a few years later, however, when the oil economy in Texas collapsed. As donations dried up, astronomers watched their telescope project go into hibernation for several years.

By late 1981, only two mirror options remained as serious contenders for the 15-meter NNTT design. Jerry Nelson and astronomers from the University of California proposed the first. Nelson's vision of the NNTT was taken from the California 10-meter telescope's segmented mirror design. The mirror's parabolic surface would be constructed from several dozen small, lightweight pieces of glass made and polished by a commercial firm. Despite the university's ample funding for technology development, Nelson and his colleagues still faced the challenge of building a prototype that demonstrated an inexpensive means of polishing and shaping several dozen mirror segments precisely, joining them into a parabolic surface, and controlling them predictably and reliably.

The second design came from astronomers at the University of Arizona. As early as 1979, Arizona astronomers expressed a preference for an even bigger version of the new and innovative telescope on Mount Hopkins they co-owned with the Harvard-Smithsonian Center for Astrophysics.[14] At the 1980 telescope conference, astronomers from the University of Arizona described a giant multiple-mirror telescope that would combine eight 5-meter mirrors. The result, they claimed, would be a giant telescope that would cost far less than any other comparable design.[15] Therefore, it surprised few people when Arizona scientists and engineers advocated a scaled-up rendering of the Multiple Mirror Telescope for the NNTT a few years later.

The University of Arizona operated an astronomy department as well as its research division, Steward Observatory. Both were led at that time by Peter Strittmatter. A British-Swiss astronomer, Strittmatter did his doctoral work in the early 1960s at Cambridge University. As a postdoctoral student, he came to the United States, first visiting Princeton and then the University of California. He did his first significant nighttime observing with Margaret Burbidge at Lick Observatory and published several papers on quasars and white dwarfs. Strittmatter found the Burbidges fun, unpretentious, and even verging on rebellious. His pioneering spirit also emerged in his enthusiastic support for the original Multiple Mirror Telescope and his avocation of larger versions of it.[16]

Strittmatter's political skills and his ambitious—some even said arrogant—management style helped him become Steward Observatory's director at the young age of 35. Goldberg soon asked him to join the scientific committee responsible for articulating the research possibilities for the Next Generation Telescope. After the committee's report appeared in 1980, Strittmatter became frustrated with Kitt Peak's progress toward a large telescope. He was familiar with the national observatory, having been offered its directorship in 1977.[17] He likened

Kitt Peak's role to that of the Duke of Plaza-Toro, who led his troops from the rear in Gilbert and Sullivan's *The Gondoliers*. Strittmatter doubted the ability of the national observatory to design and build a giant telescope that was a leap beyond anything it (or anyone else) had done before. In his view, the cooperative technology development program that Kitt Peak had established with the three universities gave the national observatory the semblance of leadership while university-based scientists and engineers did the actual design and development.[18]

Like Jerry Nelson, the Arizona contingent planned to expand on a familiar design with which they had personal experience. The original Multiple Mirror Telescope was made possible because Aden Meinel obtained valuable military-surplus mirrors. Such good luck was not going to happen twice. As a result, the biggest challenge facing the University of Arizona coalition in 1981 was how to obtain several lightweight mirror blanks, each as big or bigger than any in the world. Turning their backs on exotic and expensive low-expansion glasses, Strittmatter and the Arizona astronomers chose to return to the techniques and materials Corning had used to cast the mirror for the 200-inch telescope decades earlier. Instead of relying on industry, the Arizona group proposed the much more daring strategy of making its own mirrors.

The Great Leap Backward

In 1980, when Strittmatter returned to Tucson after a sabbatical, he heard that one of his department's astronomers was excited about ideas for making large mirrors. He asked Strittmatter for permission to use a loading dock near Steward Observatory for glass-melting experiments. "I asked him what kind of mirrors he wanted to make," Strittmatter said, "and he replied, '8-meter mirrors.' I managed to contain my mirth and that's how it moved from his backyard to Steward."[19]

The astronomer was James Roger Prior Angel and he would do as much as any single person to change the postwar concept of giant telescopes. Roger Angel grew up in a small town near London in wartime England. From an early age he tinkered with electronics, fixed radios, and built gadgets like a primitive acoustical radar. Aside from national pride in the famous Jodrell Bank radio telescope, he didn't have any significant interest in astronomy as a child.[20] When it came time for university studies, Angel had the choice between studying engineering at Cambridge or physics at Oxford. Physics was more appealing so he headed to St. Peter's College at Oxford. After graduating in 1963, he spent a year at Caltech before returning to Oxford for his doctorate. While in England, he built experimental equipment for atomic spectroscopy.

After finishing his doctoral degree in 1967, Angel accepted a postdoctoral position at Columbia University. He began to help build rocket-borne experiments to do x-ray astronomy that were launched from White Sands Missile Range. He enjoyed the pace of the work. "You could conceive of a new experiment, build the equipment, integrate it into some rocket with the telemetry, take it to New Mexico, get your data, analyze it, and write it up. The whole cycle might take 18 months. It was on a very human scale."[21] Angel began to combine results from optical observations with his rocket data. He and his collaborators used telescopes at the McDonald Observatory in Texas and Palomar, often bringing equipment they had built themselves. Angel, trained as an experimental physicist, also brought a different attitude from that of traditional astronomers. "We treated observing almost like a physics experiment," he said. "The telescope was just something to bring light to your detectors."

Angel's research attracted the attention of senior astronomers like Jesse Greenstein, and he moved from Columbia to a new post at the University of Arizona. During that time, Angel became increasingly involved in designing astronomical instrumentation. The plethora of telescopes near the University of Arizona made it convenient to build a new gadget and try it out quickly. The design and commissioning of the MMT provided yet another inspiration for Angel.

One of Angel's close collaborators was Neville J. Woolf. Like Angel, he was British with a background in experimental physics. After Woolf moved into theoretical and observational astronomy, they published over 40 papers together. Woolf and Angel regarded instrument building as an entirely appropriate activity for astronomers. After coming to Arizona in 1977, Woolf became an advocate of the MMT's design and encouraged Angel to imagine a bigger version of it. They were impressed with the performance of the MMT's mirrors, made lightweight by their honeycomb design. Angel soon plunged into the same area of investigation that George Ritchey had pursued 50 years earlier—making very large, yet lightweight, honeycomb mirrors.

After attending the 1980 Kitt Peak conference on large telescopes, Angel used some insulating bricks and an electric heater coil to build a small kiln at home. He melted glass from a fluorescent lighting tube first before graduating to Pyrex custard cups. It was around this time that Angel began seeking lab space, partly because his family had noticed that the backyard experiments were diminishing their supply of kitchenware.[22] For years, Angel kept chunks of fused glass interspersed with bits of firebrick in a storage closet near his office as a reminder of his early experiments.

At the end of that summer, John M. Hill, a graduate student in his early twenties working with Angel, returned to Tucson. Hill chose the University of Arizona because of its reputation for developing astronomical instrumentation. His dissertation research, on which Angel was an advisor, described techniques for using optical fibers to take the spectra of many objects at once—multiobject spectroscopy, in astronomers' parlance. Angel showed Hill the Pyrex cups he had fused together and told him that they could make telescope mirrors this way. Despite the occasional black widow spider in their equipment, the whole endeavor stirred Hill's enthusiasm, and the two scientists began to scour craft shops in Tucson for useful materials.[23]

One of their early ideas was to make mirror blanks in the same way that commercial firms had fabricated the lightweight mirrors used in the MMT and the Hubble Space Telescope. Angel and Hill delved into the technical literature about mirror making. Angel's files from this period contain papers dating to the 1930s on glass-melting technology, unclassified reports on lightweight mirrors for spy satellites, and technical publications from Corning and Kodak.

Hill and Angel began to wonder whether, instead of joining many pieces of glass as Ritchey had done, casting a mirror blank from molten glass was the best route. Unlike many commercial mirrors, which were made from expensive, low-expansion specialty glasses, Angel and Hill opted to use ordinary borosilicate glass. Similar to Pyrex, it was much cheaper than the fancier low-expansion glasses. It also melted at a much lower temperature, making it easier to cast into complex shapes.[24] Because they had chosen to make mirrors using techniques and materials similar to those employed by the workmen who cast the

200-inch mirror, some astronomers joked that Angel's approach was the "great leap backwards."[25]

Initial funding for Angel's work came from the University of Arizona, but there was comparatively little of it. After the NSF funded Kitt Peak's technology development proposal for the NNTT in 1981, Geoffrey Burbidge directed much of the money to support Arizona's mirror-making experiments. Angel's "mirror group," as it was called, received $60,000 the first year, $150,000 the next year, and more afterward.[26]

The mirror group soon grew to about half a dozen people. Angel relocated its operations to the basement of the Optical Sciences Building a few hundred yards away across campus. In their new home, they built a furnace big enough to melt enough glass for a 2-meter mirror blank. At a conference in Germany in 1981, Angel and Hill presented their first paper that described detailed plans to make honeycomb mirrors of borosilicate glass.[27] Acknowledging their debt to Ritchey, the MMT, and spy satellite mirrors, they showed a test casting 60 centimeters in diameter and reported plans to build a bigger furnace for making mirror blanks as large as 8 meters. While this paper was an early demonstration of their technique's possibilities, the unrealistic schedule they outlined would bedevil Arizona mirror-making efforts for years.

The classic way of making a mirror blank was to cast it in a stationary furnace with a flat surface and then have a team of opticians laboriously grind the glass away until it had the proper curvature. Angel proposed doing away with this onerous intermediate step by melting the glass in a specially designed rotating furnace. When a liquid is rotated at a constant velocity, its surface assumes the shape of a parabola, which was exactly the curvature astronomers needed.[28]

Angel's basic plan was as follows: Place small chunks of borosilicate glass, which melts at about 1200° C, in the bottom half of the rotating furnace. Stack the glass on a mold of ceramic cores held in place with silicon carbide pins. Lower the dome of the furnace, lock it in place, and begin heating. As the temperature slowly rose, the furnace would begin to rotate. The glass would gradually soften and flow through the carefully spaced gaps around the solid ceramic cores, while a smooth surface of glass formed on top of them. The furnace's rotation would force the surface of the glass into a parabolic shape and this top layer would become the mirror's faceplate. After a long cooling period, the furnace would be opened. The glass disk would be lifted out and technicians using high-pressure water jets could wash away the mold material. What was left would be a glass mirror blank with a roughly parabolic surface and mostly hollow honeycombed interior.

Angel, Hill, and the other members of the Arizona mirror group developed and refined this basic mirror-casting process over many months. It entailed solving a whole host of mechanical and materials problems in a cut-and-try fashion along the way. The mirror group's logbook entries say a great deal about the technical problems they contended with daily and their feverish pace of work. On August 14, 1982, after pulling several late shifts babysitting their furnace, John Hill scrawled that he was finally "going home to sleep in a real bed."[29]

Throughout 1982 and into 1983, Angel and the mirror group continued to practice mirror casting. Their work in the basement of the Optical Sciences Building often wreaked havoc when soot from the furnace interfered with more precise optics research. Supplied with another $250,000 from Kitt Peak, they gradually moved their operation to an abandoned Jewish temple on the fringe of the campus. Engineers held meetings in the former rabbi's office and the staff painted a big window in the building to resemble the hexagonal pattern of the molds in which they melted their glass. In its early days, the Steward Observatory Mirror Lab, as it became known, relied on student volunteers who formed a tightly knit group, dating each other and, in one case, marrying. Articles about their activities began to appear with increasing frequency in astronomy publications and popular science magazines.[30]

One of the Mirror Lab's most challenging tasks was building a prototype rotating furnace. Daniel Watson was one of the first students hired as a technician and he worked at the lab for almost 20 years. He and the other staff helped build the first rotating furnace from scavenged parts. They mounted the furnace structure on top of a large ball bearing taken from a decommissioned radio telescope.[31] An "oven pilot" rode on the rotating furnace—sitting in a seat liberated from a 1966 Ford Mustang—and monitored the casting process by keeping a close eye on the furnace temperature and the rotation speed. Watching the computer monitors while spinning up to 15 times per minute made more than one person sick. Dan Watson was one of the few who mastered the art of staring straight ahead for several hours with stereo speakers strategically placed to either side of his head while he oversaw the casting. Part of his success, Watson claimed, was due to a diet of dark beer and Mexican food.[32]

Pursuing a technological solution for bigger telescope mirrors, however, was not enough. Like Jerry Nelson with his segmented mirror, Arizona astronomers also had to convince the science community that honeycomb borosilicate mirrors were the right path to bigger telescopes. This meant getting mirrors out of the lab and into the observatory. Establishing credibility for their technology was a time-consuming, but necessary, task for Angel, Woolf, and Strittmatter. They worked extensively to convince colleagues planning new telescopes to base their designs around the lab's forthcoming giant mirrors.[33]

Unlike Jerry Nelson's mirror technology program, Angel and the Arizona operation did not have the benefit of a large amount of seed money. By 1983, the NSF, through Kitt Peak's technology development plan, had given the Mirror Lab almost $500,000. This funding was incremental and much less than the resources the University of California made available to Nelson. The Mirror Lab slowly inched its way toward ever-bigger castings. The first mirror blank to come out of the rotating furnace at the former temple was only 75 centimeters in diameter. Angel's lab had a long way to go before one of his mirrors became the foundation for a world-class giant telescope.

Dueling Designs

The competition between segmented and honeycomb mirror technologies for the NNTT became an exciting race. Since the 1980 telescope conference, Jerry Nelson and his mirror team had labored to design and build a full-scale prototype that demonstrated their ability to precisely control mirror segments. They concentrated their tests on a 1.8-meter hexagonal mirror segment linked to a second reference segment. Nelson's group built the apparatus to test the ability of its sensor and actuator systems to control the two segments with respect to one another. Although Nelson said that scaling up from the prototype to a 10- or even 15-meter mirror would be "just like paving your bathroom floor," getting to this stage of technical readiness would take several years.[34]

While much of Nelson's focus remained on the University of California's 10-meter telescope, the techniques developed by his group were directly applicable to the 15-meter NNTT. Each segment in the telescope would rest on three actuators. These would apply small forces over a hundred times a second to adjust the segment's position with submicron accuracy. Nelson's team designed supports that would keep the segments centered both in side-to-side and up-and-down directions. Sensors also had to be designed that could register the relative positions of all the segments and transmit the information many times a second to maintain the global shape of the mirror. All of these parts had to work reliably in cold and harsh mountain conditions and be easily reproduced to boot. A schematic of the entire segment control system for the 10-meter telescope was tremendously complex—168 sensors, 108 actuators, dozens of segment supports all woven together by a web of wires and feedback mechanisms. Plans for the 15-meter NNTT were even more complex and, for both telescope designs, computer control and modeling simulations were essential tools. Where Roger Angel's approach harked back to a time when teams of muscled workmen ladled hot glass into mirror molds, Jerry Nelson's team was staking its reputation on microelectronics, computer software, and systems engineering.

Burbidge's support of the NNTT program brought a significant shift in the national observatory's role. Other than helping Nelson develop the hardware, computer programs, and techniques necessary to test his stressed-mirror polishing approach further, Kitt Peak engineers and astronomers were largely confined to roles as technical managers for the NNTT project. Astronomers and engineers from the Arizona and California universities took over much of the important technological development and telescope design. Each institution in the cooperative effort hoped to derive some special benefits that would further its own telescope ambitions. While this approach had payoffs for Angel, Nelson, and their colleagues, it established a relationship between Kitt Peak and the other institutions, especially the University of Arizona, that would be problematic, if not outright hostile, in coming years. The difficulty stemmed from Kitt Peak's goal of a giant national telescope, while astronomers at the University of California and the University of Arizona wished to build their own facilities.

Kitt Peak's participation in plans to build the NNTT was more limited than the leadership role that Goldberg had originally envisioned when he initiated plans for a giant national telescope in 1974. Despite the observatory's reduced role, Burbidge was a firm advocate of the NNTT program and the collaborative effort to explore the technologies necessary for it. He lamented that, unlike high-energy physics, there were few optical astronomers "inclined to make sacrifices to build big machines . . . The astronomical community is happy to support a project only so long as it doesn't affect anything they already have."[35] He recalled a time in 1982 when money for Kitt Peak was especially tight. Members of Kitt Peak's various user and visitor committees indicated they were quite willing to abandon the NNTT project to prevent any existing telescopes from being mothballed.

Years later, after Burbidge stepped down as Kitt Peak's director, he noted that he held a card up his sleeve when it came to the NNTT. Barry Goldwater, Arizona's influential senator, was a big booster of Kitt Peak and occasionally visited the observatory. Burbidge, possibly displaying some of his tendency for rhetorical overstatement, believed that if the situation for the NNTT became truly desperate, he could always make a direct pitch for the project to Goldwater, perhaps even offering to name the giant telescope after the senator.[36]

Scientists and engineers interested in the evolving NNTT project had several opportunities in 1982 and 1983 to discuss the latest plans for the 15-meter telescope. In March 1982, about 400 people attended meetings in Tucson that explored new technologies for giant telescopes. Guests were treated to tours of Kitt Peak and the MMT facilities, and Jerry Nelson and Roger Angel explained their respective mirror technologies. In light of the competition that developed later between Angel's Mirror Lab and commercial firms, one of the more noteworthy talks came from representatives of Corning Glass Works.[37] The company observed that the market for large mirrors had been traditionally small, with astronomers, the military, and reconnaissance agencies ordering custom products. The mirror market was shifting, said the representatives, to "increased sizes, increased volume, and more standardization of design." Corning's presentation made it clear that it was not going to relinquish its tradition of making large mirrors to entrepreneurial astronomers like Nelson and Angel.

Throughout 1982, the NNTT designs of the California and Arizona groups were gradually refined and made public. In June 1982, Kitt Peak invited three dozen astronomers and engineers to an NNTT Design Workshop. This three-day meeting was held in the relaxed resort atmosphere of Flagstaff, Arizona; its purpose was to make known the findings from Kitt Peak's Technology Development Program and start the process of selecting a final design for the NNTT.[38] Larry Barr, project engineer (and soon the project manager) for the NNTT, cautioned those gathered that the telescope's specifications would be "severe" and predicted that choosing a design would be neither "dainty or tidy."[39] Jerry Nelson and Roger Angel, in a scene that would be repeated many times in the next two years, presented their respective groups' designs for the NNTT.

Figure 9 shows scale models built by Kitt Peak staff of the two competing concepts. Between the two NNTT designs is a model of Kitt Peak's existing 4-meter telescope. The fact that the 4-meter telescope itself weighed some 500 tons and was covered by a dome one hundred feet in diameter suggests the absolutely gargantuan size of both NNTT designs.

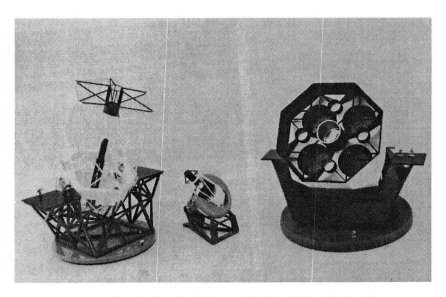

Figure 9. The two designs for the 15-meter National New Technology Telescope. California's segmented design is on the left, Arizona's "four-shooter" is on the right, and the 4-meter telescope at Kitt Peak in the middle provides a sense of scale. Courtesy of NOAO/AURA/NSF.

The multiple-mirror version of the NNTT championed by the University of Arizona is shown on the right. Angel and Woolf based their final concept, which they called the "four-shooter," on four 7.5-meter honeycomb mirrors to be created by the Mirror Lab.[40] These would be 75 percent lighter than traditional mirror blanks of the same size and arranged on a common mount. The secondary mirrors would be placed in a unique upper section that could rotate like a carousel. Two different sets of four secondary mirrors could be brought into play, depending on whether astronomers wanted to do wide-field or infrared observing. The Arizona version of the NNTT could use all four primary mirrors at once and simultaneously send all the light collected to a single instrument at a common focus. In another configuration, each 7.5-meter mirror could be used as part of four "separate" telescopes—pointed in the same direction, of course—with different instruments analyzing the photons collected.

Jerry Nelson and his colleagues from the University of California naturally recommended that the NNTT be a scaled-up rendition of their 10-meter telescope design. The 15-meter version would have 60 mirror segments and use the sensing and alignment technology Nelson's team was developing. As with Arizona's four-shooter, interchangeable secondary mirrors would provide different optical arrangements for wide-field and infrared work. Because Nelson's group had already formulated a precise set of specifications for a 10-meter telescope, his team easily scaled these up by 50 percent. His cost estimates suggested that an NNTT with a segmented primary mirror would cost about $82 million in 1982 dollars.

Despite their different approaches, the two telescope designs shared several features. Both the California segmented and the Arizona multiple-mirror concepts emerged from an established tradition of conceptualizing large telescopes and each adhered to a clearly identifiable technological style. Both designs were scaled-up versions of planned or existing telescopes but untried on the scale proposed; neither was obviously superior to the other. In terms of design, both incorporated weight-saving techniques for the primary mirror to achieve a large and stable collecting area. The two designs, like all the large telescopes astronomers were discussing, incorporated a space-saving alt-az mount like that used for the original Multiple Mirror Telescope. Angel and Nelson achieved a further reduction in the size and cost of the telescope and its enclosure by proposing faster focal ratios, in the range of $f/1.75$ for the primary mirrors. Finally, both telescope concepts were designed around research programs in which optical astronomy was increasingly coupled with observations in the infrared region.

Building any large tool for science requires tremendous resources, including but not limited to funding. The design and construction phases are lengthy and may occupy a significant part of an astronomer's or engineer's career. Moreover, the institutions and individuals advocating particular designs, such as those for the NNTT, invested their professional reputations along with considerable funding in a particular design. With two viable NNTT contenders, Geoffrey Burbidge and his staff were faced with choosing a winner.

Community Participation

To select a design for the NNTT, Kitt Peak was obliged to solicit a broad range of opinions and recommendations from the national community of astronomers. The last time astronomers were planning major new national ground-based telescopes was in the early 1960s when Kitt Peak and Cerro Tololo were developing their 4-meter facilities. The basic design of these, however, followed that of the 200-inch. The two competing concepts for the NNTT diverged from anything built before and depended on unproven technologies.

Dozens of astronomers and engineers from outside the national observatory participated in shaping the NNTT's design and debated its scientific usefulness. This inter-institutional cooperation was a relatively novel experience for the American optical astronomy community, but entirely understandable given that the NNTT was expected to be the most expensive national facility for ground-based astronomy ever built.

Breaking the "Palomar paradigm" wasn't the only act that brought astronomers together. In the two decades since AURA won approval for its 4-meter telescopes in Arizona and Chile, the nature of long-range planning in the American astronomy community had changed. The National Academy of Sciences had already commissioned and released two decadal surveys of astronomy. Preparing these reports required astronomers to negotiate and prioritize their wish lists for new telescopes. By 1980, astronomers no longer questioned the value of this activity, and securing the blessing of the Academy was essential for any major new national initiative in astronomy to go forward.

In April 1978, after the NAS picked George Field to chair its third decadal review, the Harvard-based astrophysicist assembled a committee of prominent astronomers to consider the options. Joseph Wampler chaired the Panel on Ultraviolet, Optical, and Infrared Astronomy, which included Roger Angel and Jesse Greenstein as consultants as well as Geoffrey Burbidge and Fred Gillett from Kitt Peak.

Wampler's panel wanted, of course, to recommend that the NSF fund some type of much larger optical telescope, 15 to 25 meters in size. They were faced with the challenge of making this new telescope appear complementary to NASA's Hubble Space Telescope. The Hubble launch had been pushed back to 1985, but this date was still long before any giant ground based telescope would be built. Some members of the astronomy community, however, opposed more ground-based telescopes and instead argued that the astronomy community's goal should be even larger space telescopes.[41] By this time, some astronomers were already proposing telescopes in space and on the moon with mirrors as large as 20 meters.[42]

In response, Wampler identified three basic conundrums faced by ground-based astronomers. First, the absence of a clearly superior solution for big telescopes created divisiveness in the community. Second, because scientists could use existing ground-based telescopes to address such a wide range of research problems, some of astronomy's patrons assumed that these scientists did not think a giant new telescope was necessary. Finally, the new telescopes under consideration were so gigantic and expensive that there were no preexisting models to guide the equitable management of the proposed facilities. For example, if a giant new telescope were built with public funds, how would time on it be awarded? Any suggestion to award observing time to only three or four dozen astronomers annually for their personal research promised only more community divisiveness over the project.[43]

When the NAS released the Field committee's decadal survey in January 1982, it contained four major recommendations for the next decade of astronomy. Two were for ground-based astronomy. As was the case in the previous survey, radio astronomers' wishes received higher priority than those of optical astronomers. The report's first ground-based recommendation was that the NSF fund and build the

Very Long Baseline Array, a connected network of ten 25-meter radio telescopes spread over the United States with an estimated cost of $50 million. Second on the list of ground-based initiatives (and third overall) was what the report called a New Technology Telescope—a 15-meter facility with a $100 million price tag for making optical and infrared observations. The word "National" was omitted from the project's name, perhaps in light of the fact that several groups besides Kitt Peak were pursuing large telescopes. Field's report acknowledged that the construction of such a facility would be possible only because of the recent advances made in "optical fabrication techniques, design concepts, and electronics." The 15-meter telescope was expected to be a cornerstone of astronomy for the next decade and, the report emphasized, "the design studies needed before the [telescope] can be constructed are of the highest priority and should be undertaken immediately."[44]

Despite the high ranking that the Field committee gave to the 15-meter telescope project, the NSF did not dramatically increase the NNTT's funding or visibly promote the national telescope project. In fact, the opposite happened. In March 1981, NSF representatives appeared for their annual hearing before the House of Representatives' subcommittee that controlled appropriations for the NSF's annual budget. Because this group of lawmakers had billions of discretionary dollars to allocate over such areas as housing, veterans' affairs, and space exploration, the subcommittee—called the Cardinals of Capitol Hill by one writer—wielded enormous power and influence.[45]

Edward P. Boland, a longtime Democratic representative from Massachusetts, was the subcommittee's influential chair. He was skeptical about scientists' seemingly endless demands for new instruments and, in years past, had questioned the necessity of NASA's Hubble Space Telescope. Boland queried the NSF representatives about their plans for a national 15-meter telescope. Despite the fact that the NSF was asking for only $400,000 to continue Kitt Peak's collaborative technology-development effort, this relatively minor item caught the subcommittee's attention. Boland understood that awarding seemingly insignificant design funds often led his subcommittee to fund an entire project. When Boland asked what effect deferring the design studies might have, the NSF's representative replied that, while it might disrupt efforts already underway, the building phase was far enough off that "the impact would not be that great."[46] In other words, the NSF did not push the NNTT as a major priority. Some astronomers wondered whether this brief exchange foreshadowed an unwillingness on the part of the NSF to advocate for the NNTT even though it was ranked highly in the decadal survey.

Despite his role as the initiator of Kitt Peak's new giant telescope project, Goldberg lost enthusiasm for the NNTT after stepping down as the observatory's director. He disagreed with the direction Kitt Peak had taken under Burbidge's leadership and disapproved of the collaborative program to build the NNTT. Goldberg was also disappointed with astronomers' scientific justification for the NNTT and believed its success depended on astronomers' ability to sell the NNTT as an essential complement to the Hubble Space Telescope.[47]

In July 1982, Goldberg sent a long, confidential letter to Stuart A. Rice, a chemist from the University of Chicago and a member of the National Science Board, the NSF's governing body. That board, composed of Presidential appointees, had authority over the agency's budget and plans. Goldberg told Rice how Kitt Peak's partners in the collaborative effort had their own ambitions to build giant telescopes. "Once these private telescopes are funded," Goldberg asked Rice, "can you imagine the NSF and Congress putting up $100 million for a National Telescope that would increase the aperture only by 50 percent? Nonsense! . . . My feeling is that Kitt Peak ought to leave the field of large ground-based telescopes to the universities that want them." The national astronomy facilities of the future, Goldberg predicted, would be space telescopes. He concluded by suggesting the National Science Board ask "searching questions" about the NNTT, such as "whether the present scenario for two private telescopes followed by a somewhat larger national version is realistic or even desirable." Rice thanked Goldberg for his candor, noting that it was only with information such as Goldberg's that "the National Science Board can intelligently formulate optimal policy."[48]

Despite Goldberg's pessimism and the lack of serious support from the NSF, Kitt Peak's staff remained optimistic that the NNTT would be built, especially with the impetus from the Academy's endorsement. Nelson and Angel continued their technology development efforts and Burbidge assembled a Scientific Advisory Committee (SAC) in March 1983 to select a winning design from the two contenders. Its responsibility was to identify and prioritize the major areas of research that the NNTT might do, evaluate the two competing designs, and, most importantly, recommend one for further development.[49]

Burbidge carefully balanced his panel geographically and in terms of the types of telescopes and astronomical research its members favored. The SAC included astronomers from Caltech; the Universities of Arizona, Texas, and California; and Kitt Peak. Panel members spanned a broad range of seniority, from veterans like Allan Sandage to protégés such as Robert Kirshner and Jeremy Mould. Burbidge asked Robert D. Gehrz to be the group's chair. Gehrz was then a 39-year-old astronomer from the University of Wyoming and a rising star in infrared astronomy. He had worked as a student under Nick Woolf before serving on committees for AURA and the NSF. Gehrz also had recently helped plan an infrared-optimized telescope for the University of Wyoming and was familiar with the technical and scientific issues of telescope building.

When the SAC met for the first time at the end of March 1983, Gehrz explained that his was not a technical evaluation committee but one that would give a scientific justification for why one design was better than the other for doing research. Despite this compartmentalization, his committee was obliged to consider the NNTT's basic specifications and performance requirements. Responding to the possibility that his committee members' own institutional affiliations might cause conflicts of interest, Gehrz said, "We all probably have our prejudices—we should set these aside . . . to develop an unbiased recommendation."[50] He acknowledged that many of the SAC's challenges were political. "There were a lot of scientific arguments," Gehrz said, "but they were all marginal. Picking the right design came down to whether you believed your calculations and what was most important to you."[51]

One of the youngest members on Gehrz's committee was Roger L. Davies, an English astronomer who had recently finished his Ph.D. at Cambridge. After coming to the United States to work at Lick Observatory, Davies was advised by Sandra Faber, who was then closely involved in developing the scientific justification for California's 10-meter telescope project. In 1982, Davies accepted an offer to join the staff at Kitt Peak and the next year Burbidge asked him to join the SAC. For a young scientist like Davies, serving on the panel was an exciting and challenging assignment that helped him think about telescopes "in a more quantitative way."

One of the main topics the SAC addressed was how the NNTT would compete with and complement the Hubble Space Telescope. "You didn't want to put emphasis on the telescope's design," Davies said, "in an area where the space telescope was going to beat you."[52] The Hubble Telescope, once it was launched and orbiting above the earth's turbulent atmosphere, would have exquisite resolution of faint objects and it could make observations in the ultraviolet spectrum. For high-resolution spectroscopy, infrared observing, and sheer light-gathering power, the NNTT offered enough potential advantages to make it desirable to the astronomy community. Promising embryonic technologies like adaptive optics and infrared arrays as well as its much greater field of view made the NNTT an attractive tool for science. Besides, astronomers estimated the NNTT would cost less than a tenth of what NASA was spending on the Hubble Space Telescope.

Serving on the SAC was a politically sensitive and time-consuming task. Throughout 1983 and into 1984, it met over a dozen times. The SAC held workshops to assess current telescope technology, and Kitt Peak invited over 40 consultants from universities and industry to talk to the SAC about specific technical areas, including CCD detectors, image processing, and optics manufacturing. The members also heard presentations from astronomers and engineers planning other giant telescopes to assess the state of the art.

In November 1983, Gehrz's committee held a meeting of special importance in Tucson. After much debate, the twelve members of the committee formally determined the scientific research agenda for the NNTT by ranking its main science priorities. Each member of the SAC submitted votes to Gehrz describing the most important observing techniques possible with the new telescope and the types of research associated with each technique. When the votes were tallied, SAC members had selected spectroscopy (including multiobject, infrared, and high-spectral resolution techniques) and infrared imaging.[53] It was in these areas, the SAC explained, that the NNTT would have unsurpassed advantages over both smaller ground-based telescopes as well as future space telescopes.

For example, astronomers interested in infrared imaging expected that gains in sensitivity, which are proportional to the square of a telescope's diameter, would give the NNTT a massive advantage over any existing ground-based telescopes. They looked forward to using this capability to study areas of the Milky Way galaxy where stars were obscured by clouds of dust and gas. Because these emit radiation primarily in the infrared, the increased power and resolution would help astronomers distinguish active star formation regions in our own as well as neighboring galaxies. Meanwhile, multiobject spectroscopy would enable scientists to undertake research programs on the large-scale structure of the universe that depended on voluminous amounts of information to produce a meaningful data set.

The telescope's basic performance and design specifications followed from this list of crucial research techniques. Over the next several months, the committee negotiated and refined a set of performance requirements for the telescope. As Gehrz described, "It was like the Army deciding what specs it wants for a Sherman tank. You write down what you want it to do and then you look at your design alternatives."[54] For example, scientists gave the technique of multiobject spectroscopy highest priority for the NNTT. To collect the spectra from several dozen galaxies simultaneously, Gehrz's "Sherman Tank" would need a fairly wide field of view of about of one-half to one degree.

Throughout their deliberations, the SAC heard numerous presentations from Angel, Nelson, and other Arizona and California astronomers. Each side made fervent pitches concerning why the segmented or multiple-mirror design would be best for a giant national telescope. Throughout these tense meetings, Nelson, Angel, and other participants maintained decorum but remained staunchly adamant about the superiority of their own approaches and the flaws in their competitor's. These exchanges were often quite pointed. When Nelson and Terry Mast presented a final report in June 1984 advocating their mirror design, Angel and Woolf responded with a running commentary in the paper's margins. At the end of the report, Nelson and Mast stated, "There is broad and intensive support in this country for segmented mirror design and development . . . It is widely believed in the national technical community that future large mirrors will be segmented." Arizona's scientists retorted by attacking the cost and technical difficulty of Nelson's design: "The segmented mirror telescope would be in danger of being another Sugar Grove [a much-publicized Navy failure in the 1960s to build an enormous radio telescope] . . .

Somewhere there must be 36 million good reasons for building a segmented mirror telescope. We are waiting."[55]

Because the NNTT was to be a national telescope, Gehrz's committee received opinions and comments from astronomers all over the United States and abroad who favored a particular design (or no telescope at all). In June 1984, Kitt Peak organized a two-day open meeting in Baltimore that coincided with the American Astronomical Society's annual meeting. Dozens of astronomers heard Nelson and Angel describe their concepts and offered their own views on the project. A letter Geoffrey Burbidge received after the Baltimore meeting is representative of the concerns astronomers expressed. "What effect will this development [the NNTT] have on the high quality support we have experienced at KPNO?" asked the scientist, "With the Ten Meter Telescope, Texas, and other projects, is the effort in Astronomy as a science properly balanced?" After considering the options available, the astronomer told Burbidge that he favored Nelson's design "partly out of conservatism, partly intuition, and to a lesser extent technical questions."[56]

After reviewing what they had learned from these workshops and conferences, the SAC still had not found any showstoppers for either Nelson's or Angel's design. While each concept still had many engineering problems, no single technological problem existed that would force the SAC to choose one design over the other.[57] In short, the two telescope designs were in a dead heat.

The extent to which astronomers participated in the design selection process for the NNTT illustrates the perceived importance of the project and the impact astronomers believed it might have on their community. The SAC's interactions with the larger science community served several functions. Astronomers were given the opportunity to offer input on the type of research that they wanted to do with a new national telescope. The responses of astronomers helped guide the SAC as it prepared to choose the winning design. Their comments also helped AURA and the NSF gauge overall interest in the project.

French sociologist and philosopher Bruno Latour describes how technologists create a paper world in which "machines . . . are drawn, written, argued, and calculated" before becoming part of the "messy, greasy, concrete, world."[58] The more complex the machine is, the more paper is required to bring the project into existence. A vision of the

NNTT emerged from the reports, memos, and letters prepared by the SAC and the extensive feedback other astronomers provided. As astronomers and engineers built their "paper telescope," the boundaries between the worlds of technology development, astronomical research, politics, and science funding became increasingly indistinct. Members of Gehrz's committee, Kitt Peak staff, and interested members of the astronomy community were obliged to be fluent in and negotiate all these realms as they debated the NNTT's design and purpose.

Shootout in Santa Cruz

Outside the NNTT's paper world, developments were taking place that altered the landscape of American astronomy and had the potential to affect the project's future. Throughout 1983, there were rumors and discussions about the reorganization of the national observatories. The three centers—Kitt Peak, Cerro Tololo, and the National Solar Observatory—were then managed by AURA under separate contracts with the NSF. At AURA's annual meeting in the spring of 1982, its board members approved a plan to combine these observatories under a single director in conjunction with three associate directors. Not everyone agreed with this move. Critics claimed the NSF wished to avoid difficult decisions about allocating resources between the centers and that the reorganization would not make the centers' management much more efficient.[59] Despite these complaints, in June 1983, the National Optical Astronomy Observatories (NOAO) was established and headquartered in Tucson. At about the same time, AURA moved its corporate offices from Tucson to Washington, D.C. to allow it to interact more closely with the NSF, NASA, and other scientific organizations.

AURA chose John T. Jefferies to be NOAO's first director. Jefferies was an Australian solar astronomer and theorist with extensive experience in managing science organizations. For almost two decades, Jefferies ran the Institute for Astronomy at the University of Hawaii. He helped found the institute and built it up from insignificance to a world-class organization. His biggest coup while in Hawaii was persuading several institutions to put their telescopes on Mauna Kea. In return for providing sites on top of the extinct volcano and helping manage the telescopes, the University of Hawaii's astronomers got a guaranteed share of observing time on the new facilities. This powerful lure helped Hawaii become a powerhouse in astronomy and made Mauna Kea an astronomical mecca.[60] In 1978, AURA invited the University of Hawaii to join its ranks.

Geoffrey Burbidge remained as Kitt Peak's director and oversaw the NNTT project while the NSF reorganized the national centers. When this process was complete, in May 1984, Kitt Peak announced that Burbidge would step down the following September to make room for Sidney C. Wolff, the first woman to occupy Kitt Peak's director's office.[61] An observational astronomer, Wolff received her Ph.D. from Berkeley in 1966. After spending time as a postdoctoral student at Lick Observatory, she accepted a post at the University of Hawaii. Both Wolff and Jefferies actively promoted Mauna Kea as the best location for new telescopes in the northern hemisphere. Some astronomers found her appointment unusual; her research specialty was stellar evolution and formation, while Kitt Peak's focus continued to shift to extragalactic topics. Woolf brought skills she had learned while planning and managing the first major telescope owned by the Institute for Astronomy, an 88-inch instrument on Mauna Kea, and became a staunch advocate for national astronomy centers.

When the NSF combined the three national astronomy centers under one director, the agency also added a fourth division to NOAO. The Advanced Development Program was formed to help the national optical observatory promote novel techniques and instrumentation for its own telescopes and the broader astronomy community. Jefferies and other astronomers also expected the Advanced Development Program to take a prominent role in designing and engineering the NNTT.

Jefferies recruited Jacques M. Beckers to lead the Advanced Development Program. Like Jefferies, he was originally a solar physicist. In 1979, after almost two decades in that field, the Dutch-born Beckers shifted to optical astronomy when he became the first director of the Multiple Mirror Telescope Observatory. Encountering the culture of nighttime astronomers while working with what was then the world's most innovative telescope thrilled Beckers. He devoted himself to debugging the Multiple Mirror Telescope and making it ready for operation. When he joined NOAO in March 1984, Beckers recalled having a "certain attachment" for multiple-mirror telescopes, but he wasn't asked to take an active part in recommending a design for the NNTT.[62]

Beckers' technical successes using the MMT, with its complex and demanding optical and electronic systems, made him an especially attractive candidate to lead the Advanced Development Program. In the mid-1980s, most astronomers perceived the national observatory as not being at quite the same level of technological sophistication as university and private institutions. Developing frontier instrumentation and building a new giant telescope was vital, Jefferies believed, for the future health and status of the national observatory.[63]

In May 1984, the astronomy community received another major announcement. David P. Gardner, the University of California's president, announced that the school had just finalized details on the largest donation in its history. Marion O. Hoffman, the widow of a wealthy car importer, had given the university $36 million worth of securities, real estate, and artwork to help build the largest telescope in the world. As Gardner explained to his faculty members and the press, Hoffman's bequest was generous but still not enough to finance an entire 10-meter telescope. The University of California would therefore seek other partners willing to put in money in exchange for observing time; and in a controversial move, Gardner invited Caltech to contribute $25 million for a quarter share of that time. For over three decades, Caltech astronomers had had bragging rights to the 200-inch, still the world's best and biggest telescope. The possibility of playing second fiddle was galling.[64]

Hoffman's donation, however, made the 10-meter telescope's future appear secure and guaranteed that Jerry Nelson's work on his segmented mirror technology would continue with even more intensity. Hoffman's gift must have piqued the envy of astronomers at places such as Arizona and Texas, who had to fund their telescope projects in a piecemeal and insecure fashion.

In July 1984, with these developments in mind, the NNTT's Scientific Advisory Committee held its final meeting in Santa Cruz—some participants called it the Great Telescope Shootout. On the morning of July 13, members of the SAC met in executive committee. One of the questions they considered was how to announce their recommendation to the community. For instance, if the committee selected the multiple-mirror style, the announcement had to be made in a way that would not damage California's 10-meter telescope project.

Roger Angel and Jerry Nelson made a final pitch for their designs to the SAC. According to Larry Barr, the NNTT's chief engineer, Angel was pessimistic about the chances for Arizona's multiple-mirror concept being selected. Angel believed that the Berkeley group had articulated their case exceptionally well, while their technology development efforts (unlike his own) had been lavishly funded for several years. He also rightly believed that the extensive work the Californians had done for their 10-meter project had directly benefited their 15-meter NNTT design.[65]

The committee deliberated all day and identified a long list of technical criteria to help it choose a winner. The most important requirements were that the NNTT would have a wide field of view, excellent image quality, and be able to accommodate the rapid changeover of scientific instruments. Astronomers on the panel interested in infrared observing with a future NNTT also pushed for a winning design that would emit as little heat as possible. If there were too much emissivity, the telescope's own background would swamp faint stellar signatures and degrade its performance.

In addition to the technical aspects of the design, there were scheduling and political factors. The likely construction of California's 10-meter telescope, for example, meant that any segmented version of the NNTT might be postponed while California's facility was being built. The multiple-mirror version of the NNTT was not immune to similar considerations. The committee recognized that a working model of the Arizona design already existed in the original Multiple Mirror Telescope, but some members questioned the merits of building a larger version of a telescope that had yet to demonstrate unequivocally the multiple-mirror concept.[66]

The debate continued the next day; Robert Gerhz remembered the sessions as agonizing. "Everyone was soul-searching. All the cases were eloquent. Everyone had good reasons for one over the other."[67] Astronomers agreed that a choice had to be made soon as it was becoming too expensive for their community to support technology development efforts for the NNTT at both Arizona and California. The segmented-mirror design, Gehrz's notes say, was superior: "If you could get a segmented mirror telescope to work, why would you go the other way?"[68] Some SAC members were comforted by Nelson's plan to have industrial contractors fabricate and polish the essential mirror segments once his team had demonstrated the basic process. Angel's approach, in comparison, was for his university-based lab to make what was traditionally a commercial product.

Angel predicted that the Mirror Lab could start producing 7.5-meter honeycomb mirrors in only a few years. Astronomers were inclined to believe that Nelson's precise, large-scale fabrication and control of segments would take longer to realize. The majority of the scientists opted, in Gehrz's view, for "conservatism"—this meant using the large, single mirrors they were used to. "It basically boiled down to a gut-level feeling," he said, "that controlling 30 or 40 segments flying in close formation was going to be very tough."[69] At the SAC's final executive session, the committee complimented Nelson's careful and innovative design for California's 10-meter project. Scaling this up to a grander scale, however, made some worry about "building a mediocre monster."[70] That evening, Gehrz asked his committee to assess the risks and merits of each design and called for an open ballot vote. When the numbers were added, Angel's multiple-mirror design edged out Nelson's telescope by a tiny margin. The giant new national telescope was going to be an Arizona "four-shooter."

In choosing the Arizona design, the SAC avoided investing too much of its community's political and monetary assets on one untested technology. If they had chosen the segmented design and it failed, the national telescope project would have suffered along with California's private project. In August, a week before he stepped down as Kitt Peak's director, Geoffrey Burbidge wrote the SAC and thanked them. He noted that the task of building a new national telescope had just begun. Roger Angel still had to demonstrate that his honeycomb mirror casting technology could be scaled up and Kitt Peak didn't expect to submit a final proposal for the NNTT's funding until at least 1986. But with the right amount of technical and political skill, Burbidge believed it was possible for American astronomers to be doing research with a giant new national telescope by the end of the century.[71] While Burbidge's prediction was right, American astronomers had little idea of the bumpy road they would travel to get there.

CHAPTER 5

Growing Pains

AURA's decision to build Arizona's design for the NNTT breathed new life into Angel's Mirror Lab. In February 1985, after the AURA board ratified the plan to build the multiple-mirror version of the NNTT, the Mirror Lab received a $900,000 infusion of NSF funds to continue its mirror-casting program. The lab then moved to a specially designed facility underneath the east wing of the university's football stadium. The staff's goal was simple: Cast and polish the world's largest mirrors for a new generation of giant telescopes.

The heart of the Mirror Lab facility was its new giant rotating furnace, which Angel and his team of engineers and technicians had designed to cast mirrors as large as 8 meters. The entire furnace, 40 feet wide and 22 feet high, weighed as much as 100 tons when loaded with chunks of glass. It rested on a single ten-foot roller bearing in a central pedestal that also contained the microelectronics and computers to monitor the temperature at several hundred points inside the furnace. Electric heating elements in the furnace slowly brought the temperature up to 1200° C while cameras inside the furnace allowed the oven pilot, now no longer dizzily spinning onboard, to watch the operation from a control room.

In 1934, Pathé film clips of the 200-inch mirror's casting featured teams of Corning workmen muscling hot glass from the furnace to the mold. A large mirror casting at Angel's lab was not as immediately dramatic. At one casting, a local news photographer standing next to the hot spinning furnace asked, "Where is all the hot glass?" The absence of ladles and synchronized teams of workmen did not rob the operation entirely of drama. Warmth from the furnace was perceptible as soon as one entered the casting room; it generated the equivalent of 2,000 household space heaters. The furnace, steadily and swiftly spinning, created its own wind as it whirled like an amusement park carousel on steroids. Time-lapse photographs of glass inside the furnace, showing it gradually softening to a glowing, smooth vitreous form, were hypnotic evidence of the long process.

The cost of casting a large mirror was equally impressive. The Mirror Lab estimated in 1987 that glass and mold materials for making a single 8-meter mirror would cost over $1.2 million.[1] In the event of a casting failure, neither the mold pieces that lab technicians hand-machined nor the pieces of borosilicate glass specially ordered from Japan could be reused. Labor and other capital equipment added to the cost. If the Mirror Lab were to succeed, there could be no mishaps. One ruined casting would bankrupt the operation and launch doubts about the lab's competence.

In March 1985, as construction of the giant new furnace and facilities continued, the Mirror Lab successfully made a 1.8-meter mirror blank using its old prototype rotating oven at the abandoned temple. This was cast for a telescope to be jointly operated by the Vatican Observatory and the University of Arizona. Angel described the project to Pope John Paul II when the pontiff visited Arizona in September 1987, causing some to refer to the observatory as the "Pope 'scope." Angel was not joking when he suggested the Mirror Lab would not only cast the mirror, but also polish it. More shocking, the mirror would have a deeply curved surface, giving it the radically fast focal ratio of $f/1$. The telescope would focus starlight at a point only 1.8 meters above the primary mirror, resulting in what one astronomer called the "Fastest Mirror in the West."[2]

As the lab's new facilities under the stadium took shape, the builders set aside space to permit the polishing of mirrors as large as 8 meters. A dedicated optical testing tower over one hundred feet high was devised. Mirror blanks were moved underneath the tower with a special hover-cart while lasers at the top of the test tower illuminated the mirror's surface. By comparing the laser light's interaction with the mirror's surface and that of a reference surface, opticians could evaluate the progress of polishing and determine what areas of the mirror still needed work.

Arizona's foray into mirror polishing was a bold expansion of the Mirror Lab's activities. Instead of just casting a mirror and sending it to a commercial firm to be ground and polished, the Mirror Lab proposed to do all of these operations in-house. Its final product would be a finished mirror blank, ready to be aluminized with a reflective coating and integrated into a waiting telescope.

In April 1986, a special subcommittee organized by the NSF released a report reviewing the current state of large telescope projects. It paid special attention to technology development for the NNTT. Mirrors, it said, still remained "the most important technical problem to be solved." The report described progress on the segmented and honeycomb mirror techniques, noting that a third option, thin meniscus mirrors, had received no comparable attention. It concluded by endorsing the work at Angel's Mirror Lab and reaffirming support for the NNTT project.[3]

The NSF report also acknowledged potential problems facing the NNTT project. NOAO was across Cherry Avenue from Steward Observatory on the Arizona campus. Naturally, there was healthy scientific competition between the staff at the university and the national observatory, but institutional relations between the two astronomy centers were often so problematic that scientists referred to the street as Cherry Canyon. At times, this competition manifested itself in jealousy and a lack of cooperation. Some at the University of Arizona resented the relatively lavish federal funding NOAO received and believed the national observatory failed to make the best use of it. Meanwhile, NOAO staff complained that the university astronomers were arrogant and self-serving. A 1985 memo from the NSF noted that "there is little evidence of cooperative collaboration in the actions of the Arizona-Steward group toward NOAO."[4]

One cause of friction was the question of ownership of the NNTT project. Angel's mirror techniques, essential for the project, were developed at the University of Arizona, while its astronomers and engineers had largely designed the "four-shooter" concept. It is not surprising that some Arizona astronomers viewed the NNTT as their own as much as a national facility. These resentments were exacerbated when Peter Strittmatter and his colleagues pursued other telescope projects with different institutions. In 1985, *Sky & Telescope* revealed that Ohio State University and Arizona were considering building the Large Binocular Telescope, two 8-meter telescopes on a common mount. Besides the NNTT, Strittmatter had discussions with half a dozen institutions contemplating giant telescopes, all based on mirrors from Angel's spinning furnace.[5]

In the opinion of some staff at NOAO and the NSF, Arizona's ambitions could delay the national telescope project. Each mirror casting was a time-consuming operation and Arizona's other telescope plans seemed a distraction from what NOAO saw as the Mirror Lab's first order of business—making mirrors for the NNTT.[6] Arizona, in turn, countered that it could go ahead with its own plans as the NNTT had no large-scale funding yet.

Disagreement about where to build the NNTT created additional stress between NOAO and Steward Observatory. As early as 1983, astronomers from both institutions identified two possible locations for the telescope. John Jefferies, NOAO's new director, favored building the NNTT high atop Mauna Kea in Hawaii, the site he helped establish. Not surprisingly, Arizona astronomers wanted it closer to home. They lobbied NOAO and the NSF to build on Arizona's Mount Graham. Located only 140 miles from Tucson, Mount Graham was a lush pine-covered peak over 10,700 feet high where, not coincidentally, the University of Arizona planned to build its next generation of astronomy facilities, including the Large Binocular Telescope and the Vatican Telescope.

Astronomers had long recognized the critical relationship between a telescope's location and the quality of images it produced. An ideal site had minimal cloud cover at night and low humidity and was relatively accessible to vehicles and staff. The relative darkness of the site, away from urban light pollution, was another crucial factor. Finally, the stability of the atmosphere above the telescope, what astronomers called "seeing," was most important.

Random temperature changes high in the earth's atmosphere are the primary culprit for bad seeing. What Isaac Newton once called "Tremors of the Atmosphere" distort incoming wave fronts of light, making images of stars subtly jiggle or, when seen by the naked eye, twinkle. As a result, light collected by a telescope's mirror is spread out over a larger area, resulting in fuzzier images. Astronomers measure seeing by referring to the angular diameter of images produced by the telescope. Reducing the angular diameter of any object imaged with a telescope means more photons being put into a smaller area; this has the same positive effect of increasing the collecting area of its primary mirror. Choosing a site with the best seeing for a new telescope offered astronomers yet another way to collect light more efficiently and produce better images.

In the late 1970s and 1980s, with the commissioning of the original Multiple Mirror Telescope and the new telescopes on Mauna Kea, astronomers developed a better appreciation of the importance of seeing and worked to improve it. They found that their telescopes could produce better images than they had imagined possible when local conditions were controlled. Astronomers using the MMT, for example, watched the quality of images improve when they stopped warm air from leaking out of the control room to the cold telescope dome.

Astronomers began to specify that their next generation of telescopes have even more accurately and smoothly polished mirrors so the telescope's location and the seeing conditions, not its optics, would be the limiting factor for image quality. The NNTT's specifications required that its optics produce images only 0.25 arc-seconds in diameter, a significant improvement over the image quality at older telescopes, such as the 200-inch, where images of only 1 arc-second might be achieved on the best nights.[7]

Characterizing a telescope site's quality is an inexact science. In the spring of 1983, staff from Kitt Peak began a lengthy effort to compare Mount Graham with Mauna Kea. Heavy equipment had to be lugged to the tops of the peaks to record data for months at a time. Snow and wind often damaged the apparatus, making travel difficult or impossible, and the data gathered were subject to multiple interpretations. Each mountain also had its own vocal advocates.

Mount Graham supporters soon faced serious political obstacles that went deeper than their internecine struggles with NOAO. The University of Arizona's plan to build several telescopes, including the NNTT, on Mount Graham attracted fierce opposition. The mountain was home to an endangered squirrel and local Apache groups claimed the mountain was sacred land. Traditionally, astronomers and environmental groups shared common cause in fighting urban sprawl and the light pollution that threatened mountain sites. Astronomers' plans for Mount Graham and threats of vandalism by radical groups such as Earth First! fractured this partnership. Peter Strittmatter and his colleagues made a point of highlighting the scientific prestige and economic benefits new telescopes would bring to Arizona, but they were caught off guard. Mount Graham became a public relations predicament. For years, the U.S. Forest Service delayed any telescope construction pending further environmental studies.[8]

Kitt Peak's evaluation of the two mountain sites continued months behind schedule. A site selection committee met in late 1986 to recommend a location. After reviewing the data and hearing presentations from Arizona and Hawaiian astronomers, it chose Mauna Kea as the best location for the NNTT, citing its outstanding characteristics for infrared observing and image quality.[9] While the final report ignored the political difficulties associated with building on Mount Graham, these certainly affected the decision. The choice to build the NNTT in Hawaii disappointed Arizona astronomers and may have diminished their enthusiasm for the project. Local reaction to the decision in Tucson was muted; but in Hawaii, local papers blared, "Hawaii Bags Big Telescope" and "Mauna Kea Gets the Big One," and applauded John Jefferies for bringing the NNTT to the island.

As debate over where to put the NNTT simmered, the Mirror Lab was ramping up from a small experimental operation to a production facility. For scientists like Angel, larger mirrors meant new management challenges. Unlike a commercial venture, in which money and labor might be shifted quickly to address crucial tasks, the Mirror Lab had a limited number of staff it could hire. "We were undertaking brand new things on a very large scale," Angel said, "It was much more like heavy-duty experimental physics. The scale of our operation was a completely new thing for astronomers."[10]

Angel and his group slowly began casting ever-bigger pieces of glass while they finished the new spinning furnace. After successfully making the 1.8-meter mirror for the Vatican telescope, the lab's next step was a series of 3.5-meter mirrors. As mirror castings became bigger, they required more staff, thus making any delays more costly. The Mirror Lab tried a succession of management strategies to address these challenges. Difficulties occurred as the lab tried to shed its research image and transform into a production facility without actually hiring professional managers. Some engineers and technicians at the lab resented the astronomers' reluctance to cede responsibility on quotidian technical matters and disliked the observatory's traditional caste system that placed astronomers' scientific knowledge ahead of technicians' practical experience.

In 1988, Angel's lab finally cast its first 3.5-meter mirror for a telescope project organized by the Astrophysical Research Consortium (ARC), a group of astronomers from Princeton and the Universities of Chicago, New Mexico, and Washington. At the time, this telescope would have been the fourth largest in the United States. With the NSF providing an initial $3.75 million, ARC astronomers hoped to use their new telescope, located at Apache Point in New Mexico, to make observations remotely from their home institutions. Angel saw the ARC mirror as proof his spin-casting technique gave a mirror as good as he promised. "Perhaps the tensest moment for me over the whole history of the Mirror Lab was when we lifted the ARC mirror off the furnace and it stayed in one piece," Angel said, "You don't know you've made a sound structure until you see it actually live on its own."[11]

In 1985, Angel had estimated that his lab would cast its first 3.5 meter mirror by September 1986 and move on to 7.5-meter castings a few years later. Budget and scheduling setbacks made these projections overly optimistic and some astronomers gossiped that the Mirror Lab was always a "day late and a dollar short." Frustrated by delays, one ARC scientist vowed not to shave until their mirror blank was done. For years, a photograph—taken in 1988—hung in the Mirror Lab, showing him gazing into the finished mirror, razor finally in hand.

Upping the Ante

Throughout the early 1980s, Caltech had largely been a bystander as other groups rushed ahead in pursuit of giant new telescopes. Caltech astronomers discussed their options for remaining at the forefront of astronomical research and considered several different collaborative schemes. One was to join forces with the University of Arizona. Difficulties associated with the Mount Graham site, as well as differences in research styles, made them hesitant to partner with Arizona scientists, whom they saw as relative newcomers on the frontier of large-telescope astronomy.[12] Another possibility was to join with the University of California, an option that looked even more attractive after the 1984 announcement of the $36 million donation from the Hoffman estate.

During 1984, Caltech executives and Howard Keck, son of Standard Oil tycoon William M. Keck, held extensive confidential negotiations about a possible large donation. In August, University of California administrators were informed of these discussions. David Gardner, president of the University of California, quickly found himself in the middle of a high-stakes game.[13] His school still had the $36 million from the Hoffman bequest that would have positioned Caltech as a junior partner in a 10-meter project. Suddenly, Caltech was flush enough to build a 10-meter telescope by itself, using the University of California's design. There was a chance that the University of California, after having invested years of effort, would be left behind.

After weeks of discussion, during which Caltech made it clear that it was prepared to proceed alone, a solution was reached. Caltech and the University of California formed a joint corporation, composed of representatives from the two universities, to manage the telescope project. Keck's donation would pay for most of the telescope, estimated now to cost about $85 million, and Caltech would legally own the facility. The University of California's contribution would defray some of the annual operating costs, about $3.5 million a year for 25 years; it also quietly returned the Hoffman donation.

On January 3, 1985, Marvin L. Goldberger, Caltech's president, shocked the science and philanthropy communities when he announced the largest single gift ever made for a scientific project. Keck had offered Caltech $70 million to build the 10-meter Keck Telescope on Mauna Kea. Caltech's private telescope would use the segmented-mirror technology developed by Jerry Nelson and his colleagues. Personally, Jerry Nelson cared little about who paid for the telescope and was happy that Keck's donation meant his design would be realized. Instead of being a lesser partner with the University of California, Caltech became the principal owner of the 10-meter telescope project. There was speculation that Howard Keck, who was also a Caltech trustee, might fund a second 10-meter instrument later. Speaking at the January 1985 press conference, Keck said that he was "making a significant gift to further the interests of science and the interest in mankind in learning why we exist." Upon hearing the news, Ronald Reagan offered the presidents of Caltech and the University of California his "best wishes for a New Year that has truly started off with a Big Bang."[14]

Astronomers from Caltech and the University of California, who represented fewer than 5 percent of the astronomy community, were elated by the January announcement and the news that they would share observing time on the giant new telescope. University of Hawaii astronomers also saw their scientific capital surge with the news; they were guaranteed 10 percent time on the Keck Telescope in exchange for providing a site on Mauna Kea. Jesse Greenstein, now 76 years old, rejoiced that this massive donation revitalized astronomy at Caltech. Greenstein lauded the cultural importance of Keck's philanthropy, calling it practical and ennobling. The construction of the Keck Telescope signaled that private telescopes and university-based astronomy were not just alive; they were surpassing any plans for a new national facility.[15]

In September 1985, Howard Keck joined university officials to break ground for the telescope atop Mauna Kea. There were still many technical intricacies to solve before the telescope could be built. Jerry Nelson knew, for example, that his "bend-and-polish" technique could make mirror segments from circular pieces of glass, but the Keck Telescope's design called for hexagonal segments. When Nelson's team cut the circular pieces of glass into hexagons, the shape of the segments changed unacceptably as the glass relaxed slightly.

Unlike Arizona's Mirror Lab, where all the casting and polishing were done at one university-run facility, the Keck Telescope's management decided to let industry solve the final problems with the mirror segments. In late 1985, Caltech awarded a $10.8 million contract to Itek Optical Systems, a Massachusetts company with extensive defense-related experience. Delays and budget overruns soon frustrated scientists and engineers from the Keck project. Itek was more accustomed to large defense contracts rather than working closely with individual university-based scientists. Besides cost and scheduling issues, staff from the Keck project complained that the contractor ignored their hard-won experience. "We knew more than Itek did in this area, and they weren't willing to accept that," recalled one senior manager.[16] Eventually the Keck project took the contract away from Itek and gave it to Tinsley Laboratories, a California-based company that soon showed better performance.

Nelson later admitted that fixing the distortions caused by cutting segments into hexagons was "a bigger problem than we had first thought."[17] To control these distortions, his group employed sets of small aluminum springs called "warping harnesses." These forced the mirror segments to take the correct figure. Another technique the project relied on was "ion figuring." Technicians at Kodak gathered information about each segment's profile and an ion gun bombarded the high spots to create a more accurate optical surface.

As money from Keck's massive donation began to make itself felt, the organization of the 10-meter telescope project changed. Until about 1985, work on the telescope was largely the province of a relatively small group of astronomers and engineers, their efforts still directed toward research and development. Once construction began and expenditures soared, the project's management became more complex. Nelson was still the 10-meter telescope's chief scientist, but he and the other astronomers had to share power with a wider array of persons. The Keck project had a board of directors who gave overall guidance for the telescope. There was also a Science Steering Committee, chaired by Sandra Faber and Wallace Sargent, which took responsibility for developing the instruments that would make the telescope scientifically useful. Other committees took responsibility for the telescope's enclosure, its software, and electronics. The Keck Telescope's organization began to more closely resemble a medium-sized space astronomy mission rather than the small university-based operation it once was.

The project manager who oversaw the daily progress and problems was Gerald M. Smith, a former electrical engineer. Before joining the Keck project, Smith had worked on space programs for NASA's Jet Propulsion Laboratory and had managed large projects like NASA's lunar Ranger program and major satellite for infrared astronomy. In the 1970s, he also guided construction of a NASA-funded infrared telescope on Mauna Kea. This experience familiarized him with the mountain, local politics, and the difficulties of building a large facility under harsh conditions.

Despite his experience, Smith's appointment to the Keck Telescope project caused some astronomers, especially Nelson, a good deal of grief. The two did not develop a close or comfortable relationship. Smith tended to view the telescope project as a massive engineering undertaking rather than a personal dream nurtured for more than decade.[18] According to Nelson, Smith was an efficient, sometimes coldhearted, manager who drove the project forward relentlessly. Nelson disliked Smith's autocratic style, disagreed with his management techniques, and complained of being left out of important decisions.

Staff at Arizona's Mirror Lab experienced similar growing pains. The Mirror Lab was founded largely on Strittmatter's entrepreneurial leadership and Angel's ideas. When the lab expanded and telescope projects began to anticipate receiving their mirrors, management and scheduling issues became increasingly important. The NSF continued to make large grants to Angel's operation. By 1987 its annual budget was over $1 million. The agency, in turn, required progress reports and more active management from NAOA and the University of Arizona.

From 1985 onward, as it scaled up to bigger mirror castings, the Mirror Lab comprised two distinct facilities under one roof. One was an experimental operation where almost every procedure was innovative; equipment was often hand-built, and the scientists were in charge. The other emerged as the Mirror Lab transformed itself into a small production facility where managers and engineers tried to run daily operations with the reluctant acquiescence of the astronomers. Reconciling these dual identities was difficult at times and took several years to achieve successfully.

As it scaled up, the Mirror Lab developed the potential to compete with larger commercial firms specializing in large optics. In the mid-1980s, companies such as Corning in New York and Germany's Schott Glasswerke displayed increasing interest in making large mirrors for the next generation of telescopes. Eschewing segmented and honeycomb-mirror technologies, these glass companies developed the capability to fabricate thin meniscus mirrors as large as 8 meters. The scale of these industrial operations and the resources at their disposal were fundamentally different from the "cut-and-try" approach and incremental progress that prevailed at Angel's relatively small lab. These differences set the stage for a fierce contest between competing mirror technologies—Angel's honeycomb versus Corning's meniscus—which nearly destroyed the Gemini Telescopes project a few years later.

Dark Technologies

Ever since Galileo showed his telescope to the Venetian senate, military leaders recognized the potential value of astronomers and their technologies. For a long time, this relationship was indirect. After World War II, however, defense interests began to influence and reshape astronomy. Radio astronomy, of course, benefited enormously from scientists' wartime experiences with radar, while captured German V-2 rockets were the nucleus around which the entire field of space science coalesced. The postwar relationship between astronomers and the military was synergistic and often benefited both communities. Astronomers gradually availed themselves of new devices (like CCDs) developed originally for the military. They also took advantage of declassified military studies for their research, such as the Air Force's Infrared Sky Survey completed in the early 1970s. At the same time, the military valued scientists' knowledge in the areas of optics, solar phenomena, and atmospheric seeing.

Not surprisingly, civilian research was subordinate to military interests. Astronomers often found themselves working with equipment inferior to that available to colleagues who worked on classified projects. This did not mean that they had no inkling of technologies developed in the world of classified defense research. As one astronomer quipped, "One of the most infuriating things is to know that a detector far better . . . is orbiting the earth, attached to a large optical telescope, looking downwards."[19]

During the Reagan years, scientists and engineers continued to improve an array of classified military technologies that had some value to astronomy. A few eventually migrated from the secret world of classified research to become part of the scientists' toolkit. In the 1980s and 1990s, astronomers worked feverishly to integrate a new technique called "adaptive optics" with the new generation of large telescopes and their ever more complex systems.

Almost as soon as the telescope was invented, astronomers noted the deleterious effect of bad seeing on their viewing. In time, they identified the problem: Distant stars and galaxies emit light as spherical wave fronts that are scrambled upon entering the earth's atmosphere. Correcting for this effect required equipment that could restore the original wave front pattern before the light was brought to the telescope's focus.

Horace Babcock, before he became director of Palomar, published an article that described a primitive method for "compensating astronomical seeing."[20] Babcock's system began with an Eidophor, a device used in European theaters to project television pictures onto a movie screen. An Eidophor incorporated a mirror coated with a thin film of oil encased in a vacuum. Babcock proposed that light from a telescope's primary mirror be reflected to the much smaller Eidophor mirror. A beam of electrons would scan the Eidophor mirror, causing small changes in the oil film's shape. If alterations to the mirror were made rapidly enough, the light's wave front could be corrected for the poor seeing conditions before it was reflected to the focal plane.

While Babcock believed the technical effort needed to integrate his hypothetical system with existing telescopes was too extensive, he identified three basic elements that astronomers would need: a sensor to detect the incoming wave front, a deformable mirror, and a controller that used the sensor readings to adjust the mirror many times a second. Throughout the 1960s and 1970s, new technologies such as improved computers, microelectronics, and lasers enabled the military to explore the adaptive optics concept and develop it into a practical tool.

Realization of the potential of adaptive optics benefited from parallel improvements in infrared detectors. The areas of atmospheric turbulence that cause bad seeing become bigger at longer infrared wavelengths. Therefore the telescope sees fewer "isoplanatic patches," as astronomers call them, at these wavelengths. At infrared wavelengths, the deformable mirror would have to make fewer corrections per second and consequently speed and accuracy requirements for adaptive optics systems would be less severe. As a result, applying adaptive optics to infrared observing was a logical first step.

In the early 1970s, the United States military was concerned with identifying newly launched Soviet satellites. While high-resolution photographs could be taken from the ground, the images were blurred by atmospheric turbulence. Throughout the 1970s, scientists and engineers from Itek Optical Systems and the Air Force explored ways to compensate. MIT's Lincoln Laboratory also joined the effort as an outgrowth of its interest in propagating high-energy laser beams up through the atmosphere without undue energy loss.[21] Optical engineers designed small glass mirrors with surfaces that could be accurately nudged by actuators a few millionths of a meter, thousands of times a second.[22] Because their shape was flexible at that scale, optical engineers began to call these "rubber mirrors."

In the mid-1970s, after Itek's engineers demonstrated that their prototype worked, the Department of Defense agreed to scale it up for use on a real telescope. The military operated telescopes on a dormant volcanic peak on Maui called Haleakala that were not available to civilian astronomers. By 1982, a traditional 1.6-meter telescope on Haleakala was reserved for the Air Force's satellite-tracking program. The telescope featured a complex adaptive optics system whose construction demanded the integration of complex mechanical, optical, electrical, and computer systems. Its heart was a mirror with 168 separate actuators. Initial tests done by military scientists showed (although the results would not be released for years) that the system could easily resolve blurs, transforming them into images of closely spaced double stars. While many astronomers would have given a great deal for this capability, to the Air Force the stars were merely engineering targets to help fine-tune the adaptive optics systems before tracking fast-moving satellites.

When Horace Babcock wrote his seminal 1953 paper on adaptive optics, he recognized one obstacle scientists needed to overcome if they wished to use adaptive optics on a regular basis. To send correct signals to a deformable mirror, they needed to know exactly how a wave front is distorted. For them to acquire that information, the wave front sensing device needed a strong signal from a bright "reference star" very close to whatever the Air Force satellite watcher (or astronomer) wished to observe. Unfortunately, there were not enough bright stars in the sky to provide sufficient reference beacons. This posed a serious limitation to the usefulness of adaptive optics. However, a possible solution existed—what if astronomers could make "artificial stars"?

One scientist who played an important role in developing what became known as "laser-beacon guide stars" was Robert Q. Fugate. Fugate grew up in Dayton, Ohio, and, from an early age, was interested in both electronics and telescopes.[23] In 1970, after completing a doctorate in physics at Iowa State University, Fugate faced a dismal market for academicians. Trained as an experimental physicist, he took a position at Wright-Patterson Air Force Base. Fugate soon began to appreciate the resources the military offered and found his interaction with scientists from universities, industry, and the military to have its own rewards. He devoted much of his career in the 1970s to the detection of laser beams, a topic that interested the military due to their potential role in aiming missiles launched at American planes.

In 1979, Fugate accepted a position at Sandia Optical Range on sprawling Kirtland Air Force Base in Albuquerque, New Mexico. His classified research investigated how high-power laser beams were distorted as they traveled from the ground through the atmosphere to a target. This was the reverse analog of astronomers' attempts to compensate for the distortion of starlight as it travels down through the atmosphere.

In 1981, Julius Feinleib, a scientist working for an MIT spin-off firm, suggested a way to get around the problematic lack of bright reference stars for adaptive optics. One night at Haleakala, he watched a series of laser pulses shot into the night sky. What would happen, Feinleib asked Fugate and others, if they used a laser beam fired into the lower atmosphere as their reference star? The beam would bounce off oxygen and nitrogen molecules—a process called Rayleigh scattering—and create a bright reference beacon. Feinleib, Fugate, and others presented this idea to JASON, a group of prominent scientists who advised government agencies on defense-related issues. The JASON group agreed the technique was promising and recommended the Air Force provide support.[24]

In 1983, Fugate's lab (eventually renamed the Starfire Optical Range) was one of two groups the Air Force selected to do laser-beacon experiments. This was the same year Ronald Reagan made his famous Star Wars speech; soon the military plunged ahead with a massive campaign to research technologies for the Strategic Defense Initiative.[25] SDI proponents talked of shooting high-powered laser beams from earth stations to satellites. This scheme required the laser beams to be transmitted with minimal energy loss from atmospheric turbulence. Adaptive optics offered a solution and, as a result, it benefited enormously from the SDI program, receiving an estimated $600 million over the next ten years.[26]

Fugate was limited to presenting his lab's progress at classified conferences. Technicians at the Starfire Optical Range began to adapt their prototype machines and techniques to a new, $800,000, 1.5-meter telescope. Fugate and his crew were soon working 80-hour weeks to develop the equipment and software that would link the telescope to their adaptive optics system. Once they had their deformable mirror in sync with the electronics and the telescope, they used it first with a bright natural star as a beacon. The next step was to switch to an artificial beacon produced by their laser system.

Shortly after midnight on February 13, 1989, after a decade of late nights, Fugate and his team won the race to correct atmospheric blurring using a laser-guide star. "You can develop the theory about how an experiment should work. But there is nothing as exciting as actually doing it," Fugate recalled. By adjusting their deformable mirror system more than 30,000 times a second, they reduced blurry, 2 arc-second smears to extra-sharp images of stars only 0.18 arc second in diameter. In the ensuing days, Fugate's group used their system to look at objects like the Trapezium, a set of four stars in the Orion Nebula, and Saturn. "We were astonished at what we were seeing. This was before Hubble and we were getting images considerably better than almost anything previously seen from the ground on a regular basis. We just couldn't tell anyone about them."[27]

Excited by his lab's success, Fugate lobbied the Air Force for an even bigger telescope. The new telescope would be used primarily to track and image satellites and continue Fugate's adaptive optics experiments. Fugate persisted until the Air Force agreed to build a specially designed 3.5-meter facility. Because satellites move very fast across the sky, Fugate specified that his new telescope be able to move more quickly and point more accurately than any civilian telescope in existence. Achieving this goal called for a very light mirror and, in October 1988, the Air Force and Steward Observatory signed a contract for a 3.5-meter honeycomb mirror blank from Roger Angel's lab.[28]

This mirror, estimated to cost well over $3 million, was only part of an attractive package for the Arizona facility. The Air Force also wanted Roger Angel's lab to polish its mirror and offered to fund further development of the innovative polishing techniques the Mirror Lab was exploring. Casting the Starfire mirror was a substantial departure for the lab, which was used to dealing with university astronomers. It was obliged to keep more detailed documentation, a requirement that helped the lab's efforts to transform itself into a quasi-production facility. Moreover, the Air Force's money kept the ever-struggling lab afloat and enabled it to make significant progress in mirror-polishing techniques that it otherwise could not have afforded.

Reagan's SDI program divided the American science community. Hundreds of physicists glommed onto missile-defense money while hundreds more pledged not to touch it. Astronomers, one popular science writer cautioned, should refrain from "crouching beneath the table at which arms builders feast, hoping to catch a few crumbs" lest they become enlisted in the SDI program's "blind stagger." Hawkish pundits predicted that some day almost all adaptive optics hardware that trickled down to the civilian science community would carry "a 'Made by Star Warriors' label." Jerry Nelson reminded his colleagues that, while SDI might be spending vast sums money on technology that was somehow related to astronomy, their community had little to show for it.[29]

The military remained close-lipped about the performance and technical specifications of its adaptive optics program. Civilian astronomers were not unaware of the work Fugate was doing with adaptive optics. As early as the summer of 1985, some astronomers published papers describing their own variant of the laser-beacon concept developed independently of the military. While civilian scientists succeeded in reinventing military techniques and equipment, their efforts were piecemeal. NOAO's Advanced Development Program, led by Jacques Beckers, was one group of civilian scientists that explored the potential of adaptive optics for astronomy. Beckers became convinced that adaptive optics offered a bonanza for astronomy. He hired several experienced scientists to develop expertise and equipment for getting better resolution from ground-based telescopes. By 1987, almost half of his budget went for adaptive optics.[30]

Beckers and his staff recognized the relative ease of developing adaptive optics for use at infrared wavelengths. This complemented related efforts by NOAO to improve infrared detector arrays and to use adaptive optics for solar observing. The funding Beckers had, about $500,000 annually, was a pittance compared to what the military was lavishing on Fugate's adaptive optics program. Despite this, Beckers and his group achieved some successes and they optimistically planned to integrate their nascent system with Kitt Peak's telescopes. Hampered by insufficient resources and military secrecy, however, significant applications of adaptive optics for astronomical research had to wait until the Cold War ended.

Taking Whatever They Could Get

In 1985, Richard N. Malow had what many in Washington, D.C. considered one of the city's most powerful jobs. Holding the unassuming title of "clerk," Malow was the chief staff member for the House appropriations subcommittee that controlled the NSF's budget. "King Malow" to some, he wielded substantial behind-the-scenes power on Capitol Hill. While Edward Boland and the other members of Congress met the press and appeared on talk shows, clerks like Malow worked discreetly to monitor and control the appropriations process. "Congress is charged with the power of the purse," Malow explained, "We looked at every nit."[31]

Tall, lanky, and bespectacled, Malow was a science and space buff with a manner that led at least one interviewer to conclude he would be an unrelenting poker player. In 1972, after earning his degree in political science, he moved to Washington and started working on the Hill.[32] Malow especially monitored federal programs and agencies pertaining to science and technology. As one senior representative said, "Dick is my eyes, my ears, my antenna." Malow made recommendations to Representative Boland who, after approving them, took them to his appropriations subcommittee for discussion. Malow also prepared the probing questions Boland asked witnesses during annual hearings. In the mid-1970s, Malow gained notoriety among scientists for questioning the necessity of NASA's Hubble Space Telescope. With Malow's assistance, Boland challenged the project and the two men earned the respect, if not the love, of the astronomy community.[33]

One of Malow's concerns was astronomers' plans to build the NNTT. "In 1985, I was looking," he recalled, "as was Boland's committee, at what were shocking deficits." In December 1985, President Reagan signed the Gramm-Rudman-Hollings Act into law, compounding Malow's worries. This law set targets for deficit reduction that lawmakers were obliged to meet to avoid automatic spending cuts. Malow knew that true enforcement of the act "meant that agencies like the NSF were going to get hit hard." The NSF had already committed itself to the Very Long Baseline Array (VLBA), astronomers' first priority according to the most recent decadal survey. Malow also recalled that scientists, including astronomers, lamented how the NSF's expenditures on large projects shortchanged funding for individual research grants. "I was very concerned, when it came to the NNTT, that we not get ourselves hooked on something that was going to cost $200 million or more and not be able to support it with the budget problems."[34]

Even the Very Long Baseline Array, which ranked higher than the

NNTT in the Field decadal survey, ran into major political problems. In 1983, Malow and other House staffers prepared a report criticizing the rising costs of the survey's recommendations. In March 1984, armed with this report, Boland attacked the VLBA and, more broadly, the NSF. His congressional experience told him that if his committee permitted initial design work for new science facilities, scientists would respond by requesting more money to continue the project. Boland also complained that the NSF was not budgeting enough money for science education. In 1984, he won a political victory when he persuaded his colleagues that 40 percent of the money the NSF requested for the VLBA should go to science education instead.[35] Recognizing an exceptionally poor climate for requesting money for yet another big project, the NSF wisely avoided discussing the NNTT at the 1984 hearings.

In 1985, Reagan nominated Erich Bloch to be the NSF's new director. Bloch, formerly a vice-president and industrial engineer at IBM, was an unlikely choice for the top slot at the NSF. Unlike his predecessors, Bloch was neither an academic with a Ph.D. nor a scientist. His lack of traditional credentials, as well as his combative managerial style, stirred controversy among scientists. Bloch emphasized industrial competitiveness and engineering instead of the basic science research for which the agency was noted.[36] When he arrived at the NSF, Bloch recalled that George Keyworth, Reagan's science advisor, ordered him to rejuvenate the agency and make it pull its weight by helping American industry.[37] Astronomy and other basic science disciplines did not have the same priority.

Maarten Schmidt, the Caltech astronomer renowned for discovering quasars, was president of the American Astronomical Society from 1984 to 1986. He recalled being asked to testify before a congressional subcommittee about astronomy in 1986. "It may be an original view," Schmidt speculated, "but I think Mr. Bloch hated astronomy . . . I'm still not sure that Mr. Boland didn't get to him."[38] One possible motive for the NSF director's dislike of astronomy came soon after Bloch took his new post.

In February 1985, Bloch appeared before Boland's subcommittee for the first time as the new NSF director. When it came to astronomy, the honeymoon did not last long. Dick Malow was well aware of the record-setting donation the Keck Foundation had made for a 10-meter telescope. Two weeks before the hearing, the *New York Times* ran a story describing the Keck Telescope. The article quoted Jesse Greenstein saying that the new facility would "be able to conduct observations far beyond the reach of the Space Telescope" at a fraction of the cost.[39] The article, which Malow provided to Boland, also described the NSF's plan for a new 15-meter national telescope that would be even more powerful than the Keck Telescope.

Boland, long opposed to NASA's space telescope, was incensed by the article. He recalled that when NASA officials testified before his subcommittee, they promised the Hubble Telescope would make all other optical telescopes virtually obsolete. "What Dr. Greenstein is saying, in effect, is that we have spent $1,175,000,000 to date on a telescope that is not even launched as yet," Boland said, "but will be far less capable than the $70 million to be spent on this ground-based telescope."[40] After Boland patronizingly offered copies of the *New York Times* article to the NSF representatives, no one from the agency dared bring up funding the NNTT.

Bloch's already less-than-favorable views of astronomers suffered another blow when a few lawmakers took the NSF to task for a managerial misstep made by John Jefferies. As NOAO's new director, Jefferies proposed that the national observatory alleviate its budget woes by closing the solar telescopes it operated at Sacramento Peak, New Mexico. Solar astronomers rapidly mobilized in protest and Bloch found himself being politely grilled by Senator Pete Domenici of New Mexico. Under the watchful eyes of New Mexico's elected officials, the facilities remained open. Jefferies, along with AURA and NOAO, suffered considerable political fallout from the debacle, and his authority as NOAO's director was compromised.

As Leo Goldberg had predicted, news of the Keck Telescope's funding diminished the likelihood that Congress would authorize money for the NNTT. Why, Malow asked, would the federal government invest $100 million or more in a telescope when philanthropists were already providing the same service?[41] The federal government, some lawmakers reasoned, had met its obligations to astronomy by funding the Hubble Space Telescope and the NSF's new radio telescope array. Meanwhile, x-ray astronomers were clamoring for their own billion-dollar space telescope. When it came to asking for more telescopes, Boland wryly observed, "astronomers will take whatever they can get and more."[42]

The *Challenger* explosion in January 1986 grounded NASA's space shuttle fleet and brought another lengthy setback for space astronomy. Optical astronomers had been anticipating the Hubble Telescope for more than a decade and were understandably disappointed. George Field, after reviewing how astronomy was faring after his decadal survey, reminded colleagues that delays and cost increases reflected badly on all of them. If astronomers wished to be rewarded with bigger and more expensive facilities, Field argued they had to demonstrate that they could handle management as well as scientific challenges.[43]

The average Congress member seemed not to understand that both NASA and the NSF funded astronomy and tended instead to see all money spent on telescopes as "supporting astronomy" regardless of where the money went. Some ground-based optical astronomers believed they had been unfairly neglected while radio and space-based astronomers enjoyed relatively ample funding for grander projects. Moreover, the European Southern Observatory and Japanese astronomers had made public their bold plans to build giant telescope facilities. America's astronomers, some in Washington reasoned, could just as well do research at facilities paid for by some other government.[44] Malow, as yet unconvinced of astronomers' need for another national telescope facility, continued to keep a watchful eye on their long-range plans.

Despite imposing political and financial obstacles, Jacques Beckers, Larry Barr, and other advocates for the NNTT worked to build support for a new national telescope in the astronomy community. They promoted the telescope project through newspaper articles and press releases and gave talks to astronomy departments all over the country about plans for a new national telescope. Community interest in the project failed to reach a critical mass, however. From Beckers's perspective, the fight for the NNTT needed victories on at least three fronts. One, of course, was the political battle for funding in Washington. There could be no progress unless the astronomy community spoke with a clear and unified voice in support of the telescope. Scientists at schools with their own telescope facilities—astronomy's "haves" —had their own plans for bigger instruments like the Keck Telescope. Amongst the have-nots, there was "a big fear that this grandiose project would push the little guy out of the way."[45] The have nots were the majority of the astronomy community and they worried that the NNTT's cost would force the closure of smaller national telescopes on which their careers depended. Meanwhile, even if the NNTT was built, observing time on this cutting-edge research instrument would be limited while astronomers' research proposals would undergo fierce peer review for the time available.

Poor relations across Cherry Canyon between NOAO and Steward Observatory frustrated Beckers and hurt progress on the NNTT. He had worked closely with many of the Steward scientists while directing the Multiple Mirror Telescope Observatory. While he personally got along with Roger Angel and Peter Strittmatter, there was a larger institutional conflict. The slow progress of the Mirror Lab in scaling up to bigger castings added to his misgivings. He drew graphs showing when the lab had predicted it would make its first 8-meter mirror versus actual progress and "they didn't converge for many years."[46] In fairness to the Mirror Lab, incremental funding from the NSF coupled with its annual uncertainty made it difficult to develop and keep realistic long-range plans.

John Jefferies also disliked the arrangement he inherited from Burbidge for supporting the Mirror Lab financially. Under Burbidge's leadership, money from the NSF went to the national observatory, which then subcontracted the work to Steward Observatory. Jefferies found it difficult to monitor the Mirror Lab's progress and expenditures, an arrangement that contributed to bad feelings between Steward and NOAO. The NSF was concerned enough to request monthly progress reports about the Mirror Lab from the national observatory.[47]

The NSF, perhaps sensing a lack of community consensus for the NNTT, refused to commit firmly to the project either to Congress or to scientists. The NSF's Division of Astronomical Sciences told John Jefferies that the science agency enthusiastically advocated NOAO's development of "new telescope technology," but did not support "a specific instrument called the National New Technology Telescope project," which was "neither proposed nor funded."[48] Jefferies, struggling in the face of NOAO's declining budget, did not enthusiastically support the new national telescope in the same manner as Goldberg or Burbidge had.

By the end of 1986, the *New York Times* noted that "budget constraints have stalled this country's most ambitious astronomical project."[49] The NNTT project had become unstable. Community support for it was not overwhelming, and its future challenges considerable.

While the NNTT was designed bigger than any other telescope project, astronomers and administrators questioned its scientific justification when private and international groups were building roughly comparable facilities. At the same time, community dissension about the future of NOAO as a scientific institution was growing.

Future Directions

While the skies darkened for the national observatory and its giant telescope project, they were bright for astronomy as a science. In 1987, the sudden appearance of a supernovae visible to the naked eye inspired astronomers' research. A year earlier, Margaret J. Geller and John P. Huchra, two astronomers from the Harvard-Smithsonian Center for Astrophysics, released the first results from an extensive survey of galaxy redshifts. Working with other astronomers and using data obtained mostly with modest-sized telescopes, Geller and Huchra mapped the position and recession speed of tens of thousands of galaxies. Instead of a uniform distribution, Geller and Huchra found the galaxies stretched out in space in a massive sheet-like structure over 800 million light years long but only 20 million light years tall. One of the first images released from the survey showed that the points on their map, each representing the position of an entire galaxy, formed a human-like stick figure. This eventually became an icon of late-twentieth century science; Geller and Huchra dubbed this galactic arrangement "The Great Wall." Its discovery encouraged astrophysicists' continuing reappraisal of the universe's large-scale structure.[50]

Another team of observational astronomers and theorists, led in part by Sandra Faber, was discovering an additional example of large-scale structure in the universe. Using data obtained from studying the velocities of several hundred elliptical galaxies, a team of seven scientists announced that the Milky Way galaxy, along with many others, were streaming toward what they called the "Great Attractor."[51] This was a huge concentration of mass, equivalent to a half a million galaxies or more, that stretched over hundreds of millions of light years. The detection of the Great Wall and the Great Attractor shook up cosmologists' understanding of the universe and helped stimulate astronomers' demand for more and larger telescopes.

1987 was also a watershed year for NOAO. Lack of support for its 15-

meter telescope project had reached a critical state. More broadly, many in the astronomy community perceived a crisis regarding the ambitions and future of the entire national observatory system. With universities already planning and building the next generation of large telescopes, what was NOAO's purpose? Some astronomers proposed reducing the role of the national observatory and giving more federal resources to private and state-operated institutions. Others favored the opposite—a stronger and more prominent position for NOAO in American astronomy.

In 1986, the AURA board selected Goetz K. Oertel as its second president and asked that he help resolve these dilemmas. Oertel grew up in the harsh conditions of wartime Germany. After the war, he studied physics at Christian Albrechts Universitaet in Kiel. In 1957, he received a Fulbright scholarship and emigrated to the United States, where he finished his doctorate in the University of Maryland's physics program. After doing research in solar physics, Oertel gradually moved into science administration at NASA and the NSF. He quickly earned the respect of Leo Goldberg and saw the older astronomer as a mentor. In 1978, Oertel took a new position in the Department of Energy where he oversaw large-scale programs at Savannah River and Los Alamos. Out of astronomy's mainstream for years, Oertel was pleased and somewhat surprised to be recruited by AURA. He was also well aware of the traditional divisions in the community between the haves and have-nots.

Soon after Oertel became AURA's president, John Jefferies stepped down as the director of NOAO. Sidney Wolff, Jefferies' protégé, took over in early 1987 as NOAO's first woman director. Oertel and Wolff both began to question the future of the NNTT and, more broadly, how the national observatory fit into America's science portfolio.

Oertel believed the NNTT project was flawed politically and technically. "Four 8-meter telescopes on a common mount sounded itself like a tremendous challenge," Oertel recalled. "If that system had somehow not worked, you would have had four telescopes forever condemned to look in the same direction."[52] While Wolff was Kitt Peak's director, she witnessed the difficulties facing the NNTT and saw that NOAO's efforts to sell the idea to the community, Congress, and the NSF were flagging. "Building that telescope would have probably cost twice as much as the most expensive project that NSF had ever undertaken," Wolff said, "and the astronomy community was not very excited about the science it could do."[53]

Oertel, working at AURA's headquarters in Washington, D.C., also began to hear unpleasant sounds from Capitol Hill. "The NNTT didn't fall on very good ears in some parts in Congress and also at the NSF." Oertel soon got a phone call from Dick Malow, "who really lambasted me for even talking about the telescope because he said, 'You're not supposed to be spending any money on it!' It was made very clear to me this wasn't the way to go or else there was a long way to go to convince the system that this was the thing to do."[54]

In response, Oertel commissioned the Future Directions for NOAO Committee. Like the soul-searching exercise initiated by Goldberg a decade earlier, AURA asked the committee to articulate the appropriate role for the national observatory. Steve Strom, an astronomer at the University of Massachusetts and a former Kitt Peak scientist, chaired the diverse group of sixteen prominent astronomers from private, state, and federal institutions. In August 1987, the AURA board convened a retreat for its members in the mountain resort of Keystone, Colorado. Members of Strom's committee attended along with NSF representatives. The crucial issues on the agenda were the future role of NOAO, the needs of the American astronomy community, and the types of facilities the national observatory should develop.[55]

As he listened to participants debate, Strom saw the national observatory and the NNTT project simultaneously under pressure from different factions in the community. "There were two competitions. Between the elitist institutions who didn't want NOAO to succeed and the smaller institutions that didn't want NOAO to succeed but for very different reasons." Frank Low, Arizona's renowned infrared astronomer, was also on the committee, and he agreed that the fundamental battle here was between astronomy's haves and have-nots.[56]

One option they discussed was for NOAO to be America's premiere institution for ground-based optical and infrared astronomy, analogous to the United Kingdom's Royal Observatories. Achieving this "flagship NOAO" meant giving the 15-meter NNTT highest priority. Those in favor said this could increase the programmatic coherence, political unity, and competitive strength of American astronomy. Others countered that the plan was impossible to implement given the limited federal funding available. And, just as Leo Goldberg found in

1976, no one realistically expected that the top-tier private and state-funded observatories would cede their prestige (and the accompanying federal funding) to NOAO voluntarily.

At the other extreme, some on Strom's committee proposed that NOAO should only supplement the research programs of private and state-run observatories. NOAO under this service model would be analogous to the NSF's Antarctic Program, which operated facilities for other scientists to use. The service model would consume minimal federal funds and leave the lion's share of money to the other observatories. NOAO would not be expected to contribute much scientifically to astronomy and future national facilities such as the NNTT would not be built. Critics pointed out that a national observatory operated this way would become a government-maintained lab perpetually fending off mediocrity and elimination.

In the end, the committee chose a middle strategy described as "first among equals." NOAO would complement the private and state observatories, leading in some areas and providing support in others. This meant NOAO would build new telescopes to the extent that they were not being pursued by other private groups, concentrate on unique areas of instrumentation and research, and still continue to provide support for American astronomers lacking their own resources. In other words, NOAO still had to meet the needs of practically the entire astronomy community.[57]

Participants at the Keystone meeting didn't take a vote to kill the seven-year-old NNTT project even though its flaws were now clear. Low recalled, "Although on paper the NNTT would produce grand results—lots of collecting area and lots of spatial resolution—it was going to be very expensive. It wasn't going to be quick. It wasn't going to have any predictable payoff. It had a lot of lofty goals but it wasn't sufficiently practical."[58]

Strom's committee still believed the national observatory needed a big project. "It needed to be more realistic, something that could be achieved relatively quickly," Frank Low explained, "It had to be technologically innovative, offer scientifically compelling justification, and be a good illustration of the kinds of things that the national telescope could do."[59] As they examined their options, participants at the meeting kept returning to stratagems based on building 8-meter telescopes.

In September 1987 the Future Directions Committee released its final report. It recommended that NOAO "build as rapidly as feasible two 8-meter class telescopes (one in each the Northern and Southern hemispheres) with superlative image quality and located at excellent sites" instead of the NNTT.[60] Shortly after the Keystone meeting, AURA's board affirmed this advice and voted to "construct and operate 8-meter telescopes to provide outstanding research facilities for the U.S. optical and infrared astronomy program during the early 1990s."[61] Officially, these 8-meter telescopes were pitched as stepping-stones to AURA's long-term goal of building the NNTT in the mid-1990s. It is unlikely that this mollified any NNTT supporters, who surely knew that building even one 8-meter telescope, let alone a pair, would take a decade or more.

Roger Davies had come to Kitt Peak in 1982 as a young scientist and helped promote the NNTT; he was still at the national observatory when the Keystone meeting took place. Davies was one of the astronomers in the midst of discovering the Great Attractor and in an especially productive time in his career. He observed the NNTT's cancellation jar the staff at NOAO. "It was a catastrophe. When the AURA board decided to build the two 8-meters, morale went through the floor." After the NNTT was canceled, several experienced astronomers and engineers left the national observatory. With NOAO's plans scaled back and astronomers in the United Kingdom talking about building their own 8-meter telescope, the differences between the two science communities was lessening. Davies, despite having tenure at NOAO, returned to England to help jump-start a large telescope project there. To Davies, the events in the summer of 1987 were part of a larger pattern. "The national observatory can have as large a telescope as they like as long as its mirror is not bigger than 80 percent of what Caltech has!"[62]

There were other casualties. Since 1984, inflation and budget stagnation had seriously eroded NOAO's purchasing power by as much as 20 percent. To save money, Sidney Wolff recommended that the entire Advanced Development Program be closed.[63] Jacques Beckers's hopes for the national observatory to develop cutting-edge technology collapsed and most of the astronomers he had recruited dispersed to develop their techniques elsewhere. Beckers himself left NOAO to take a position with the European Southern Observatory.

In the end, American astronomers and science managers decided that telescopes less ambitious than the NNTT, but still very powerful, would enhance the health of American astronomy and enable the continued existence of NOAO. The introspective examination that Oertel, Wolff, and the Future Directions Committee undertook kept NOAO alive as an institution that offered more than services. It also kept NOAO as a player in the quest for bigger telescopes, something that was especially important for NOAO at this time. By late 1987, several different large telescope projects were already underway. Not to be left behind by the Keck project, the Carnegie Institution of Washington announced plans for its own 8-meter telescope. The University of Arizona had its unique Large Binocular Telescope as well as a plan to convert the Multiple Mirror Telescope it jointly operated with the Harvard-Smithsonian Center for Astrophysics to a single 6.5-meter mirror. The Japanese were planning an 8-meter national facility for Mauna Kea that spared no expense. Trumping all of these was the ambitious plan the European Southern Observatory formally announced in 1987 for the Very Large Telescope, an array of four 8-meter telescopes in Chile estimated to cost more than a quarter of a billion dollars.

NOAO and AURA's institutional security came at a price. For over a decade, the national observatory had nurtured plans for a giant new telescope that would serve the entire science community. Budgetary, technical, and political obstacles forced NOAO's astronomers and engineers to scale back their ambitions twice—from the 25-meter NGT to the 15-meter NNTT and then to two 8-meter telescopes. Given the big projects now envisioned by many groups worldwide, NOAO might soon find itself less competitive and operating smaller telescopes. If this happened, NOAO's position on the frontline of astronomy technology, already seen as precarious by some, would sink lower.

Oertel saw the NNTT simply "fade away like MacArthur, an old soldier" after the August meeting in Keystone. Afterwards, many astronomers saw the NNTT as a technological dead end or simply a misguided program. In spite of this, the NNTT was not a complete failure. It nurtured the successful development of Nelson's segmented mirrors and Angel's innovative spin-casting and polishing techniques. These technologies helped astronomers break away from the Palomar paradigm that had dominated telescope design for decades. The effort also kept the national observatory involved in a technologically challenging project that was highly visible to (if not fully supported by) the scientific community and funding agencies.

Goldberg had accurately predicted the challenges the NNTT and the national observatory would confront. By the late 1980s, some astronomers looked forward to new large telescopes funded by generous private and state monies. University astronomy had not withered away and Jesse Greenstein's beloved "aristocratic astronomy" was more powerful than ever. As George H. W. Bush campaigned for President, Oertel and Wolff began the daunting task of creating broad support and finding funding for the 8-meter national telescopes. They both believed the twin 8-meter project was important for America's scientific competitiveness and essential for the national observatory's future. While the telescopes might not claim bragging rights to the world's largest mirrors, Oertel and Wolff believed they would still be excellent, even essential, tools for astronomers.

On November 1, 1987, a few months after AURA abandoned the NNTT, Leo Goldberg died. Reflecting on his former classmate's life for a memorial, Greenstein said, "all the seeds of his personality were already in place when [we were] struggling at Harvard . . . Leo was both an important scientist and leader, but not a saint." Nevertheless, Greenstein urged the writer to "emphasize the positive and not look for spots on the sun."[64]

CHAPTER 6

Astropolitics

Goetz Oertel was familiar with Otto von Bismarck's saying, "He who has his thumb on the purse has the power." After AURA's 1987 retreat in Colorado, Oertel and NOAO staff committed themselves to building two 8-meter telescopes all American scientists could use. They soon watched as political and fiscal realities transformed their ambitions into an undertaking few might have imagined a decade earlier. Instead of a strictly American project, the Gemini 8-Meter Telescopes project became an international endeavor involving the United States, Canada, and the United Kingdom and one in which the U.S. had only a 50 percent share. Bismarck also said, "Politics is the art of the possible," a maxim that astronomers were obliged to heed as they planned their next move.

Government support of British astronomy began in 1675 when Charles II chartered the Royal Greenwich Observatory and appointed John Flamsteed as the first Astronomer Royal. Professional astronomers' reliance on the crown's largesse for telescopes and research support continued into the twentieth century, and the long tradition of philanthropic and university support that American astronomers enjoyed was absent in the British science community.

After World War II, Great Britain did not emphasize ground-based optical astronomy. While the clear nights and high mountain peaks of the western United States helped make America preeminent in optical astronomy, England's climate encouraged research and instruments at radio astronomy's frontier as well as excellence in theoretical astronomy. When Great Britain did build a new optical telescope, internal politics led to its location outside of London near what was then the Royal Greenwich Observatory's headquarters. With skies overhead often obscured by clouds, the 98-inch Isaac Newton Telescope never met scientists' hopes that it would be Britain's "greatest single contribution to the development of observational astronomy."[1]

While English astronomers waited for the Isaac Newton Telescope—its planning and construction process took 21 years—the British government reorganized its scientists. In 1965, hoping to establish greater control over scientific research and enhance economic prosperity, the government established the Science Research Council (later called the Science and Engineering Research Council or SERC).[2] In the process, the Royal Greenwich Observatory and the Royal Observatory, Edinburgh, were placed under SERC's control.

Over the next twenty years, SERC's support of astronomy in general decreased by more than a third; optical astronomy fared even worse. The loss of basic science funding coupled with a lack of suitable observing sites and first-class telescopes resulted in what one senior scientist called a "missing generation of astronomers."[3] The optical astronomy that was done tended to be very traditional—"brass telescopes and celestial mechanics" as one astronomer derisively recalled.[4] Britain's refusal to join the European Southern Observatory in 1962 exacerbated the situation. Unhappy with the short shrift optical astronomy was receiving, many British scientists took their talents and research programs to the United States, where they made important research contributions and won leadership roles in the American community.

In the 1970s, the fortunes if not the funding of British observational astronomy began to improve. In 1967, SERC authorized a joint British-Australian project to build a 3.9-meter telescope in New South Wales. The Anglo-Australian Telescope was similar to the successful 4-meter designs used for AURA's telescopes at Kitt Peak and Cerro Tololo and astronomers began observing with it in 1975.

The Anglo-Australian Telescope served as a rallying point for Britain's moribund optical astronomy community and demonstrated to British astronomers that international collaborations were a viable way to obtain access to well-sited bigger telescopes. Michael G. Edmunds, a graduate student at Cambridge when the AAT opened for business, recalled, "The AAT opened up the possibility that younger people could get access to a large telescope." Roger Davies, another young British astronomer in the late 1970s, called the AAT "the major development that brought U.K. optical and infrared astronomy into the modern era."[5]

Astronomers using the AAT took advantage of its sophisticated instrumentation. One of these devices was the Image Photon Counting System (IPCS). It was built in the early 1970s by a young British scientist named Alec Boksenberg. He developed the IPCS to be analogous to the human eye so it both recorded and processed images. The complex electronic system had an image tube at one end to intensify light collected by the telescope. Boksenberg coupled this to a television camera so the IPCS could both record and show what the telescope was observing. The result was a very sensitive light detector that offered astronomers two-dimensional images in real time (unlike a CCD, which is read out only after a period of time has elapsed.) Boksenberg's device had a loudspeaker attached to it so, as Edmunds recalled, "you could actually hear the photons coming in—bip, bip, bip—and watch the image build up. It was very exciting to use."[6]

Copies of the IPCS were built and used at the Anglo-Australian Observatory and at Palomar. Boksenberg himself co-authored several papers with Caltech astronomers after they collected data with his IPCS system on the 200-inch. These successes won Boksenberg the director's office of the Royal Greenwich Observatory (RGO) in 1981. He quickly developed ambitious plans for optical astronomy in the United Kingdom. As the new director of a centuries-old institution, Boksenberg later explained, "it was necessary to re-direct or re-emphasize the work of RGO away from some of its more traditional activities . . ."[7] Recognizing that a new facility could give new life to British optical astronomy, he encouraged Great Britain to develop a new observatory on San Miguel de la Palma, an 8,000-foot mountain in the Canary Islands. The project was another successful international partnership for British astronomers, who shared management tasks and financial responsibilities with Spain, Sweden, and Denmark in exchange for observing time.

In 1985, British scientists inaugurated their new facility and they soon transferred the formerly hapless Isaac Newton Telescope to La Palma's sunny skies. The observatory continued to grow as funding from SERC enabled completion of the 4.2-meter William Herschel

Telescope (the world's third largest at that time) in 1987. On Boksenberg's watch, RGO became a major player in optical astronomy; much of that vitality was centered on La Palma and the new telescopes there.

La Palma was not the only site for telescopes British astronomers were developing, nor was RGO the only royal observatory. The Royal Observatory, Edinburgh (ROE) was founded in 1834 by royal decree and directed by the Astronomer Royal for Scotland. In the mid-1980s, Malcolm Longair was ROE's director. A Scotsman who received his doctorate degree at Cambridge's famous Cavendish Laboratory (of which he later became director), Longair first specialized in radio astronomy. He later was the only non-American selected by the Space Telescope Science Institute to give scientific advice to NASA's Hubble Space Telescope project. Both he and Alec Boksenberg were appointed directors of royal observatories while unusually young, something Longair attributed to the missing cohort of talented British optical astronomers who had left England for lack of good telescopes.[8]

While Boksenberg's observatory developed La Palma, ROE built its new telescopes on Mauna Kea. In 1979, ROE began managing the United Kingdom Infrared Telescope, the world's biggest telescope dedicated exclusively to infrared observing. One of the special features of what astronomers called the UKIRT was its 3.8-meter mirror. Only half as thick as the mirrors of conventional telescopes at that time, its design foreshadowed the commercial development of thin meniscus mirrors by commercial firms like Corning.

Astronomers, including many in Britain, were increasingly interested in exploring the sky at infrared wavelengths, thanks in part to the improving capabilities of infrared detectors and instrumentation. They also looked forward to combining infrared data with information from traditional optical wavelengths. One of these scientists was Charles Mattias (Matt) Mountain, a sturdy-looking man with curly blond hair whose rapid talking was interspersed with a wry sense of humor. Mountain came from what he called a traditional English family "who did sensible things." His mother's family was Swedish and artistically inclined. As a young boy, Mountain struggled in school—he later believed a mild dyslexia hindered his early learning—and his parents had limited career aspirations for their son.

Nonetheless, Mountain did well in math and science courses and originally considered a career in high-energy physics. During a summer at CERN, Europe's center for particle physics, Mountain worked for a large collaborative project when "it dawned on me that particle physics wasn't what I was looking for."[9] His graduate advisor at Imperial College steered him toward astronomy and he finished his doctoral degree there in 1983. Mountain discovered that infrared observing was the "physicist's astronomy with all the detectors and dewars. The infrared lab made sense to me as a physicist." His talent for building instruments—he designed an infrared spectrometer for his dissertation—caught Malcolm Longair's attention and he was recruited personally by the Astronomer Royal of Scotland to join the ROE's staff.

At ROE, Mountain and others were assigned to particular projects to build new instruments. He found a "creative tension" there as Longair encouraged scientists to work alongside engineers and technicians. As ROE's director, Longair reorganized the institution so the chief engineer reported directly to him rather than having this person's input and opinions filter up to him through the science staff. At ROE, it was not uncommon for instrument projects to have one scientist for every ten engineers or technicians. This style, Mountain believed, contrasted with astronomy's traditional "principal investigator model" in which work was led by a single scientist (perhaps helped by a student or two) with the engineers in a subordinate role. By 1985, the twenty-nine-year-old Mountain was the project scientist for a new $1.2 million spectrograph ROE was building in conjunction with a large team of engineers, optical designers, and software specialists. Given his initial distaste for the large-scale and impersonal nature of high-energy physics, Mountain found the irony rich. "You need professional engineers so things get done right. The scientist helps keep that large team going," he explained. "The single investigator approach wasn't going to cut it at the front rank in the U.K."

By the mid-1980s, British astronomers saw their community divided along several axes. Boksenberg's RGO was allied with European partners at its La Palma site and its staff was predominantly doing astronomy at visible wavelengths. Longair and the ROE were more acquainted with the American astronomy community via their facilities on Mauna Kea and were especially committed to infrared astronomy. Some astronomers also perceived a generation gap as younger scientists often collaborated with American colleagues and accepted jetting off to Hawaii or Australia to collect data for a few nights as a normal part of their job.[10]

Richard S. Ellis was one of the young British scientists who was used to observing on Mauna Kea. He had wanted to be an astronomer ever since he was six years old. He built his own telescopes and read shelves of popular astronomy books as a teenager. While a graduate student at Oxford, Ellis traveled to Australia to do research on the Anglo-Australian Telescope. He made good use of Boksenberg's new IPCS detector system and saw how a new telescope equipped with state-of-the art instruments had revitalized optical astronomy in the United Kingdom.

In the mid-1980s, British astronomers were considering what their next big telescope project should be. SERC's studies recommended different options.[11] To chart priorities, SERC established a Large Telescope Panel in early 1987 and asked Ellis to chair it. At this time, Ellis was a thirty-six-year-old astronomer building his career at the University of Durham. Besides Ellis, the panel had only two other members: Michael Edmunds from the University of Wales and James H. Hough, a physicist at Hatfield Polytechnic. The panel members were all relatively young and they harbored ambitions to put British optical astronomy at the leading edge. This meant securing access to the next generation of big telescopes. Ellis and his panel realized SERC would balk at funding another British-only telescope project so soon after the completion of new telescopes on Mauna Kea and La Palma. International collaboration, therefore, was the path most likely to lead to a larger telescope for their community. Their main dilemma was finding a suitable partner.

Members of the Large Telescope Panel began traveling all over the world to court potential collaborators. Sometimes they even began their presentations with a slide reading "U.K. Large Telescope Marriage Brokers." They sought out Japan's astronomy community (then planning an 8-meter national telescope on Mauna Kea) and negotiated with German scientists as well as representatives from the European Southern Observatory.

In March 1987, Ellis met with John Jefferies, then still the head of the National Optical Astronomy Observatories, in Tucson. With support for the NNTT project waning, Jefferies believed this was "an opportune time to move" on a joint American-British effort to build an 8-meter telescope on Mauna Kea. Jefferies soon resigned from NOAO, but his successor, Sidney Wolff, kept lines of communication open and supported the idea of a collaborative effort.[12]

After innumerable presentations to astronomers and science managers, Ellis and his panel concluded they had two options.[13] One was to partner with Spain to build an 8-meter telescope at La Palma. Ian F. Corbett, head of SERC's astronomy program in the late 1980s, described this as "a political statement about Europe and building on our prior investment in La Palma."[14] La Palma was relatively more accessible than Hawaii, and some astronomers favored the alliance because Spain would be a minor partner, giving British scientists greater control.

The other option was a partnership with the Americans through NOAO. This would be, as some saw it, a more equal alliance, a comforting fact to some British astronomers who questioned their ability to procure all the necessary high-tech components for a large telescope. Goetz Oertel encouraged negotiations between Wolff and British astronomers and signaled AURA's general support for some form of partnership. Optimistic about the prospects for "one or more 8-meter telescopes," Oertel wrote Ellis on the heels of the NNTT's cancellation that a collaboration offered British and American astronomers a convenient way "to get more and better science done at lower cost."[15]

In October 1987, Ellis's Large Telescope Panel sent SERC its interim report. The panel's first priority was securing at least a 50 percent share in a new 8-meter telescope project by either collaborating with United States or Spain. The report was incorporated into a more formal strategy issued by SERC the following year. The Ground Based Plan, as it was called, set British strategy and priorities to the twenty-first century. This included participation in an international 8-meter telescope project. Newspapers in England reported that SERC was prepared to commit £15 million to a "new-generation ground-based telescope."[16] What these stories did not spell out was who the United Kingdom's partner would be nor how astronomers in the United States might react to news that their hope for a new national telescope was taking an international flavor.

A Multiplicity of Telescopes

Astronomers and politicians established the European Southern Observatory in 1962 with the goal of providing unparalleled observing opportunities in the Southern Hemisphere. After an extensive site survey, ESO astronomers chose a remote mountain ridge locally known as

La Silla ("the saddle") in Chile's Atacama Desert. ESO purchased land around the mountain and signed a treaty with Chile's government that made its observatory the world's largest diplomatic enclave and gave its staff diplomatic status. ESO then boldly began to build telescopes; by 1988, it operated 15 different telescopes on La Silla, including two 4-meter class facilities.

Astronomers based the European Southern Observatory's management on CERN. This organization had enabled European physicists to pool their resources in order to construct increasingly powerful particle accelerators that were beyond the capabilities of any single country. ESO took the same strategy and applied it to astronomy. Compared to the United States' pluralistic approach toward funding and long-term planning, which brought together philanthropic, university, and federal support into a diverse but unwieldy combination, ESO's organization was much more top-down and monolithic. While this made it somewhat less flexible, ESO could directly forge or force consensus among the 1,400 or so scientists from its member states. American optical astronomers were simply not able to achieve (or have imposed upon them) the Europeans' unified undertaking of a few major projects.

ESO's commitment to bigger telescopes was most visible in its plans for what one journalist called "Europe's Astronomy Machine." The Very Large Telescope, despite its prosaic name, was an incredibly ambitious undertaking. Similar in scale to the ill-fated NNTT, the Very Large Telescope project had a collecting area equivalent to a 16-meter telescope. Instead of placing four separate 8-meter mirrors on a common mount, ESO opted to build an array of four separate 8-meter telescopes in Chile.

On December 8, 1987, less than four months after the AURA board laid the NNTT to rest, delegates from ESO's eight member countries—Belgium, Denmark, Germany, France, Italy, the Netherlands, Sweden, and Switzerland—formally approved the Very Large Telescope. ESO's science managers estimated their long-term commitment would cost at least a quarter of a billion dollars, with millions more needed annually once it became operational in the late 1990s.[17]

ESO's press release for the Very Large Telescope, not surprisingly, announced that the project would make Europe "second to none in the exploration of the Universe for a long time to come."[18] Not conceived of as four separate telescopes but rather as an integrated research facility, completion of the massive project would enable European astronomers to collect more light than all the other existing telescopes in the world combined, a point Sidney Wolff brought to the attention of Congress.[19] Citing increasing international competition is a common strategy in scientists' pitches to Congress for more funding. In this case, it is difficult to accuse Wolff of hyperbole. When built, the Very Large Telescope would give ESO astronomers eight times more glass to collect photons with than all the telescopes NOAO currently controlled.

Compared to the relatively piecemeal efforts underway in the United States, ESO proceeded along a more deliberate path toward the Very Large Telescope. For example, in 1982, Italian and Swiss astronomers gained access to ESO's telescopes after their governments paid several million dollars in membership costs. With money from these new members, ESO began building a new telescope on La Silla called the New Technology Telescope. ESO astronomers and engineers specifically planned this modest-sized 3.6-meter telescope as a test bed for technologies they would use later for the much larger Very Large Telescope facility. Besides giving ESO astronomers yet another new telescope, the New Technology Telescope would be a tool to gauge how new design features worked during actual observing runs.[20]

ESO based its New Technology Telescope neither on Jerry Nelson's segmented mirror technology nor on Roger Angel's lightweight honeycomb mirrors. Instead, ESO astronomers opted for one of the thin meniscus mirrors that commercial firms like Corning and Schott Glasswerke were ramping up to produce. Since the early 1980s, these firms had invested considerably in facilities to make thin mirrors for research and defense applications. A few years later, their work began to pay off as commercially fabricated meniscus mirrors became a third path that telescope designers took with increasing frequency.

Anticipating the expertise and knowledge they would accrue with the New Technology Telescope's meniscus mirror, ESO engineers specified that the mirrors for the four separate telescopes of the Very Large Telescope facility be only 17.5 centimeters thick. This meant the 8-meter pieces of glass would have the relative dimensions of a contact lens. This was a bold and radical departure from traditional telescope mirrors. Japanese astronomers made a similar design choice for their

8-meter national telescope, choosing a meniscus mirror from Corning over one of Angel's honeycomb mirrors.

In order for the meniscus mirror to work, however, ESO astronomers needed to demonstrate the workability of one key technology. Because meniscus mirrors are so thin, they flex slightly, but unacceptably, under their own weight as the telescope moves. ESO adopted a preventive solution called active optics. After the mirror was polished, it would be transferred to a mirror cell, where it rested on several dozen supports. Sensors analyzed the shape of the primary mirror about once an hour while astronomers used the telescope (infrequent analyses compared to how adaptive optics systems and Nelson's segmented mirror system functioned). Distortions detected in the mirror's shape were smoothed out by a system of computer-controlled actuators. While the active optics system would merely enhance the New Technology Telescope's performance, it was absolutely critical to the success of the entire Very Large Telescope project. Without this technology, the giant 8.2 meter mirrors would produce ruinously poor images.

ESO chose Schott to make the mirrors for all of its new telescope projects. The German company also provided the raw glass segments for the Keck Telescope project. Schott made its mirrors from its proprietary low-expansion material called Zerodur. Initially, however, there were major difficulties in fabricating the 8.2-meter mirror blanks for ESO's Very Large Telescope—Schott's first three attempts cracked in its giant annealing oven. Handling the thin and fragile 23-ton mirror blanks was another engineering challenge and Schott did not successfully make its first 8-meter mirror until 1991.

In March 1989, ESO astronomers made their first scientific observations with the 3.6-meter New Technology Telescope. Real-time pictures transmitted back to ESO headquarters in Germany showed the active optics system performing flawlessly.[21] The telescope's razor-sharp images proved it was possible to make moderately-sized thin mirrors perform as well as the thicker, rigid mirrors Roger Angel was promoting. The question that lingered in everyone's mind was whether much larger meniscus mirrors could be reliably made and, more importantly, controlled to astronomers' demanding specifications without costing a small fortune.

Sidney Wolff and staff at NOAO in Tucson monitored ESO's methodical progress toward its Very Large Telescope and saw the excellent "first light" results ESO scientists obtained with their smaller prototype telescope.[22] Wolff, with AURA's support, opted to follow ESO's cue by first building an intermediate telescope at Kitt Peak. This, Wolff reasoned, would accomplish several goals. Basing a new telescope on one of Angel's recent 3.5-meter castings was an opportunity to test the honeycomb mirror technology. It had been almost twenty years since the national observatory had built a new telescope, and an intermediate project could provide experience and feedback the NOAO staff could use later for the twin 8-meter telescopes. Just as important, astronomers were placing increasing demands on NOAO's facilities, and a new modest-sized telescope would relieve some pressure for observing time.

NOAO still continued to confront severe budget pressures. These cutbacks persuaded Arthur Walker, chair of the NSF's advisory panel for astronomy, to tell a congressional subcommittee that national centers and the ground-based astronomers who relied on them were "literally starving to death." The Stanford physicist drew imaginatively upon his experience helping investigate the *Challenger* explosion to warn of the risks the NSF and Erich Bloch were taking by not supporting more basic science.[23] Appearing before the same subcommittee, Wolff said the recent collapse of a 300-foot radio telescope at Greenbank, West Virginia, symbolized America's eroding infrastructure for science. Just as threatening, she said, was the possibility that, unless Erich Bloch revitalized funding for astronomy, "we'll be going back to the situation we had in the 1950s, when optical astronomy was dominated by Caltech and the University of California and most astronomers had no access to the best observing facilities."[24]

In response to NOAO's declining budget and increased demands for telescope time, Wolff and Oertel explored alternative strategies for building the observatory's proposed telescope prototype.[25] In 1989, NOAO announced a partnership with the Universities of Wisconsin and Indiana to fund a 3.5-meter telescope on Kitt Peak based on a honeycomb mirror from Angel's lab. Yale later joined the effort to build what became the modest-sized WIYN (for Wisconsin, Indiana, Yale, and NOAO) telescope. This was the first public-private consortium to build a major new telescope, a noteworthy step for NOAO and university astronomy programs eager to obtain new observing facilities they otherwise might not be able to afford. A few years later, NOAO announced a similar joint venture with Columbia University and the University of North Carolina to build the Southern Observatory for Astrophysical Research (SOAR), a 4-meter telescope in Chile.

NOAO pursued the WIYN and SOAR telescopes in part to test new telescope designs, innovative instruments, and new techniques for using them such as remote observing and multiobject spectroscopy. But, by the late 1980s, building 4-meter class telescopes was no longer a cutting-edge endeavor. The technological frontier had shifted now to 8- to 10-meter instruments such as the Keck and ESO projects.

Partnerships of one kind or another were an increasingly common strategy astronomers adopted to get at least partial access to their own telescopes. By 1990, almost all new telescopes under construction were part of some type of collaborative effort, a major shift in the organizational landscape of American astronomy. Collaborative partnerships were not without risks, however, as the University of Arizona discovered when Ohio State University suddenly pulled out of a joint project to build the Large Binocular Telescope on the controversial site of Mount Graham. Ohio State's withdrawal surprised many scientists. For years, Peter Strittmatter and his Arizona colleagues had maneuvered around environmentalists' objections to telescope construction on Mount Graham. This news, coupled with the funding uncertainties that threatened Angel's Mirror Lab on an annual basis, encouraged Strittmatter to adopt what he wryly called a "rigid policy of flexibility."[26]

American astronomers were still divided over the best role for their national observatories. In response to a community survey by AURA that asked scientists to "define the goals of the national observatory," an overwhelming majority of replies were conservative. Many astronomers entreated NOAO to forego future telescopes and simply keep the ones it currently had in operation.[27] Both Oertel and Wolff believed NOAO should build new facilities to avoid being left behind. Twin 8-meter telescopes were central to their plans. Yet, they could not completely disregard their constituency's fears that older telescopes would be closed to pay for new ones in what many saw as a zero-sum game.

In September 1989, AURA submitted its plan for the "NOAO 8-M Telescopes" to the NSF.[28] Unlike Kitt Peak's earlier requests for new telescopes in the 1970s, which were relatively cursory, the four-volume proposal was comprehensive. Knowing many American astronomers did not fully believe NOAO should build anything, let alone two giant new telescopes, Wolff and Oertel worked to sell the ambitious project to scientists, the NSF, and Capitol Hill. NOAO newsletters featured articles about the project's progress and enlisted other astronomers' advice and support. AURA itself circulated at least three drafts of its multivolume proposal through the science community, revising it and shoring up weak areas.

AURA asked for almost $144 million of federal money to fund the twin-telescope project. The heart of each telescope would be an 8-meter honeycomb mirror supplied by Angel's Mirror Lab. Despite years of research and development by people like Angel and Jerry Nelson, astronomers still viewed the mirrors as the project's riskiest component. The mirrors were certainly the most expensive component. Two 8-meter mirrors cast by Angel's lab would cost an estimated $7 million; polishing, ancillary hardware, and shipping added another $14 million. In the event that Angel's lab failed, meniscus mirrors would be the fallback technology, albeit a more expensive one.

NOAO engineers designed the twin telescopes as versatile general-purpose facilities capable of collecting images and spectra at both visible and infrared wavelengths.[29] It was important for project advocates to suggest areas in which the NOAO telescopes might outperform other telescopes under development. For example, the Keck Telescope's 10-meter segmented mirror would probably have some limitations in terms of the quality of images it produced. Consequently, the AURA proposal stressed the telescopes' ability to produce superb image quality—0.25 arc-second resolution was their goal—as one area in which they might beat the other telescopes under construction. Adaptive optics, incorporated into the design from the beginning, would enhance image quality even more.

NOAO anticipated taking advantage of what its engineers had learned in the last fifteen years about telescope design by using a primary mirror with a short focal length, a compact alt-azimuth mount, extensive computer modeling of the telescope and its subsystems, and a building design and site chosen to provide best possible seeing conditions. Technicians would quickly swap among three different secondary mirrors at each telescope to permit a wide range of observing techniques from wide-field imaging to infrared observing and enable astronomers to execute several research programs on the same night. NOAO's new facilities, the proposal claimed, would be the first large telescopes "designed from the outset to meet the requirements of infrared astronomy," another aspect that might distinguish the project from other telescopes.[30]

Instead of presenting a cosmic shopping list of potential research areas the telescopes could attack—everything from planets and asteroids to extragalactic quasars—AURA adopted a strategy NASA had employed to win support for some of its space telescope missions. The Space Telescope Science Institute, for example, anchored science objectives for the Hubble Space Telescope around a few core research topics called Key Projects. A certain fraction of observing time was devoted to major research initiatives (such as determining the Hubble constant, a number representing the current rate of the universe's expansion) that the science community as a whole agreed were most important. Astronomers who helped write AURA's proposal identified three "astrophysically significant problems"—star formation, galactic structure, and galaxy evolution.[31] Just to accomplish specific research programs in these three areas, the proposal estimated astronomers would need over 7,000 hours of observing time, along with at least seven major instruments costing well over $25 million. The concept of key projects, with NOAO reserving large blocks of telescope time, represented a degree of top-down science planning that many American astronomers might have found objectionable fifteen years earlier.

AURA's proposal reflected many of the changes in American astronomy since 1974 when Leo Goldberg first initiated Kitt Peak's exploratory Next Generation Telescope project. Consider, for example, AURA's vision for the operation of the twin 8-meter telescopes. National facilities, the proposal said, should not be just "data collection sites" but rather "complete scientific systems."[32] Like all NOAO telescopes, astronomers would get time on the twin 8-meters by successfully submitting a peer-reviewed proposal. Astronomers working on the telescopes' core research programs would be obliged to reduce their data "in such a way that it is suitable for archiving and can be made available to the whole community."[33] In other words, not only would the telescopes be nationally available but, in many cases, the data they collected would be a national science resource. This strategy, familiar to NASA's space astronomy missions, was a departure for more traditional ground-based astronomers. Telescopes were rapidly being viewed both as research tools and as components in a larger system, reflecting the growing complexity, cost, and scale of astronomical observing.

In 1974, Jesse Greenstein told colleagues at a large telescope conference that he didn't trust a photographic plate "unless I take it myself."[34] At the time, many astronomers shared his view. Fifteen years later, AURA's proposal declared "the astronomy community has become increasingly accustomed to the acquisition of observations without the scientist being present at the telescope." Part of this shift in practice was brought about by NASA missions that relied on remote data collection. Astronomers also recognized that seeing conditions at telescope sites varied widely and quickly throughout the night. By incorporating some form of flexible telescope scheduling, they could take advantage of the best seeing. Careful monitoring of the local environment, including seeing conditions, moonlight, and cloud cover, was necessary in conjunction with rapid instrument switching and complex scheduling aided by computer algorithms similar to what engineers had written for the Hubble mission. While many astronomers were initially suspicious about the "queue scheduling" of their observing programs, the payoff would be data that took advantage of the best seeing conditions without astronomers having to wait at the telescope for those circumstances to emerge by chance.

Since the national observatory had completed its last big telescope project in 1974, ground-based telescopes had become nodes in a larger network of astronomers, time allocation committees, and scheduling routines, all linked by satellite, phone lines, and fiber optics and coordinated to collect photons more efficiently. On the basis of the cost and size of its instruments, optical astronomy joined the leagues of Big Science decades earlier. Now the actual practice of doing astronomy, long limited to scientists operating on their own initiative, was beginning to be reshaped as well by the growing scale and sophistication of new telescopes astronomers were building.

Gemini Emerges

Thirteen months after AURA sent its proposal for two 8-meter telescopes to the NSF, Congress passed H.R. 5158. Tucked away on page

153 of this appropriations report was the following statement: "The committee recommends the $4,000,000 requested for engineering design studies and the purchase of glass for two new 8-meter telescopes." This was good news for American astronomers. After years of waiting and fruitless effort, astronomers had finally obtained funding for a new national telescope.

The bill also stipulated that, in order "to protect the U.S. investment," the National Science Foundation could only initiate work on the northern telescope after a "firm, fixed, cost-sharing arrangement was concluded with all foreign partners." Moreover, the "U.S. contribution to this project [may] not exceed 50 percent of the total cost" and even if "foreign participation is ultimately agreed to or not . . . the U.S. share of this project will not exceed $88,000,000."[35] In other words, Congress told astronomers they could have one telescope or half of two telescopes. In either case, they were forbidden to spend more than $88 million. The proscriptive language was an unwelcome shock to many American scientists. How did their telescope project, initially conceived as an American undertaking, become a tightly budgeted international endeavor?

The NSF was familiar with using international partnerships to pursue ambitious science projects. By 1988, for instance, science agencies from over fifteen countries were contributing money to an NSF-funded and managed ocean-drilling consortium that paid for specially designed ships and teams of scientists to collect deep-sea cores of rock and sediment. In March 1984, Lewis M. Branscomb, a science advisor to President Johnson and later the National Science Board's chair, told the House appropriations subcommittee that "international dimensions" should be integrated as an "organic aspect of the scientific enterprise" at all levels of the NSF's programs.[36]

As the NSF's director, Erich Bloch heeded Branscomb's advice. Bloch, as noted earlier, had reasons for not being overly fond of astronomers and their requests for new telescopes. Bloch himself was more concerned with issues such as America's economic competitiveness and directed NSF money toward projects that offered some promise of financial gain. He was not averse, however, to telling a somewhat skeptical Congress that investment in astronomy and other basic sciences might create industrial payoffs.[37] As Bloch had a personal interest as well as a mandate from the National Science Board to foster internationalism in future science projects, astronomy became a convenient vehicle to pursue his larger goal. He gradually let it be known that he desired "cooperation between ourselves and some of the trading partners in the area of astronomy centers and astronomy equipment."[38]

Bloch saw astronomy as a "good test case" for other collaborative efforts and a way to get advanced facilities for United States scientists when federal funding for astronomy was stagnant and national science policy dictated other priorities.[39] Bloch's top-down imposition of international collaboration also appealed to congressional interests. Dick Malow monitored AURA's evolving plan for two 8-meter telescopes. He was concerned that the NSF's commitment to large, complex facilities might cut into its more traditional mission of supporting individual researchers. Malow worried, he said, that if the nation's deficit problems continued and the NSF embarked on another big project, "we were going to start down a trail, once started, we could never shut off." Cost-sharing via international collaboration was a sensible strategy he favored.[40]

The NSF's top-down encouragement of international collaboration differed from what was happening in the science communities of its potential partners. In both Canada and the United Kingdom, support for a joint project existed both at the grassroots level and at the top levels of science management. Canada and the United Kingdom already participated in international telescope collaborations and their scientists accepted these as part of the reality of building ever-more complex and expensive tools under tight budget constraints. At the same time, their respective science agencies were experienced and comfortable with managing collaborative international projects.

In May 1988, the Canadian Astronomical Society recommended that the National Research Council, Canada's equivalent to the NSF, pursue collaborative efforts. Canadian astronomy was funded in much the same way as in the United Kingdom with almost all money coming from the government and no tradition of private funding. In September, after further evaluation, Canadian representatives met formally for the first time with Wolff and other NOAO staff in Tucson. To few people's surprise, Wolff told Canadian representatives from the National Research Council that the United States was unlikely to provide funds to build an American-only telescope project. Moreover, Bloch had confided to her that, as the NSF had built new national radio telescopes for close to twenty years, it "was now optical astronomy's turn."[41]

What could a partner such as Canada offer a telescope collaboration? Simon J. Lilly was an English astronomer who became prominent in Canada's 8-meter efforts when he moved to the University of Toronto. He later represented Canada on different science committees for Gemini. He believed American scientists could take advantage of Canada's considerable experience operating international telescope facilities on Mauna Kea. Astronomers using the Canada-France-Hawaii Telescope, a 3.6-meter telescope dedicated in 1979, routinely took images with exceptionally fine quality. "We had a heritage with the CFHT. It showed the world what type of image quality you could get on Mauna Kea. The American optical community as a whole had very little experience on Mauna Kea and most of them had not observed there."[42]

On June 1989, the Canadian Astronomical Society unanimously voted to join the collaboration with the United States and the United Kingdom for two 8-meter telescopes. Less than a month later, the pace to put a formal agreement in place quickened as representatives from the United Kingdom and Canada held their first tripartite gathering with American scientists in Ottawa. Their plans for the twin telescope project assumed that the United States would pay 50 percent of the costs, with the United Kingdom and Canada each contributing 25 percent.

After the potential partners met, Goetz Oertel faced the delicate task of informing AURA's constituency that international collaboration was likely. Rumors this might occur, of course, circulated among scientists. At the summer meeting of the American Astronomical Society, NOAO described the international collaboration that would make its 8-meter project possible. Five months after AURA submitted its 8-meter telescope proposal to the NSF, Oertel sent a formal "Dear Colleague" letter to American astronomers letting them know that international collaboration was planned. In April 1990, AURA's board gave its official support to the venture.[43]

American astronomers expressed a mixed reaction to the internationalization of "their" telescope project. "The grassroots opinion was for the 8-meter telescopes," Oertel said, "but they wanted them to be national. Erich Bloch gave absolutely no encouragement to this idea."[44] Their reticence surfaced publicly when *Science,* the weekly journal of record for American science, published an article about the twin 8-meter project. After describing its potential capabilities, the article reported that the NSF would not commit to building the telescopes without foreign contributors paying half.[45]

The U.S. astronomy community was split over whether it would be better to build only one national telescope in Hawaii or share half of two international telescopes that would give researchers a view of the entire sky. Many scientists were angry that the NSF would not provide all of the funding, some later saying that Bloch "threw American astronomy to the dogs of international collaboration."[46] Rumors that Caltech and University of California astronomers might be getting a second 10-meter telescope from the Keck Foundation increased their resentment. Some also worried that an international effort would take longer to build and be harder to manage. George Field, chair of the previous decadal survey, noted that the lack of clear progress was disappointing given that his report had advocated a new, national telescope more than eight years previously. Others, though, shared the view of Edward Stone, a Caltech physicist and later the director of the mission-oriented Jet Propulsion Laboratory, that "half a telescope is better than none."[47]

On the other side of the Atlantic, British astronomers accepted international collaboration but were still divided on the question of a partner. In January 1989, after receiving Richard Ellis's report recommending the United Kingdom invest in an 8-meter project, SERC established a two-year study to specify the design details. Roger Davies left Kitt Peak and returned to the United Kingdom to lead the design study at Oxford.[48]

While Davies and a small team of engineers carried out technical studies, Richard Ellis and the other two members of the Large Telescope Panel navigated the rocky waters of British science politics. Despite recent progress toward a collaboration with NOAO and Canada, many in the United Kingdom still favored partnering with Spain to build an 8-meter machine on La Palma. These efforts were largely led by Alec Boksenberg, who wished to ensure the continued health of the Royal Greenwich Observatory and its facilities on La Palma. The debate received considerable coverage in the British press. An article entitled "Spoilt for Choice" in *The Economist* included a cartoon that fairly summarized the position in which British astronomers found themselves. It showed a circle of arguing scientists while, above them in the pans of a giant balance, sat two telescopes.[49]

To make the case for La Palma and a European-based partnership, Boksenberg deployed several arguments. The United Kingdom would be firmly in control—scientifically, financially, and managerially—with the European option. The necessary infrastructure was already in place. An 8-meter telescope on La Palma could be based on the design of RGO's recently completed 4.2-meter telescope there while NOAO was still hashing out its plans. Building on La Palma would give about 55 percent of the available telescope observing time to British astronomers, while the American-Canadian option, Boksenberg argued, gave only a quarter share on two different telescopes, both at less accessible sites.[50] This option was soon complicated by Spain's announcement that it would fund only 25 percent of the project, making a third partner necessary for the La Palma option. The Soviet Union even considered entering the project, a move that probably did not help the case for La Palma.

Malcolm Longair and other staff members at the Royal Observatory, Edinburgh, countered Boksenberg's arguments throughout 1989 and 1990. The high and dry site of Mauna Kea, they argued, was superior for a telescope project with a high priority on infrared observing.[51] Partnering with the American astronomy community also offered British scientists the possibility of a surprise technological windfall. For several months, the NSF had been negotiating with the Air Force to declassify its adaptive optics technology, plans for which Wayne Van Citters, the NSF's director of Astronomical Instrumentation and Development, had revealed to British astronomers and science managers at SERC.[52]

As British astronomers dithered throughout 1990, American scientists and administrators continued to apply pressure on the British astronomy establishment to choose the NOAO/Canadian option. Erich Bloch kept the advantages of international collaboration at the forefront in his correspondence with Sir William Mitchell, his counterpart in the United Kingdom.[53] After months of meetings and frustrating delays, on December 19, 1990, SERC managers finally decided to partner with the United States and Canada. Ellis, who had worked hard on getting his country into an 8-meter project for over four years, was delighted. "I was always a fan of collaboration with America. I liked the country, the vibrancy, all the exciting things happening. I thought the U.K. would benefit by having its eyes opened and the U.K. was a match for the U.S. intellectually.[54]

AURA and the NSF, meanwhile, lobbied to ensure that the initial funding for the twin 8-meter project came through. In early 1990, NOAO's newsletter announced that President Bush's fiscal 1991 budget included start-up money for the two telescopes. Bloch soon appeared before the House appropriations committee to justify the agency's plans for the project. That same year, however, the NSF was seeking funds for several large-scale science projects. This put Bloch in the uncomfortable position of defending the NSF's ability to build new equipment while still supporting individual research and education initiatives.

Three of these big projects were astronomy-related, a situation that posed a threat to the 8-meter telescopes project and Bloch's hope for an international venture. One project was to build a 300-foot radio telescope at the National Radio Astronomy Observatory in West Virginia to replace one that had collapsed in 1988. Robert C. Byrd, a powerful senator from West Virginia, earmarked $75 million of the NSF's budget to make sure this quickly became a top agency priority. Bloch could either acquiesce or risk angering Byrd, the Senate Appropriations Committee's ranking member. Another large project wending its way through the NSF was the Laser Interferometer Gravitational Wave Observatory (LIGO). LIGO was a proposed collaborative venture between Caltech and MIT to build two separate facilities in the United States, each with an L-shaped configuration of underground tunnels two and a half miles long equipped with exquisitely sensitive detectors. A small but persuasive group of physicists wanted these two facilities to detect gravitational waves, the perturbations of space-time predicted (but not as yet conclusively observed) to result from events such as the collision of black holes.[55] In 1990, the NSF asked for $47 million as the first installment on LIGO's estimated $192 million price tag.

Members of Congress interrogated Bloch about his agency's request. Was there a need for so many colossal projects at a time when, as Massachusetts representative Chester Atkins said, there were "immense needs" in education and industrial competitiveness?[56] Bloch tried to point out that, compared to truly expensive projects like the multibillion dollar Superconducting Super Collider under construction in Texas, the NSF's requests were really quite moderate. After Atkins quickly asked whether the NSF wanted to use a (soon-to-be-cancelled) physics facility that was over budget as its new benchmark for big projects, Bloch and his colleagues quickly back-pedaled from this comparison.

Bloch defended the NSF's request on the grounds that, while projects such as LIGO might benefit only a small cadre of scientists, new facilities would help draw young people to science and engineering careers while fostering industrial spin-offs. The NSF, Bloch testified, had done "a lot of soul searching" before requesting funds for LIGO. Atkins' response to the agency was that it was a "lot cheaper to search your soul than to search for a black hole."[57]

LIGO was neither a telescope nor a traditional observatory. In fact, there was considerable resentment among astronomers about LIGO. They saw it as equipment for a branch of physics that was not yet established and resisted attempts to link it in any way with astronomy's more traditional tools. Dick Malow personally opposed LIGO and gave the subcommittee probing questions to ask about the project. When Congress made the final appropriations later that year, LIGO's supporters saw their multimillion request forestalled until 1992.

Compared to the budget requests for LIGO and the Greenbank radio telescope, the $4 million needed to begin building the twin 8-meter telescopes was a trifle. Malow knew the telescopes would serve a much broader community of scientists than LIGO. Still, the NSF needed to affirm the project was a high priority. Bob Traxler, a Democrat from Michigan and the subcommittee's new chair, told Bloch, "We know they are all your children, you love them equally, and you want every one of them. You've got to make the choice."[58]

The most important condition imposed on the twin telescope project, according to Malow, was not its final cost but his determination to see the participation of the United States limited to 50 percent. Without this, Malow believed American astronomers would have less incentive to find international partners. In May 1990, Malow visited NOAO headquarters in Tucson to meet the staff and gather facts about the telescope project. They estimated the cost of two 8-meter telescopes to be $176 million, a number that the skeptical Malow took back to Washington with him.

Malow found the project's cost estimate low and was surprised when no one came to him asking to increase the project's budget before Congress made it official. "I suspect that if NOAO had gone and talked to the Europeans, who were building four 8-meter telescopes, they would have found a number substantially different than $176 million multiplied by two."[59] Ian Corbett, a science manager for SERC who monitored the budget negotiations, concurred. "It was certainly never a number that was subjected to any degree of scrutiny through international peer review . . . it bore no relation to what we expected the final design of the telescopes to cost."[60] Consequently, the twin telescope project was budget-constrained even before engineers and astronomers agreed on its final design and developed an appropriate cost estimate. America's optical astronomers had their first real NSF funding in over twenty years to start building a major national telescope, albeit one now internationalized and facing a challenging cost cap.

Planning a Decade of Discovery

With money now in hand, astronomers took a suggestion from Canadian scientists and adopted the moniker "Gemini" for their project, an astronomically appropriate name for twin telescopes. They still had several hurdles to overcome to move the project forward. For example, they had to convince the science community that Gemini was vital to its future. This was an increasingly important priority as the number of large ground-based telescope projects grew. In the United States, approval from the astronomy community meant securing a top recommendation from the National Academy of Sciences.

Since 1964, the Academy's decadal reviews of astronomy had become the tool elite scientists used to select strategic priorities and make recommendations for the next decade's big projects. Getting a high ranking from a decadal survey was no guarantee a project would be built. In 1982, the Field report had called the defunct 15-meter National New Technology Telescope a high priority. On the other hand, not getting the blessing of the NAS survey could create serious obstacles for a major new project. Greenstein's 1972 report, for example, gave the forerunner of the Hubble Space Telescope only secondary priority. Two years later, Edward Boland used this to attack the project. A small group of astronomers, led by space-astronomy enthusiasts from Princeton, lobbied Greenstein and members of his panel, a task that required delicacy given Greenstein's predilection for private science facilities and his equivocal attitude toward space astronomy.[61]

John N. Bahcall was one Princeton scientist who successfully helped sell NASA's space telescope to Boland and Congress. As a boy growing up in Louisiana, Bahcall planned to be a rabbi. He first studied philosophy at the University of California at Berkeley but later switched to physics. He got his Ph.D. from Harvard in 1961. This was soon followed by a professorship at Caltech, where he studied quasars using data taken with the 200-inch, before he joined Princeton's Institute for Advanced Study in 1971. He and Lyman Spitzer, a Princeton astronomer and statesman of science from Greenstein's generation, made an effective team as they fought and ultimately won funding for the Hubble Space Telescope.

In February 1989, the National Academy of Sciences announced that the 54-year old Bahcall would lead its fourth decadal survey. Bahcall first interviewed dozens of senior astronomers, including previous survey chairs such as Greenstein and Field, and inquired about their colleagues' judgment and vision. The comments he received were so frank and sensitive that he later destroyed the notebook containing them. Bahcall, on the advice of Greenstein, also solicited advice from Washington insiders like Malow and staff from the Office of Management and Budget. In the midst of his polling, American astronomers elected Bahcall president of the American Astronomical Society, a timely match that raised his profile in their community and gave him a wider base of scientists on which to call.

"I didn't look for representation," Bahcall noted. "I looked for excellence of scientific achievement and judgment. A sense of who the community trusted and respected." By the end of 1989, Bahcall had selected fifteen astronomers, including two Nobel laureates, to serve on his main panel. Sidney Wolff represented both the optical astronomy community and the national observatory. There were also several specialized panels that reported to the main group. Over 300 astronomers participated directly, more than in previous surveys. "We want to make sure we don't miss anything," Bahcall told a reporter. "In the end, the strength of the report we issue will depend on consensus."[62]

Forming a consensus, however, meant that Bahcall and his panel had to deal with the most contentious and divided group of astronomers—the ground-based optical community. As Bahcall saw it, "The era of robber barons in the U.S. was tame compared to the relations in the optical and infrared community. In the previous surveys, they had really shot themselves in the foot. As a result, there was no major optical or infrared facility built. The intensity, the feelings, were therefore much higher." There was, he recalled, a "cesspool of distractions" because the optical community was not used to working in groups or to helping each other achieve priorities.[63]

Bahcall picked Steve Strom to lead the subpanel for ground-based optical and infrared astronomy. Strom was a former Kitt Peak astronomer who chaired AURA's Future Directions committee in 1987. Sidney Wolff served with Wallace Sargent from Caltech as a co-chair of this group. Strom, Wolff, Sargent, and twenty-two other optical astronomers on the subpanel knew that as many as fourteen new telescopes bigger than the venerable 200-inch Hale Telescope might enter service in the next decade. This would be astronomers' single biggest surge in light-collecting capability ever and almost all of these telescopes would be operated by European or Japanese agencies or be private facilities off-limits to the majority of American astronomers. This possibility disturbed some who were familiar with the continuing growth of the American astronomy community and its demographics. The number of American astronomers had increased more than 40 percent since 1980 and the community remained relatively young (the median age was 42).[64] In short, ground-based astronomers were on the verge of witnessing an unprecedented number of new telescopes and most American scientists might spend their careers looking through observatory gates rather than gazing at stars.

The Bahcall Committee's deliberations were shaped by changes the discipline experienced in the last ten years as well as its anticipation of new technologies to come. A closer look at astronomers' publications provides an indication of what was happening. During the 1980s, it became increasingly common for American scientists to collaborate with international partners when publishing papers. Foreign scientists contributed with greater regularity to American science journals, reflecting the growing strength of other countries in astronomy. There was a noticeable tendency for publications to include data collected from more than one spectral region—optical observations combined with radio data, for instance. Astronomers were also combining data from space and ground-based facilities more often. The average number of authors on a typical peer-reviewed astronomy paper rose, a continuing transition from the days when a single researcher wrote most papers. Observational (as opposed to theoretical) papers reflected these trends even more.[65]

During the 1980s, there was also a major shift among astronomers' patrons. Historically the NSF provided most research grants to individual scientists and supported the majority of federally funded, ground-based astronomy facilities.[66] In 1980, the NSF provided almost two-thirds of astronomers' grant money. By 1989, NASA had become the dominant agency for researchers' funding. This transformation had occurred for several reasons. For years, the NSF's overall budget for astronomy had steadily declined in real dollar amounts. Construction funds for some major projects like the Very Long Baseline Array had depleted resources for individual research grants, while the size of the average research grant had shrunk by 35 percent to $55,000. Meanwhile, NASA was not only putting more money into grants but also was changing how it dispensed them. Researchers awarded time on NASA's space astronomy missions typically received money to cover their observing expenses and to help analyze the data they collected. In contrast, America's oversubscribed ground-based observatories in Arizona and Chile only offered astronomers telescope access and left it to individual scientists to get funds for data analysis.

Despite all of these changes in demographics, publishing trends, and patronage, papers by astronomers using data obtained from ground-based optical telescopes still represented the majority of all astronomical publications.[67] Members of Bahcall's committee anticipated traditional ground-based telescopes would continue as astronomers' workhorses, in part because of two important technological developments.

The first of these was the imminent declassification of adaptive optics. Earlier in the 1980s, Wayne Van Citters, head of the NSF's Advanced Technologies and Instrumentation in the NSF's Astronomy program, took part in a classified war scenario sponsored by the Strategic Defense Initiative. This helped him learn how far the techniques and technologies of adaptive optics had progressed in military labs. During the late 1980s, Van Citters also began to see more proposals from astronomers requesting NSF funding to develop adaptive optics for their civilian telescopes. Van Citters contacted Robert Fugate, the Air Force's adaptive optics guru, for help evaluating these proposals. The two scientists wanted to avoid a situation in which astronomers duplicated the adaptive optics techniques the Air Force already had developed. European scientists were increasingly interested in developing adaptive optics technology as well and Van Citters believed there was a chance that American astronomers might fall behind in a scientifically valuable area if the Air Force developments remained classified.[68]

Declassifying adaptive optics was a long and complex process. Fugate and Van Citters had to first convince the Pentagon that declassification would not harm national security. They also had to show a scientific rationale for declassification. Their case was helped when, in November 1989, people flooded through the newly opened Berlin Wall. The wall and the Soviet Union soon collapsed, calming military fears that declassification of Star Wars technology posed a security risk.

After years of working in the isolated world of classified laboratories, Fugate was eager to show astronomers how adaptive optics could improve the performance of their ground-based telescopes. In May 1991, Fugate was scheduled to describe his lab's work on adaptive optics and laser guide stars to an excited audience at the American Astronomical Society meeting in Seattle. At first, Fugate wasn't even sure the Air Force would allow his presentation, and he later heard rumors that his request went all the way to the White House. He had never addressed a large crowd of astronomers and was nervous about how they would receive his work. After his talk, dozens of astronomers came up to him. "There were a million questions," he recalled. "Where can I get one of these? How does this really work? How much does it cost? Can I put this on my telescope? It was just pandemonium."[69]

Bahcall's committee was aware of the impending declassification and assumed military-derived technologies would be available to civilian scientists soon. Their final report, *The Decade of Discovery*, identified improvements in spatial resolution—the ability of a telescope to separate features and show small details—as a promising area in which astronomers could reap many scientific payoffs. Adaptive optics, the report argued, would enable new 8- to 10-meter telescopes to compete with space-based telescopes, helping ensure the ground-based telescope would remain a powerful research tool for decades to come.[70]

The second major development Bahcall and his committee saw as critical to astronomical research was the continued improvement of infrared detectors. Until the mid-1980s, astronomers who studied the sky at infrared wavelengths had access to only single-element detectors like the ones Frank Low developed in the 1960s. They had to create two-dimensional pictures of the infrared sky by laboriously combining hundreds of separate observations.

The picture literally was very different for the military and intelligence communities. As early as 1955, the Air Force and the CIA demanded satellites that could detect Soviet missile launches and do surveillance from space.[71] Industrial contractors responded with machines such as the Missile Defense Alarm System. The classified facilities depended primarily on lead-sulphide detectors that astronomers also used. When infrared radiation (in the 1- to 4-micron range) falls on a lead-sulphide detector cell, its resistance changes. This can be measured and related to the amount of infrared radiation. Military and surveillance detectors, unlike those used by civilian scientists, soon featured hundreds of detector cells combined into more sensitive instruments.

In the 1970s, the Air Force and the CIA encouraged the development of more sensitive infrared detectors. A slew of new infrared devices appeared that were similar to the semiconductor-based CCDs astronomers and spy satellites observed with at visible wavelengths. Scientists experimented with combinations of mercury-cadmium, indium-antimony, or doped silicon that were sensitive to different wavelengths of infrared light. These new devices were hybrids—thumbnail-sized silicon chips carefully attached with small blobs of indium to another layer containing the individual detector elements. Hundreds or thousands of sensors could be arranged in a rectangular array to produce two-dimensional images. Each of the individual picture elements (pixels) was extraordinarily sensitive. By the mid-1980s, classified satellites sported detectors with hundreds of pixels and could transmit real-time images, including nighttime infrared images, back to earth.[72]

Military arrays, however, were designed to produce high-contrast pictures that could quickly show "bright" events like a missile launch. Astronomers required different capabilities from infrared detectors such as the ability to accurately measure absolute light levels and faint objects. Unlike the Air Force, civilian astronomers had neither the money nor the connections to persuade commercial firms to make infrared arrays specially suited for their research. Even NASA was not able to take advantage of the most sophisticated military arrays; NASA's Infrared Astronomical Satellite, launched in 1983, featured only sixty-two individual discrete detectors.

Engineers and astronomers worked to convince the military that declassification of some of its infrared technology would not present national security concerns. Gradually, commercial firms offered modest-sized infrared arrays to astronomers. For astronomers used to working with single-element detectors, even the modest-sized 32×32 arrays (in other words, 1024 individual pixels) available around 1985 represented more than a thousand-fold increase in observing efficiency. Dedicated research groups at a variety of institutions (including NOAO) worked to integrate larger and more sensitive arrays provided by industry into new instruments. Infrared astronomy was maturing from a "technique-oriented science" to one in which scientists could reliably use their tools to attack a wide range of research problems. Astronomers anticipated new scientific discoveries by looking, for example, through dust that obscured visible light to observe emerging young stars, and science magazines heralded infrared astronomy's new image.[73]

Wayne Van Citters and other astronomers knew infrared arrays would be a perfect complement to adaptive optics. The atmospheric distortion of longer wavelengths (in other words, infrared) of light is easier to correct, so it was likely that the initial civilian use of adaptive optics would be in the infrared. With the appropriate funding and facilities, Bahcall's report predicted the 1990s could be the "decade of the infrared."[74]

What new telescope projects would the decadal survey support to take advantage of greater spatial resolution and improved infrared sensitivity? Because several new telescopes would be built with private or state money in the next decade, Bahcall knew it would be difficult to advocate yet another large and general-purpose telescope, even if it were for the entire community. "I would not have personally supported a telescope which was not special," he said. "I wanted the U.S. to have some unique facilities." Initially, he was not impressed with ideas put forth by some of his committee members. Bahcall recalled one member who, after listening to a series of prosaic recommendations, told the group that, even if these were achieved, they would only enable the United States "to become a successful rival of Paraguay in optical astronomy."[75]

Frank Low was a key person who helped Bahcall identify a unique opportunity for the next decade of American astronomy. The Arizona astronomer was a member of the decadal survey's subpanel for optical and infrared astronomy. Low and his colleagues "used the word 'unique' a lot" in their discussions. "We wanted something of national significance for the 'have-not community' but we also wanted it to be unique, competitive in the world."[76] Low envisioned a large telescope especially designed for work in his favorite spectral region, the infrared. Its honeycombed 8-meter mirror, made in Roger Angel's Mirror Lab, would be silver-coated, which had better reflectivity for infrared light than traditional aluminum coatings, to improve its infrared performance. Placing the telescope on Mauna Kea, arguably the world's best observing site, would take advantage of excellent seeing conditions while the integration of adaptive optics and the best detector arrays available would allow it to produce even sharper images in the infrared.[77]

In January 1990, Bahcall talked to Low at the American Astronomical Society's meeting about the scientific possibilities a large infrared-optimized telescope might offer American astronomers. "It was a telescope to do something that we never have done before . . . It was clear the scientific frontiers were virgin and vast," Bahcall said, "I was thrilled by the idea. I did everything I could to help them develop that concept because it took my breath away."

Bahcall knew that not everyone in the astronomy community wanted one, let alone two, new national telescopes. "I think if we put it up for a vote, the community would have voted overwhelmingly not to have a big telescope. They would have said to buy a one meter telescope for every astronomy department . . . so we can all have our own facilities." Like Greenstein earlier, Bahcall was not inclined to allocate resources just so many small facilities would be available to as many scientists as possible. "That would have been the democratic way of doing it. It was clear that was the view of the community. Our job was not to represent faithfully the views of all our colleagues but to provide leadership."[78]

In April 1991, the National Academy of Sciences released the report prepared by Bahcall's committee. Like previous decadal surveys, Bahcall's panel listed recommendations for major new facilities; three of these were either ground-based tools or tailored for infrared observing. An "infrared-optimized 8-meter U.S. telescope" on Mauna Kea was the number one ground-based priority, while the third recommendation was an "8-meter optical telescope, operating from the Southern Hemisphere." LIGO appeared nowhere in the 181-page report, reflecting the view of Bahcall and others that the detection of gravity waves was a task for experimental physicists, not astronomers.[79] In terms of new NSF programs, the report blessed the NSF's plan for new nationally available 8-meter telescopes with a modicum of official support from the science community to go along with their first installment of funding.

John Bahcall acquired his knowledge of how the science community determines priorities early in his career when he helped lobby for the Hubble Space Telescope. His enthusiasm for an infrared-optimized telescope followed. Both the Hubble and the 8-meter telescopes would be national facilities open to all members of the American science community. The Hubble Space Telescope was absolutely unique, however. The same could not be said for a new 8-meter telescope project. Securing community support for Hubble required forming a diverse coalition of astronomers who wanted a major space telescope.[80] Astronomers and engineers incrementally altered Hubble's design to increase its appeal to a broad group. Gemini's advocates used a different strategy. Getting community support and federal funding for a general-purpose national facility was not going to be easy. Therefore, acting on the suggestion of Low and others, Bahcall's committee endorsed one 8-meter telescope as an instrument with unique infrared capabilities while the second, lesser-ranked telescope would observe the sky from the Southern Hemisphere. This helped establish a clear and special "discovery space" for each.

Negotiations and consensus-building over several years and in many different countries enabled Gemini to emerge as a fragile but viable project. This required assembling an international *mélange* of interested astronomers, administrators, and politicians. Their lobbying efforts for Gemini were done in part to maintain a fair distribution of resources within the community. For Canadian and British astronomers,

Gemini was an opportunity to secure at least a share in what they hoped would be a world-class science instrument. Many American scientists argued that Gemini would give everyone access to some of the world's largest telescopes at a time when almost all giant telescopes were being built as private facilities for astronomy's haves.

It would be naïve to think that the persons and committees who determined the direction American astronomy would take did so solely out of altruism. Pursuit of the twin 8-meter project also kept NOAO, temporarily at least, as a major player in the development of the next generation of large telescopes. Without its participation in a major new telescope project, more sharp questions would be raised about NOAO's viability as an important science institution. Astronomers' ideals of equity and fairness helped serve broader institutional needs such as the national observatory's health and its participation in Gemini offered a justification for its continued existence.

Jesse Greenstein, now officially retired, was one of the senior astronomers who vetted Bahcall's report before its release in 1991. Greenstein told Bahcall how astronomy had become "infinitely more difficult" since the decadal survey he led in 1971. Funding, in his view, was "fractured," "small-science grants to individuals [were] dead," and the national observatories were "on a starvation diet." But Mauna Kea was "already a flower in full bloom" and "Keck is far on its way." In fact, the Keck Foundation announced plans to fund a second 10-meter telescope in the spring of 1991. It is doubtful that many astronomers would have agreed with Greenstein's assertion that the balance between private and federal support "has combined as an organically successful one." As Bahcall prepared to defend his report to his colleagues and sell it in Washington, perhaps he was comforted by the elderly astronomer's reflection that "at first inspection, the Universe at larger redshifts seems smooth again—perhaps at some time scale scientific life will again be pleasant."[81]

CHAPTER 7

Smoke and Mirrors

On April 24, 1990, the space shuttle *Discovery* rocketed into Florida's morning sky with the Hubble Space Telescope on board. Two days later, NASA astronauts placed the $2 billion observatory into orbit 380 miles above the earth. For the project's engineers and astronomers, some of whom had worked on NASA's "discovery machine" for a decade or more, this was a glorious moment. As one historian noted, *Discovery* carried not just a 12-ton space telescope but the expectations of scientists, politicians, and a public audience stimulated by a relentless media campaign.[1]

A month after Hubble entered orbit, astronomers at the Space Telescope Science Institute (the AURA-managed research center in Baltimore where Hubble's science research was done) received long-awaited "first-light" images from the telescope. They soon learned that what NASA and many scientists had hyped as the "most fantastic telescope ever built" was, in the words of Maryland Senator Barbara Mikulski, a "techno-turkey." John Bahcall informed an angry Congress that he had visited them many times to promise "beautiful pictures and great discoveries." Now he wanted to share "the heartbreak that we all have in not being able to provide those discoveries."[2]

Instead of concentrating light from stars and galaxies into sharp images, Hubble's mirror blurred starlight into a broad halo. Closer examination revealed what opticians call spherical aberration—light rays far from the mirror's edge were focused at a different point than light from the mirror's center. NASA engineers and university astronomers quickly pinpointed the problem. In the early 1980s, technicians at Perkin-Elmer polished Hubble's 2.4-meter primary mirror to exquisite smoothness but the wrong curvature. As Senator Albert Gore told colleagues at a June 1990 hearing of his subcommittee on science, technology, and space, "It's perfectly wrong." Hubble soon proved a bonanza for cartoonists and late-night comics.[3]

Amidst mounting pressure, NASA hurriedly assembled an investigative panel to determine how Hubble's major optical component could be so defective. Their suspicions soon fell on the process Perkin-Elmer technicians used to test the primary mirror. Careful investigation by the panel revealed that the problem stemmed from a critical fault in the optical template that (mis)guided the primary mirror's polishing. Part of the error was caused by the insertion of a dollar's worth of household washers into a precision instrument for testing Hubble's multimillion-dollar mirror. Arizona astronomer and mirror maker Roger Angel served on NASA's panel and declared it the "single largest mistake that's ever been made in optics."[4]

What infuriated and disturbed astronomers and Congress more was that Hubble's mirror had not been inspected properly by either Perkin-Elmer or NASA engineers. Even a simple and inexpensive test could have caught the mistake. The investigation laid blame in many directions. One culprit cited was a "surprising" lack of participation by experts in telescope optics and a lack of oversight by astronomers and other scientists who were shut out of a process "operated in a 'closed door' environment."[5]

While blame rested most squarely on the shoulders of NASA and Perkin-Elmer, Hubble's mirror fiasco also damaged the astronomy community's reputation. When irate members of Congress and the public lambasted Hubble, they did not always distinguish between the contractor who built it and the astronomers who had long anticipated it. Optical and ultraviolet astronomers, Hubble's main constituency, saw their hopes for a world-class science facility postponed while NASA devised a daring, expensive, and ultimately successful repair plan.

Debates about Hubble's troubles fed scientists' ongoing debates over who should control large projects and how they should be managed. Scientists in other fields shared these concerns. High-energy physicists were divided over whether to build their next "ultimate instrument," the Superconducting Super Collider, and geneticists wondered about the wisdom of a Department of Energy–funded "big biology" approach for the Human Genome Project.[6]

Hubble's mirror fiasco engendered a lot of soul searching by astronomers. In advocating Hubble, astronomers had to adapt to the institutional culture of NASA and its industrial contractors. Throughout Hubble's history, there were battles for control between astronomers and NASA engineers and managers. Astronomers also accepted that Hubble would be operated as a large national science facility, distinctly different from the "little-science" style to which many of them were accustomed. Many scientists lamented that, as telescope projects became bigger and more expensive, individual scientists were relegated to passive and sometimes ineffectual roles. Some optical astronomers wondered publicly and privately if they had not traded too much of their autonomy in their quest for bigger and better machines.

Astronomers also began to question how their projects were organized and managed. Some California astronomers like Jerry Nelson resented the impersonal, efficiency-driven management of the Keck Telescope project.[7] In July 1990, when Roger Angel visited England to discuss big telescopes with colleagues, he talked about his recent investigation of Hubble's mirror problems. Warning of the dangers caused by an overly bureaucratic approach to big projects, he said there was "no shortcut to innovative technology via committees."[8] Angel voiced the view of many scientists who sensed an intensifying struggle between proponents of corporate science and the individual researcher.

Managing Twins

One scientist especially disappointed with Hubble's difficulties was Robert C. Bless. In the 1950s, while doing his Ph.D. in astronomy at the University of Michigan, Bless was accompanied by Leo Goldberg, then chair, to observe newly launched Russian and American satellites pass overhead. For Bless, this sparked a long interest in space-based research.[9] In 1958, he took his degree and moved to the newly formed Space Astronomy Laboratory at the University of Wisconsin, which was developing a series of Orbiting Astronomical Observatories championed by scientists like Goldberg. Bless stayed with Wisconsin's astronomy department for the rest of his career studying ultraviolet radiation from stars and galaxies, a subdiscipline of astronomy that depended on rocket- and space-based observing.

In 1971, Bless became involved with NASA's plans for what became the Hubble Telescope. His participation became more personal in 1977 when NASA selected his proposal to help build one of Hubble's primary instruments. Wayne Van Citters, on leave from the NSF's Astronomy Division, also collaborated with the Wisconsin team. Their final product, the High Speed Photometer, would measure the light intensity from faint stars and galaxies. Its successful use, however, depended on aiming Hubble precisely so starlight could pass through one of several fixed aperture and filter combinations.

As NASA engineers diagnosed Hubble's problems, they discovered the telescope's solar panels moved as it orbited in and out of the earth's shadow. The combination of mirror aberration and telescope jitter seriously compromised the performance of Bless's instrument. By 1991, Bless (who some astronomers saw as Hubble's elder statesman) was disgusted with the loss of a scientific instrument on which he had worked for years. When Van Citters invited him to serve on the newly formed Gemini Board of Directors, it was a welcome distraction from NASA's crippled megaproject.

The University of Wisconsin was a charter member of AURA and Bless was aware of the competition among astronomy's haves and have-nots for funding and observing time on ground-based telescopes. As a young astronomer, Bless was excited when AURA and Kitt Peak National Observatory were established. "We thought it was great," he said, "People were tired of hearing that the universe was owned by Palomar."[10]

At the same time, Bless also knew that NOAO had suffered in the 1980s when its 15-meter project was cancelled and its budget declined. "I think the precursor of Gemini may have been one of their last gasps to regain some preeminent position. I could understand that it was a hard blow for them when Bloch and the NSF said, 'No way unless it's international.'" Bless brought his experience with large public science projects to Gemini's board. Two decades of involvement with the Hubble program had also kept Bless relatively free from personal issues or conflicts of interest other ground-based astronomers might have had.

In October 1991, Bless attended his first meeting of the Gemini Board and saw the challenges facing the project. Ian Corbett, representing the United Kingdom's Science and Engineering Research Council, nominated Bless to be its chairman. Despite his relative naiveté about the project's history or the other board members' motives, Bless accepted the job.

One of his first tasks was understanding and navigating Gemini's formal structure. When Bless became chair, there were eight board members, four from the United States and two each from the United Kingdom and Canada. Gemini's board was the telescope project's ultimate authority. It was responsible for everything from approving the telescopes' science requirements and the annual budget to determining where the telescopes' myriad components would be built. Financial contributions from the partner countries were deposited into a single account that the NSF, acting as the project's executive agency, controlled. The NSF executed the decisions of the Gemini Board by communicating these to AURA, Gemini's managing organization. AURA in turn executed the board's executive decisions. Like Keck, the Very Large Telescope, and other major telescopes under construction, Gemini had a surfeit of committees and formal bodies that all reported to the board. AURA established its own committee to oversee its management of the project and look after financial matters. The Science Committee, chaired by Gemini's Project Scientist, gave scientific guidance that filtered up through Gemini's Director, to AURA and the NSF, and finally the Gemini Board.

Bless soon discovered that Gemini's management differed from AURA's previous arrangements with other national astronomy projects. AURA's own board, of course, controlled national astronomy facilities such as Kitt Peak and the Space Telescope Science Institute; but because Gemini was an international project with its own board, AURA had no direct control over it. The NSF, moreover, decided that no one associated with AURA could serve on Gemini's board. This policy, which puzzled and angered some senior astronomers in the U.S., placed American astronomers at a disadvantage as some of the most politically savvy scientists were barred from representing U.S. interests.[11] Complaints followed that American astronomers were consistently outmaneuvered by their more politically skilled British and Canadian counterparts.

AURA's board members were slow to realize that neither they nor the NSF had direct control over Gemini's top-level decisions. This was a point of considerable aggravation for members of Gemini's board, especially its two British members, Longair and Corbett. Gemini was, in Corbett's view, an international effort controlled by the Gemini Board, not an "AURA project which the NSF could not fund fully" and subsequently turned to an international partnership for funds. AURA, in his view, should stop trying "to see itself more as a 'partner' than a 'contractor.'"[12]

NOAO's relationship to Gemini was another sensitive issue. Gemini's ancestors were Aden Meinel's "X-inch" project, Goldberg's NGT program, and Kitt Peak's NNTT project. AURA's original proposal for twin 8-meter telescopes assumed that most of the management, design, and construction would be done by the national observatory staff. Many at NOAO saw Gemini as simply an international version of their original project, the fruit of years of effort and anticipation. To complicate matters, the Gemini project was based at NOAO's headquarters in Tucson and many astronomers and engineers there initially had key roles in the project throughout its formative years. For example, in January 1991, the Gemini Board selected Larry Randall, a former long-time Kitt Peak telescope engineer, to be Gemini's Project Manager. A major new project could breathe new life into the national observatory. NOAO engineers and scientists would be working at the technological forefront in developing world-class scientific facilities, and it might help resolve enduring questions in the science community about NOAO's future. When the NSF and Congress refused to fund a U.S.-only effort, many at NOAO felt "their" project had been taken away from them.[13]

Because AURA managed both NOAO and Gemini, clarifying the relationship between the two was of considerable importance. Goetz Oertel wished to avoid the appearance that either "the U.S. has lost control of 'its' project or that the U.S. has disproportionate control" over what was to be an international collaboration.[14] Meanwhile, Canada and the United Kingdom had concerns that money they contributed to the project might go to support NOAO rather than to build Gemini. Corbett, for example, worked to convince the other board members that Gemini needed to be separated from NOAO. Otherwise, "the culture, the philosophy, and basic *modus operandi* of the national institution will dominate."[15]

By early 1992, Gemini's board acted to sever the project from NOAO formally, eliminating any direct managerial control NOAO might have over the project. The separation of Gemini from NOAO further weakened the beleaguered observatory. And because NOAO staff (being AURA employees) could not serve on Gemini's board, NOAO's influence over the project was circumscribed more, even though Gemini had sprung from seeds that NOAO had laboriously tended.

At his first board meeting, Bless also learned about constraints placed on Gemini by Bahcall's decadal survey and Congress. The project appeared boxed in from a number of different directions: its cost, Bahcall's recommendation that the northern telescope be optimized for infrared use, and the need to build Gemini as a genuine international collaboration from which all partners derived benefit. The wording of Bahcall's report deliberately omitted elements central to the partnership American, British, and Canadian astronomers had established. It also cleaved the twin telescope project into two separate endorsements—one for an infrared-optimized U.S. telescope and a lesser recommendation for a second 8-meter telescope in the Southern Hemisphere to which the United States (and others presumably) might have access.

When Congress approved the American share of funding for Gemini, it included a total cost cap of $176 million; only half of which was the United States's share. Congress emphasized this point to Walter Massey, the NSF's new director, when he appeared before the House appropriations committee in March 1991. In preparation for the hearings, Dick Malow discussed the status of Gemini with Bob Traxler, the committee's chair. For Malow, the final cost of the project was not the critical element but, rather, that the project be built with half the cost coming from foreign partners. Malow was concerned that, if Congress did not make its demand for 50 percent foreign participation clear, "Uncle Sugar was going to end up paying god-knows-what percentage. That is one of the reasons why I reinforced and nailed to the wall the 50–50."[16]

Traxler subsequently reminded the NSF that Gemini could not proceed unless it had "firm foreign participation first."[17] Otherwise the NSF was to go ahead with plans for a U.S.-only telescope in Hawaii costing $88 million. Traxler's committee gave these instructions despite claims from the NSF that one telescope would, in fact, cost more than half (as much as $100 million) of Gemini's total cost because there would be no savings from building two almost-identical facilities.[18]

These restrictions angered many astronomers, especially as they watched the NSF's Laser Interferometer Gravitational-Wave Observatory win congressional approval and $212 million of funding. LIGO soon fell behind schedule and spiraled to more than $360 million while its scientists could not guarantee its first generation of detectors would be sensitive enough to detect much at all. Costing almost four times the U.S. share of Gemini, LIGO would serve a far smaller group of scientists in an unproven field of observational astronomy. New optical telescopes, skeptics said, would at least see *something,* a claim gravitational-wave physicists had yet to demonstrate.

Further complications were added when, in June 1991, Canada's National Research Council surprised the science community by announcing it would pull out of the Gemini project.[19] Canadian astronomers were shocked and embarrassed by the news and quickly mobilized to reverse this decision. By the end of the summer, the Canadians cautiously predicted they would still participate in Gemini, but at a funding level reduced to 15 percent.[20] While good news, Gemini still needed to replace the missing 10 percent. There was not much time left for this. The latest report from Congress said that, unless foreign funds were fully committed by January 1, 1992, the NSF was to go ahead with building a single infrared-optimized telescope for U.S. astronomers on Hawaii. This was not what Gemini's advocates wanted. To them, one of the project's main strengths was having two telescopes with full-sky coverage integrated as a single observatory. This was especially important for astronomers who were limited in their access to big telescopes that could see the southern night skies.

Gemini's supporters believed that lobbying by powerful American astronomers had helped create the congressional constraints, which were rather unusual for a large project still in its early stages. Bahcall's committee was adamant that the telescope in Hawaii should be for American use only. Bahcall himself opposed the idea that his recommendations might be used to support the case for an internationally owned pair of telescopes. "I felt we needed full time on an IR-optimized telescope," he said, "I thought we were giving things away too easily. I was not comfortable with ceding the running of our national observatory to other countries."[21] To make his case, Bahcall went to Washington, D.C. and briefed Malow on the extent to which Gemini would meet his report's recommendations.

Bahcall also wanted reassurance that the telescope on Mauna Kea would be infrared-optimized, as his report recommended. To do this meant including features such as a special secondary mirror that could subtract sky background, a primary mirror coated with silver instead of the traditional aluminum for better infrared performance, and an overall telescope design that improved its infrared sensitivity. He reminded Sidney Wolff that the emphasis on infrared observing was "essential to our ranking the northern 8-meter telescope so highly."[22] The U.S. delegation to Gemini, Bahcall complained, did not contain any infrared astronomers while the NSF continued to downplay the infrared optimization that led his committee to support the project in the first place.[23]

Other vocal American scientists also favored an American-only infrared telescope on Mauna Kea. The University of Hawaii's Institute for Astronomy, for example, had a vested interest in a large infrared telescope in Hawaii. Donald Hall, the Institute for Astronomy's director, was disappointed that the NSF and AURA would have no direct control over the Gemini telescope on Mauna Kea. He was surprised that American astronomers had acquiesced to an international arrangement despite its failure to meet the Bahcall committee's recommendation for a national facility. Hall wanted the University of Hawaii to have some influence over Gemini, even a seat on the Gemini Board if possible and a guaranteed share of observing time.[24] Hall did not lack bargaining power. His institution controlled access to sites on Mauna Kea for new telescopes and the institute had expertise in cutting-edge infrared arrays and instruments. Hall could also appeal to Daniel K. Inouye, a powerful Hawaiian Senator who appreciated the economic benefits and prestige new telescopes brought.[25]

Hall's criticisms joined a small but vocal chorus of American scientists unhappy with Gemini's direction. To Gemini's foreign partners, this was just sour grapes. "The Americans were affronted," Longair recalled, "They said, 'Why should the greatest country in the world have to collaborate to do something that is our birthright?'"[26] Corbett became increasingly concerned that the NSF and AURA were not pushing Gemini past obstacles put up by Congress and senior American scientists. Disturbed by what he believed was Americans' failure to appreciate Gemini's international nature, Corbett complained the United States feared a "cunning foreigner was trying to take poor Uncle Sam for a ride." British scientists, Corbett insisted, were not simply buying shares in an American-led project. "We are *partners* in a *joint* project," he wrote Van Citters, "If the NSF cannot educate Congress and the Senate to recognize this and to accept that a descent into xenophobia will benefit no-one, then I personally do not wish to proceed."[27]

After the first Gemini Board meeting, Oertel lobbied Bahcall and Malow for support. Oertel argued Gemini was vital because it gave American scientists access to advanced new telescopes in both hemispheres and he assured Bahcall that it would meet expectations for an infrared-optimized telescope.[28] Oertel also updated Malow about progress Gemini was making toward the 50 percent foreign participation required by Congress. Oertel claimed that Gemini, now only shy by 10 percent, was doing much better than many science projects and requested more time to sign up additional partners.[29] At a time when Congress was accusing physicists of lacking genuine foreign participation in their far-more-expensive Superconducting Super Collider project, Oertel's point was reasonable. A week later, when Congress reviewed the NSF's budget, Traxler's appropriations committee acted on Malow's counsel and gave Gemini extra time to secure the remaining 10 percent.

In the six months after attending the first Gemini Board meeting, Bob Bless had received a crash course in the project's politics and pitfalls. By January 1992, he had serious questions as to the direction it was going. Before the next Gemini meeting in May 1992, Corbett met with Bless and Don Morton from Canada's National Research Council. The three agreed this was a "critical, even watershed" meeting. If the project could not improve its organization, Corbett and the others believed that a "hostile Congress, fanned by Bahcall, et al. and Don Hall" would kill the project.[30] They were also disappointed with how the NSF and AURA were managing the project and handling communications within the project. Larry Randall's performance as Gemini's manager became a focal point for their displeasure.

Two decades earlier, when AURA built its 4-meter telescopes at Kitt Peak and Cerro Tololo, Randall was the engineering manager. Randall later served eight years as the manager for the team building the Faint Object Spectrograph, another of the Hubble Telescope's major instruments. For Gemini, Randall had formed a solid team of engineers who trusted him, and he had extensive experience with telescope engineering. Since the early 1970s, however, the sophistication of telescopes had increased enormously. Scaling up from a 4-meter to 8-meter project did not merely entail contending with twice as much complexity.

Randall's technical ability impressed the board, but it insisted he did not yet grasp Gemini's overall organizational structure or present a sufficient strategic overview of the project's status. There were concerns Randall lacked a clear plan to divvy up construction of Gemini's myriad technical systems (which were expensive technical plums that could yield jobs and prestige) among the partner countries. In these areas and many others, Randall, as Corbett recorded in his notes, "got a roasting" at the May 1992 meeting.[31]

Afterward, Corbett told Oertel in a frank and critical letter, "Gemini is, by some margin, the worst organized and managed project that I have ever come across." He still detected an "isolationist" attitude as NOAO's staff in Tucson "consolidated their grip" on the project while displaying a "centralized, 'not invented here' cast of mind." Corbett claimed NOAO was duplicating technical work already done in the UK and expressed suspicions that Gemini was a way of keeping NOAO afloat in tough budgetary times. Corbett reminded Oertel that the British investment, relative to gross domestic product, was three times that of the Americans. Unless these and other issues were resolved, "the project carries with it an unacceptably high risk of failure."[32]

Displeased with Randall's lack of what George H. W. Bush had recently called "the vision thing," the Gemini Board proposed the telescope project hire a director to deal with strategic issues. While the project searched for a full-time director, the Gemini Board asked Sidney Wolff to serve in the interim, beginning in June 1992. Gemini, of course, was now officially separate from NOAO, but Wolff retained a strong attachment to the project, which had its origins in her Tucson headquarters. Gemini needed a capable leader to guide it past political threats, but NOAO itself also needed leadership. The official "loss" of Gemini (most of the project's tasks were still being done in Tucson) had hurt morale at the national observatory and it was bracing for yet another budget crisis. Despite her reservations, Wolff agreed to lead Gemini for the time being, a decision that, in effect, kept NOAO unofficially connected with Gemini.

The Gemini Board's wrangling over management issues was done out of sight of the science community, the media, or politicians. Moreover, as Gemini created its management plan (a somewhat organic process for projects of this type), people like Hall and Corbett maneuvered with the interests of their home institutions in mind. What appeared as a highly critical letter or harsh exchange at a board meeting also reflected elements of political posturing. This relative invisibility ended in late August 1992 when Sidney Wolff announced that Corning, not Roger Angel's Mirror Lab, would make Gemini's 8-meter mirrors. According to one observer, Gemini soon became "perhaps the most politicized and contentious project ever in the history of ground-based astronomy."[33]

Mirror, Mirror?

Before Gemini's mirror controversy ensued, the project was already struggling to resolve the management and organizational issues typical for a major international science project. As Gemini's advocates navigated these seas, the mirror controversy torpedoed the project amidships. Without widespread community agreement about Gemini's mirror, the project would founder.

When the mirror controversy became public and rancorous, Gemini's supporters had to resolve it immediately so as not to undermine the astronomy community's belief that the project was both technically correct and politically legitimate. The controversy did not just threaten Gemini. The mirror debacle surrounding Hubble and the ensuing political fallout was fresh in many astronomers' minds as they debated Gemini's options. Many believed the reputation of the American astronomy community and the NSF's ability to undertake future international collaborations were also at risk.

There were signs that a fight over Gemini's mirrors was brewing months before the project chose Corning. In June 1989, the Steward Observatory Mirror Lab completed its third successful 3.5-meter mirror casting. With funding from the Air Force, the NSF, and the University of Arizona, the lab had developed its ability to polish large mirrors, thus providing "one-stop shopping" for telescope projects. Companies like Corning and Schott, in contrast, only offered a rough mirror blank and left the lengthy polishing and testing process to other commercial firms. Knowing the Mirror Lab's capabilities, most astronomers assumed as the Gemini partnership gelled that the Mirror Lab would furnish the optics.

After the series of 3.5-meter castings, Mirror Lab's next major goal was a 6.5-meter mirror blank for the Multiple Mirror Telescope. In 1987, the University of Arizona and the Harvard-Smithsonian Center for Astrophysics decided to replace the MMT's six smaller mirrors with a single large mirror. Engineers planned to fit one of Angel's honeycomb mirrors, cast with a very short focal ratio, inside the MMT's existing enclosure. This would more than double the telescope's light-collecting power and increase its field of view, offering an economical way for astronomers to get, in effect, a much more powerful telescope.

Making MMT's new mirror promised to be a major milestone for the Mirror Lab. Angel's 6.5-meter mirror design was two-thirds bigger than the 200-inch mirror Corning had made decades previously, yet it would weigh only half as much. Progress at the lab toward ever-larger mirrors was slower than Angel and Peter Strittmatter, Steward Observatory's director, expected. The NSF's practice of funding the lab on an annual basis made future planning difficult; and in the late 1980s, the Mirror Lab struggled with a high turnover of lab managers. These difficulties occurred as the Mirror Lab tried to make the transition from a research and development facility to a commercial competitor. By the end of 1991, the Gemini Board was concerned with the lab's slow progress.

At the first meeting of Gemini's board in October 1991, there was extensive discussion about Gemini's mirror options. An Angel mirror was believed to offer the best performance, but board members were concerned about management issues. At some point, the project might need to exercise oversight over the Mirror Lab if there were staff or administration problems. Ceding control to Gemini would have been antithetical to most astronomers at the University of Arizona, given their traditional autonomy in research and instrument building as well as the history of poor institutional relations between Steward Observatory and the national observatory. But, as one Gemini board member said, "Either we control the show or we let it sink."[34]

By 1991, the NSF had spent over $12 million on the Mirror Lab while other sources—the University of Arizona and the Air Force, mainly—had invested millions more. The recent Bahcall survey called the lab a "crucial element in the U.S. astronomy program" so abandoning the enterprise outright was not something the NSF wanted to do. A preferable goal was to encourage more professional management. Wayne Van Citters of the NSF raised this issue with university officials. Beginning in 1992, the NSF would stop supporting the development of casting technology at the Mirror Lab. Instead, money from specific telescope projects would support the facility's "transition from Research and Development to a production environment." Shifting to a "production-oriented activity" meant hiring a professional manager who would report to the university administration and not to Steward Observatory's astronomers. In what was one of the opening shots in Gemini's mirror battle, the NSF warned that Gemini would not use a mirror from Angel's lab unless "major management changes" were made.[35]

In January 1992, the Mirror Lab hired its first professionally trained manager. Stephen F. Hinman was an easy-going engineer with over twenty-five years of experience in Eastman Kodak's optics manufacturing program. When he arrived at the Mirror Lab, Hinman encountered circumstances that were common to many university labs. The scientists had technical authority without the accompanying financial responsibility. At the same time, the lab was not equipped to deal with the accounting requirements for what amounted to a moderate-sized manufacturing facility. "They estimated project costs on mere sketches and ideas," Hinman later said, "but often didn't think out the details until we were building the equipment. The result was sometimes cost overruns."[36] This type of problem was exactly what the NSF wanted the Mirror Lab to correct.

When Hinman was hired, the race to build the world's biggest telescope mirrors was growing in intensity, and the Mirror Lab remained a strong contender. On November 7, 1991, Caltech and the University of California dedicated the Keck Telescope, the first of the giant new telescopes, on Mauna Kea. Only nine of the telescope's thirty-six mirror segments were installed, but these alone still collected more light than the 200-inch on Palomar. Despite some software difficulties, the partially finished telescope took its first images a few weeks later. A few months before Keck's dedication, Corning announced it would build the mirror for Japan's national telescope project. The 8.3-meter Subaru telescope, named after the Japanese word for the Pleiades star cluster, would be built on Mauna Kea next to the Keck facility. By early 1992, Schott successfully made its first 8-meter meniscus mirror for ESO's Very Large Telescope facility in Chile. All three basic mirror technologies that astronomers had advocated in the 1980s—Jerry Nelson's segmented mirror, Angel's honeycomb approach, and the commercially made meniscus mirrors—were simultaneously coming on line at major observatories. None of them had yet emerged as superior in cost or performance.

On March 29, 1992, after months of delays and frenzied work, the Mirror Lab was ready for its biggest casting yet. Underneath the university stadium, lab technicians constructed an intricate mold of 1,020 ceramic cores. Over several days, lab staff loaded ten and a half tons of carefully selected chunks of borosilicate glass, some 4,000 pieces in all, into the mold. With the glass inside, the furnace weighed over 70 tons. A crane gently placed the lid on the furnace and technicians turned the heat on.[37]

Four days later, it was 1,180° C inside the furnace, which was spinning once every eight seconds. At this point there was no turning back. The glass inside the furnace had begun to soften and fill the spaces between the mold segments to create the honeycomb structure. Dan Watson, the lab's longtime oven pilot, Strittmatter, and Angel monitored conditions via closed-circuit television with groups of reporters. A failure at this point would mean the loss of months of work and $1 million worth of glass and ceramic. As the glass chunks softened and slumped, the centrifugal force of the furnace's rotation created a 33-centimeter-deep bowl in the top surface of the hot, honeylike glass. This surface would eventually collect the light from thousands of distant stars and galaxies. Weeks later, the furnace's temperature was cool enough to permit inspection of the finished mirror blank. Angel and the staff at the Mirror Lab were ecstatic. After years of effort, they had demonstrated their operation could compete with much larger and better financed commercial firms to produce an excellent product.

Glass was not the only thing heating up at the Mirror Lab. The lab's new capability prompted executives at Corning's New York headquarters to write Steward Observatory and the NSF just as Mirror Lab began its largest casting to date. Because the Mirror Lab had invited representatives from the NSF and Gemini to attend the event, Corning expressed concern that "other activities might be planned for this occasion" that would give the University of Arizona an "unfair competitive advantage" for Gemini's mirror contract. Angel's blunt response was that the Mirror Lab did not "rely upon influence to establish our competitive advantage with any constituency; we rely upon performance."[38]

In May 1992, Gemini's management requested formal bids for two 8-meter mirrors. Corning, Schott, and the Mirror Lab (through the university) responded with their proposals. The University of Arizona encountered difficulties, however, in preparing the Mirror Lab's bid. Gemini required all bidders to submit a fixed-price proposal. Bidders had to assume the risk of any casting failures and would work until the mirrors were made successfully. As a publicly funded university, Arizona could not easily assume this risk. Arizona's solution was to present Gemini with several different bids, including a scenario in which the Mirror Lab and Gemini would share the costs if one of Angel's castings failed. This bid was $12.8 million for two unpolished mirror blanks (polishing would add about another $4.5 million). This proposal required Gemini to shoulder some risk as each failed casting would cost $3 million. After much discussion, the Mirror Lab received permission to submit a fixed-price bid similar to what a commercial firm would offer. To protect the university in the event of any failures, this bid was substantially higher than their other offer—$18.5 million, not including polishing or any other costs.[39]

Corning aggressively priced the cost of two unpolished 8-meter meniscus mirrors for Gemini at around $11 to $12 million.[40] The company's proposal was understood to be priced substantially less than what Japan was rumored to have paid Corning for Subaru's mirror.[41] Fixed-price bidding was not a problem for a commercial enterprise like Corning, which could assume the financial risk if any failures occurred. Historically, astronomers associated Corning with telescope mirrors. The New York company had made the mirrors for the Palomar 200-inch and the 120-inch at Lick Observatory at its Corning, New York, factory. In the 1960s, Corning began concentrating its classified research and development at another plant in upstate New York where it developed products such as the mirrors for spy satellites and the windows for the space shuttle. Since the early 1980s, company representatives worked to keep pace with new technologies for making large mirrors.[42] Corning business executives saw products such as telescope mirrors as a way to diversify the company's pool of technology. Niche products like big mirrors, they hoped, would become more profitable while the research and development needed to make them might lead to new merchandise. Casting giant meniscus mirrors for a high-profile project like Gemini also offered Corning excellent publicity that might translate into more lucrative contracts for classified projects.[43]

At a July 1992 meeting in Washington, D.C., the Gemini Board reviewed all the mirror bids. Larry Randall and two committees of astronomers, engineers, and administrators worked over the summer of 1992 to pick a winner. By August, both committees (neither unanimously) recommended that Corning should make Gemini's mirrors. In mid-September, Gemini awarded Corning the contract for the first mirror with an option for a second.

Gemini's selection of Corning over the Mirror Lab can, initially, be understood in terms of three factors: scientific performance, cost, and scheduling. When Gemini chose Corning over the Mirror Lab, it admitted that the meniscus technology was thought to be inferior to Angel's honeycomb mirrors. One anonymous committee member even expressed concern that, with a Corning mirror, "the scientific goals of the project will not be met."[44] However, both Europe and Japan had opted for meniscus mirrors so, in this regard, Gemini had good company. Conversely, Mirror Lab advocates also noted that astronomers' confidence in Angel's honeycomb mirrors was such that already several other telescopes underway—including the WIYN telescope in which NOAO was a partner—were based on them.

These arguments were complicated by the fact that, in 1992, neither Corning nor the Mirror Lab had successfully made an 8-meter telescope mirror. Many doubted it could be done easily or cheaply. Because no telescope in the world was using either an 8-meter honeycomb or meniscus mirror yet, astronomers' arguments about future performance depended on inconclusive predictions and calculations rather than on actual experience.

Despite the belief that Angel's mirror technology promised the better performance, Gemini's board found the University of Arizona's bids impossible to accept. The Mirror Lab's bid was much more costly than its earlier presentations had led Gemini's leaders to expect. Three years earlier, for example, the Mirror Lab estimated the cost of casting an 8-meter mirror at under $3 million.[45] The University of Arizona's proposal included stipulations that Gemini's board found unacceptable.[46] Finally, Gemini remained unconvinced that the Mirror Lab possessed what it called the "deep management structure" believed necessary to make its mirrors on schedule. Any delays could cost hundreds of thousands of dollars while Gemini kept its engineers and astronomers idling.

Corning, on the other hand, could direct enough money and people to its mirror production to ensure that costs and schedule obligations were met. As for concerns that Gemini telescopes equipped with meniscus mirrors might not perform well enough—Gemini was not expected to see first light for at least six more years and the project was confident that "further engineering work [could] produce acceptable scientific performance."[47]

Scientists and engineers focused on performance, cost, and scheduling to justify Gemini's choice of Corning over the Mirror Lab. Each of these criteria could be quantified to varying degrees to support an argument for either mirror technology. Once the decision in Corning's favor was announced, other factors less easily expressed in numbers, namely community politics and differing beliefs about the way Big Astronomy should be done, emerged.

Gemini under Siege

In the summer of 1992, Gemini hired Matt Mountain as its new project scientist. His experience with infrared instruments and telescopes made him especially attractive to Gemini, which was still parrying criticism that its telescopes fell short of the Bahcall committee's recommendations.[48] As dissatisfaction with Randall's management grew, Gemini's board demanded the project function like a high-tech industrial undertaking with a more "corporate" approach. They expected to see careful monitoring of cost and schedule goals, extensive integration of Gemini's various systems, clear communication between engineers and scientists, and use of modern design and management tools. Mountain brought familiarity with this style of doing astronomy, which he learned while at the Royal Observatory in Edinburgh. When the thirty-six-year-old astronomer arrived in Tucson, he found Gemini under siege. "I didn't have any religious faith about what mirror we should have. I had expected it to be Roger Angel's, frankly. I was there to build two telescopes, not support a mirror technology. Suddenly, boom! We were into it."[49]

Angel, Strittmatter, and their supporters were shocked and appalled by Gemini's choice. They believed the bidding process they were obliged to participate in was unfair. They had not expected a direct competition with a commercial vendor or Gemini's insistence on a fixed-price bid. Arizona objected that what it considered proprietary information from the Mirror Lab circulated for months among astronomers and engineers from the national observatory. Moreover, Angel and Strittmatter had many candid discussions with NOAO staff about its mirror technology (including its costs) long before Gemini formally existed. Many of these people now worked for Gemini. One senior engineer had even left NOAO to help Corning prepare its competing bid, prompting Arizona to cry foul more vehemently.[50]

Angel and his colleagues had a considerable emotional and professional investment in their techniques. They believed their facility offered a finished mirror—cast *and* polished—that no commercial firm could match. Gemini, however, did not ask bidders to provide a finished mirror, only a rough mirror blank that would be subsequently polished. This approach, Angel believed, did not take full advantage of the capabilities his lab had worked so hard to develop. "We felt really screwed," he said, "It was like, 'Hey, a piece of glass is a piece of glass. You just make it as cheap as you can.'"[51]

The choice of Corning also revealed broader disagreement in the astronomical community. To build the next generation of astronomical tools, some scientists believed it necessary to cede control and autonomy to project managers and systems engineers. Others pointed to the crippled Hubble Space Telescope as evidence that astronomers needed to retain their influence and authority as projects became bigger and more complex. Gemini's prioritization of cost and schedule over the anticipated better performance of Angel's technology was a vote for the "corporate" approach, which some scientists felt necessary and others detested.

That Gemini admitted Angel's technology was probably superior to Corning's was especially galling to astronomers who dreaded a Hubble-like embarrassment to astronomy. Some scientists wondered why the NSF had invested millions in Angel's lab, only to see Gemini opt for a commercial vendor. "This decision to throw away a 10-year investment in a successful technology . . . clearly doesn't make any sense at all," one supporter protested.[52] The Mirror Lab offered an opportunity to return some control back to scientists who used the telescopes. "That industrial-strength brainpower is in some sense superior to university brainpower—or that it is better managed—is entirely fallacious," said one astronomer who cited the B-2 bomber and Hubble as glaring examples of "management-induced cost overruns."[53]

Compared to the corporate style Gemini favored and that Corning promised to deliver, Angel and Strittmatter operated in a manner best described as entrepreneurial. Clever thinking and risk-taking helped Steward Observatory's rise to scientific prominence. The Multiple Mirror Telescope's bold departure from traditional design reflected Steward Observatory's unconventional and sometimes aggressive expansion of its astronomy program. Throughout the 1980s, Strittmatter and his colleagues doggedly pursued plans for more new telescopes (including the unorthodox Large Binocular Telescope). The Mirror Lab was another manifestation of Arizona's entrepreneurial ethos—a facility built against the odds where astronomers made tools for other astronomers. Likewise, many of the projects based on optics from the Mirror Lab placed great value on input and judgment from individual astronomers. The Carnegie Institution of Washington, for example, had long maintained such a tradition. In 1992, it joined forces with Arizona to build two 6.5-meter telescopes in Chile with Angel's lab supplying the mirrors.

The Mirror Lab's enterprising spirit carried over to the business aspects of its production as well. Malcolm Longair recalled, "Many of us got the impression Steward Observatory thought they had the Gemini board over a barrel . . . One story I heard was that they were actually going to build their own 8-meter telescope out of the funds from these two mirrors."[54] Like Internet startup companies in the 1990s, Angel and Strittmatter had aggressively courted many patrons—the NSF, NOAO, the Air Force, NASA—to expand the lab and keep it afloat. The university and the lab anticipated reaping a profit from Gemini's mirrors, an unusual situation for a public institution competing with a privately funded company.[55] Hinman later believed they aimed too high, partly out of a desire to maximize their return and also on the belief that Gemini would be willing to pay for mirrors superior to commercial products.[56]

The loss of the Gemini contract clearly endangered the Mirror Lab's future and reputation. Astronomers from Arizona and their supporters mobilized. In doing so, they allied themselves with Gemini's existing opponents. John Bahcall, for example, remained unconvinced his survey's recommendations were being heeded, and he opposed international collaboration for Gemini. Don Hall at the University of Hawaii was still feuding with the NSF and AURA over Gemini's site on Mauna Kea, to which his school controlled access. He preferred, as did Bahcall, that the twin-telescope plan be scrapped in favor of a single, infrared-optimized American telescope in Hawaii. These feelings were fueled by another massive donation from the Keck Foundation (almost $75 million) for a second privately operated 10-meter telescope on Mauna Kea.[57] The news prompted complaints that California astronomers would have two giant new telescopes of their own while the whole American science community would be serviced by only half of an international 8-meter project.

Not all American astronomers believed that a single telescope on Mauna Kea was the best choice for their community. Many recognized that scrapping Gemini's Southern Hemisphere telescope would curtail most American astronomers' ability to observe the skies. Both Carnegie and ESO had plans for large telescopes in Chile that would be off-limits to most U.S. scientists. "If this happens," wrote Jay Gallagher, a vice-president of AURA from the University of Wisconsin, "then those of us not in California will be left out of the big leagues." This would be "a national scientific tragedy."[58] Gallagher confided to Van Citters that he had followed the "Hall-Bahcall duo around [Washington] trying to undo their strong anti-international stance on Gemini." The unfolding controversy over Gemini was, in his view, "the perfect opportunity to confirm the opinion long held by some on the Hill that astronomers are a fractious and ungrateful bunch!"[59]

Angel defended his lab by telling astronomers and science managers that the mirror choice reminded him of how Hubble's mirror problem had occurred. Angel's experience on the NASA board of inquiry that investigated the cause of Hubble's problems reinforced his opinions. Hubble's failure, Angel wrote the NSF's director, could have been avoided by incorporating the skills of "experimental scientists with deep understanding and experience with telescopes . . . I am now concerned that history will repeat itself."[60] Gemini admitted it would probably need assistance from European and Japanese telescope projects using similar mirror technologies. "What kind of management technique is this?" one astronomer asked, "Do we have so little faith in our own capabilities?"[61] Instead of the NSF getting the best possible telescope for American astronomers, Gemini had, according to Angel, "started down the path of compromise and duplication."

In a "most unpleasant meeting" with Arizona astronomers, Wayne Van Citters was disturbed to hear that Frank Low, Arizona's well-respected infrared astronomer, now also opposed Gemini.[62] Low had advocated an infrared-optimized telescope to Bahcall's survey committee; and in 1991, he helped Gemini establish design criteria that would improve its performance. According to Low, neither Gemini's mirror choice nor the telescopes' basic design were sufficient for an infrared-optimized facility. Marcia Rieke, another Arizona infrared astronomer and a newly elected Gemini Board member, expressed doubts about Gemini's scientific potential to Bahcall. She feared "the Board will not choose in favor of [infrared] needs when the inevitable tradeoffs arise."[63] These criticisms were serious. While Angel and Strittmatter could attack a specific aspect of Gemini—its mirror—the criticism from Low and other infrared astronomers thrust at the heart of the project's primary scientific rationale for American astronomers.[64]

Rumors also circulated that Gemini's British partners had requested the meniscus mirrors from Corning in exchange for their participation. While there is no direct evidence that the U.K. demanded Gemini choose the meniscus approach, many British and Canadian scientists were favorably disposed to Corning's technology.[65] The Royal Greenwich Observatory already had begun to study how to build the complex and expensive support system for a large meniscus mirror, and there was a general belief (which events bore out) that this technological plum would be awarded to them.

As a prelude to the next Gemini Board meeting, Van Citters and the other U.S. board members met again with Arizona astronomers to discuss the mirror choice and Gemini's expected infrared performance. Before he left for the rendezvous, Van Citters discussed strategy with Bob Bless. Bless was still discouraged over the difficulties his instrument on Hubble was having; Gemini's mirror woes only added to his distress. "Science is fun!!!" Bless wrote before wishing, "If only we could do some."[66]

Second-Guessed

The next meeting between Gemini representatives and Arizona astronomers lasted for seven tense hours. Discussions focused on two main issues: Gemini's mirror choice and whether the telescope would truly be infrared optimized. At the meeting's outset, Low, an imposing man with a deliberate manner of speaking, reviewed Gemini's design goals. Gemini, he said, promised scientists high-resolution images at infrared wavelengths. The telescopes' specifications surpassed the performance that astronomers anticipated from any other large telescope and was a unique science capability Gemini could offer. Low asserted the telescopes' baseline design would not meet the goals that he had helped define. An all-purpose telescope, Low reminded them, did not meet Bahcall's recommendation either.

Arizona's assault on Gemini's design continued. Angel wobbled a piece of thin material to show the difficulty of supporting meniscus mirrors in high winds, a display that made many recall Richard Feynman's famous o-ring demonstration at hearings for the *Challenger* explosion. When Wolff and Randall acknowledged their design still needed further study, Arizona astronomers were infuriated. They demanded to know exactly how Gemini was going to address the technical questions they had raised and questioned the merit of building what some saw as a mediocre design.[67]

The Gemini Board met the next day in Tucson. British and Canadian astronomers had not been invited to the meeting with the Arizona scientists, prompting complaints that Gemini had a long way to go toward real international collaboration. When he heard about the previous day's confrontation, Ian Corbett wrote in his notes, "I came to the conclusion that the Project is not yet confident in its solutions. This is indeed worrying."[68] British and Canadian members of the Gemini Board were outraged that the project was threatened by attacks from a small but vocal group of American astronomers. Malcolm Longair, Great Britain's other representative, observed "the United States didn't have the means to cope with this sort of issue the way we do in the U.K. We would put issues like this to an advisory panel to make a judgment."[69] It appeared to many board members that Arizona and Hawaii were using technical arguments to overturn the mirror choice and undo the international collaboration.

The November meetings were Matt Mountain's formal introduction to Gemini. As the new project scientist, he was "oblivious to a lot of what was going on with the project's politics." To him, Angel's Feynmann-like demonstration appeared intellectually dishonest, but "it was a very convincing and sobering lesson."[70] To succeed, Gemini needed to convince critics that the mirror choice made sense in terms of cost and scheduling, that the project could answer the technical questions Low and others had raised, and that Gemini's overall management was sound.

News of Gemini's problems spread throughout the science community and its supporters rallied to sustain it. "I have been told that the Universities of Arizona and Hawaii are attempting, for reasons of narrow self-interest, to subvert the Gemini Project," wrote Augustus Oemler, chair of Yale University's astronomy department. "To hijack this program to satisfy the interests of a few . . . would be an outrage, and a disaster for American astronomy," which would "doom all hope for any national large telescope." Other letters warned Senator Mikulski and Bahcall that "selfish interests" would harm future international science collaborations.[71]

Gemini, Angel remarked, was becoming the first telescope to produce more heat than light. It was increasingly difficult to separate strictly technical arguments from political maneuvering. Gemini's political disputes could not be hidden from Congress, and the project achieved what all Big Science projects wish to avoid—bad publicity.

On November 16, Strittmatter, Wolff, Hall, and Bahcall convened in Washington where they were joined by Oertel and Van Citters. Dick Malow and Kevin Kelly, his Senate counterpart, met with this diverse group for two-and-a-half anxious hours. There were no representatives from Gemini's international partners (again) or Corning. However, rumors circulated that the glass company had deployed its lobbyists, including Francis B. Kapper, Corning's well-connected director of Advanced Government Programs in Washington. Malow and Kelly criticized the NSF for failing, despite their warnings, to set priorities among its biggest projects (a situation Malow had warned about for years given the on-going budget climate). Gemini, LIGO, and others, according to Wolff's version of the meeting, "would be rammed down the NSF's throat. There was no sympathy I could detect for individual investigators."[72]

Ostensibly, Malow and Kelly had called the meeting to talk about Gemini's budget situation and to explore how much one telescope might cost if that option became necessary. The Congressional staffers reiterated their two main priorities. "Kevin and I were concerned about the budget issues and getting the foreign participation resolved," Malow said, "The mirror issue came in out of the clear blue to us."[73] Kelly had a stack of astronomers' messages supporting or pillorying Gemini, and he and Malow expressed surprise that scientists were airing their differences so publicly. Malow offered the project a reprieve until May 1993 to get all interested partners signed up for Gemini. Until then, the U.S. share of Gemini's funding would be frozen, one step toward outright cancellation.[74]

The congressional staffers displayed little patience for Gemini's technical difficulties and appeared interested only in meeting Bahcall's decadal survey requirements. Bahcall, although opposed to Gemini as an international venture, cautioned against abruptly abandoning Canadian and British astronomers. Bahcall instead suggested the NSF ask eminent American astronomers to review the project, including its mirror choice. Malow and Kelly agreed to Bahcall's suggestion, which some saw as another step toward Gemini's cancellation.

In December 1992, the NSF established an independent committee to review Gemini. Bahcall insisted the committee's chair be a "card carrying infrared astronomer" and suggested James R. Houck, a scientist at Cornell University. Houck, then fifty-two years old, had earned his Ph.D. from Cornell in 1967 and worked there since on both infrared research and instrumentation. Houck's committee of highly respected scientists included: Steve Strom who six years earlier had encouraged AURA to abandon the NNTT project in favor of 8-meter telescopes; John Huchra and Marc Davis, who knew each other from a large-scale galaxy redshift survey they began in the late 1970s; and Gerry Neugebauer and Judith L. Pipher, who were experts in infrared observing and instrumentation.

Six of these scientists had participated in Bahcall's survey and several were members of the National Academy of Sciences. All of them had sympathy for an entrepreneurial approach to astronomy and had worked on influential science projects by raising funds personally, building equipment, and working alone or in small teams on which astronomers had a great deal of control. Huchra, for example, had years of experience using telescopes and was a classic example of an entrepreneurial, hands-on astronomer. In the 1980s, Huchra added thousands more data points to the redshift survey he and Margaret Geller were doing—the result was the discovery of the Great Wall—yet even as this research project grew to include theorists and more graduate students, it remained a loosely organized effort with astronomers calling the shots.[75]

Princeton astronomer Jim Gunn also joined Houck's panel. Years earlier, Gunn helped design Hubble's main instrument, the Wide-Field Planetary Camera built by the Jet Propulsion Laboratory for about $70 million. When NASA grudgingly revealed the mirror problem that compromised Hubble's performance, Gunn was furious. He wrote an agonized missive expressing his belief that scientists had "lost all control of [their] destiny." As telescope projects became bigger and more complex, astronomers, Gunn said, had yielded control to a bureaucracy "unable to handle large projects of its own." Astronomers, he said, had not been "'screwed over'—we have been exquisitely vulnerable to precisely this kind of thing happening for years . . . We are a discipline of technical incompetents, happy to let our or NASA's engineers build our tools to their desires, by and large, not ours." Hubble's mirror problem, Gunn concluded, was "an ASTRONOMICAL failure" in all senses, and he was motivated to join Houck's group by a wish to avoid another "Hubble-like fiasco."[76]

Those for and against Gemini lobbied Houck's panel. One "former skeptic and critic" of Gemini resented how community "selfishness and competitiveness" now threatened it.[77] After Oertel learned of a letter-writing effort underway by scientists opposed to Gemini's design, he wrote Houck about what he gingerly termed Gemini's "avoidable challenges." Such disputes would "certainly cast a bad light on astronomy" while public disagreement might weaken the NSF's and Congress' support for astronomy in general. As a nationally available facility, Gemini was potentially valuable to all American astronomers. "To paraphrase JFK," Oertel said, "we should not ask what Gemini can do for us, but what we can do for Gemini."[78]

Meanwhile, rumors circulated of behind-the-scenes political interference. Sherwood Boehlert, a Republican representative from a congressional district in central New York near Corning's headquarters, served on the House Committee on Science. Independently, several people claimed Boehlert or Alfonse d'Amato, a New York Senator on the appropriations committee, was contacted by Corning lobbyists. While no direct evidence supported such claims, they added to an already charged atmosphere.

In January 1993, Houck's panel met in Tucson to hear presentations from both Gemini and the University of Arizona. Matt Mountain was worried after seeing NOAO's presentation to the panel about its contribution to Gemini. NOAO didn't, he recalled, understand that Houck's committee was not reviewing just the mirror choice but also the entire project. Gemini still suffered, some believed, from an identity crisis. Officially separate from NOAO, critics still equated Gemini with the national observatory that, as one prominent astronomer noted, had "never got squarely out in front of the technical challenges mounted by the general optical community."[79] Huchra remembered that at least one astronomer on Houck's panel wanted to kill the project outright in favor of a single American telescope in Hawaii. Persuading them not to abandon the twin telescopes took, he later estimated, at least two bottles of good scotch.[80]

In evaluating the two mirror technologies, Houck's committee focused on two technical concerns. One was the mirror's thermal control. To attain the exquisite imaging performance Gemini's advocates wanted, it would be necessary to keep the mirror's surface within a fraction of a degree of the ambient air temperature. Another critical question was how Gemini's mirrors would be supported. Traditionally a telescope mirror rested inside a steel mirror cell, an approach pioneered by Lord Rosse in the early nineteenth century. The mirror "floated" on supports in the cell that were connected to the telescope structure at only three points. A large meniscus mirror, however, was not inherently stiff enough to resist buffeting by high winds. To meet this challenge, Gemini's engineers proposed supporting their mirror with a complex electromechanical system of actuators instead of a traditional mirror support system. These devices would continuously monitor the mirror and adjust its shape and position to compensate for wind and gravity effects. Supporting a 8-meter mirror less than a foot thick and weighing several tons presented a challenge different from that solved by Keck for its smaller and lighter segments. This was a feat no project had successfully accomplished for a telescope as large as Gemini. Given the project's formidable goal of giving astronomers infrared images with 0.1 arc-second resolution, precise mirror control was critical.

In its report to Congress, the Houck Committee unanimously concluded that in "choosing a meniscus mirror over a honeycomb mirror, the Project has unnecessarily exposed itself to significant additional risk of failure."[81] Gemini had traded "perceived short term financial risk" for the likely possibility that the telescopes' final performance would be compromised. While Japanese and European telescope projects had opted for meniscus mirrors, these had neither Gemini's demanding imaging requirements nor its tight budget constraints. No project had yet resolved the technical questions associated with the meniscus approach at the 8-meter scale. Gemini's plan was just "too technologically risky" while the "incorrect mirror decision" was interpreted as "proof that something is seriously wrong" with Gemini's entire management.

The Bigger Picture

While the conclusions of the Houck Committee delighted Arizona astronomers, Corning responded swiftly with outrage. The Monday after the panel released its report, Matt Mountain recalled that, when he arrived at work, "there was this big black limousine parked in front of NOAO with these guys wearing heavy Washingtonian coats." Project manager Larry Randall remembered this visit also. "They were saying, and it wasn't even a thinly veiled threat, 'We can shut this project down.' They said that not only would they sue for breach of contract, but that they would kill the project."[82]

For Corning, the mirror choice was more than a matter of money. Its reputation was at stake. There were rumors that one scientist on Houck's panel had opined (off the record) that Corning's meniscus mirror would be more suitable for a birdbath than a telescope. Corning's mirror program manager complained, "We see ourselves as victims in this situation . . . We feel there is an attempt to violate and upset the competitive process."[83] Corning executive Frank Kapper denounced the NSF for having "experts in the theoretical world of astronomy" but no mechanical or structural engineers on the reviewing committee. "Should [a] university, a publicly funded institution, compete," he asked, "with privately funded businesses whose tax dollars actually pay for the institution's very existence?"[84]

The damaging review of the mirror choice also sparked an emergency meeting of the Gemini board. Bob Bless recalled this was the emotional low point in his service to Gemini. Bless told the board of Bahcall's belief, as stated in a recent letter to Malow and Kelly, that he "could not imagine a strategy that will preserve U.S. congressional support for the project if the Board does not endorse" Houck's report.[85] For two days, the Gemini board agonized over the best step to take in a landscape littered with political landmines. Malow and Kelly would probably kill the project if Gemini rejected Houck's findings outright. Switching mirrors would further undermine the project's credibility and autonomy. The NSF feared a lawsuit from Corning if Gemini canceled its contract. Switching mirrors could also attract attention from New York's congressional delegation and "might affect the NSF in other [i.e. political] ways."[86] Meanwhile, Canada and the United Kingdom refused to participate in Gemini if there was any hint of legal action from Corning.

Emotions swung wildly throughout the two-day meeting as the Gemini board labored toward a solution that would not force it to switch mirrors or appear to reject Houck's report to Congress. After Longair reminded his colleagues that the mirror selection had been made according to established procedures, the board crafted a judiciously worded response to Houck's report. Welcoming its positive statements, it wisely avoided confronting specific criticisms of the mirror choice directly and offered assurance that all technical issues would be addressed in late 1993 at Gemini's preliminary design review. If doubts persisted after the review, Gemini said it would adopt an Angel mirror for the telescope on Mauna Kea as long as it was demonstratively superior.[87] This maneuver bought Gemini several months to address the technical criticisms leveled against the meniscus mirror as the project wagered it could override both engineering and political obstacles.

William C. Harris, the NSF's assistant director for Mathematical and Physical Sciences, assured the Congressional staffers that his agency was satisfied with Gemini's decision's not to reverse course but to forge ahead until the project's design review.[88] Harris's support came at a critical time given earlier complaints from Gemini's British and Canadian partners about the NSF's indecisive leadership. On March 1, Bahcall told Malow and Kelly he could not endorse Gemini's decision to proceed with the Corning mirror until the design review. Harris again supported Gemini's strategy, in which he expressed complete confidence. As Harris told Senator Mikulski, Gemini had strong support from its international partners and was making "great strides." Despite Malow's misgivings, Harris's support carried the day. Gemini remained alive.[89]

The possibility that the board had salvaged a cooling-off period vanished when *Sky & Telescope* hit the newsstands in late March. A lengthy article, "The Gemini Project: Twins in Trouble," did not suggest which mirror was best but concluded that what Gemini needed was "first class equipment . . . [not] another Hubble-type fiasco."[90] Peter Boyce, the American Astronomical Society's executive officer, warned colleagues not to let their dispute become any more public lest the specter of Hubble cast a pall over not just Gemini but the entire field. "If Gemini should be canceled," he wrote, "it will be a long time before another astronomy project is funded."[91]

As Gemini readied itself for the upcoming design review, Matt Mountain came to believe that the controversy was not simply about mirrors. It was also about a clash of cultures. Scientists held two conflicting views about the best way to build and use instruments that were increasingly complex and expensive. One of these was centered around the university and emphasized individualism, intuition, and a small-scale approach that gave greater authority and responsibility to the scientist. On the other hand, some scientists favored a professional systems-oriented approach—what Bless called "space culture"—in which the focus was on deliberate top-down integration of the telescope's various systems and project management. This was based on large-scale systems engineering, design review boards, and science committees that, for Gemini, would make management more transparent and accountable to the NSF and the project's partners.

Debates over Gemini's future, Mountain decided, reflected this larger conflict between the entrepreneur/scientist based at a university and the corporate, team-oriented approach he learned while at the Royal Observatory, Edinburgh. "There are two approaches," Mountain explained. "The strong principal investigator [PI] has a vision and goes forward and builds the telescope. But when you are talking about $100 million telescopes, that's quite a lot for one person. Then there is the approach where you get a competent engineering and scientific team together to tackle the problem from a systems-wide view."[92]

Gemini was not the only large astronomy project that faced these cultural differences. In the mid-1980s, similar conflicts emerged when the Keck Telescope project moved from astronomers' hands to those of professional managers and engineers. In the estimation of Gerry Smith, Keck's manager, astronomers were continually worrying about a telescope's final quality. Smith, however, was willing to compromise between the performance of the telescope and keeping cost and scheduling commitments. "The emphasis always is, from my point of view," Smith said, "let's get it done."[93]

Angel, Bahcall, and members of Houck's panel reflected a different culture in which the individual astronomer possessed great authority and autonomy. Scientists like Gunn and Huchra were extremely talented, motivated, and entrepreneurial leaders who, even when working on expensive instruments and extensive research projects, brought a personal vision to make it work. Mountain, to his surprise, believed this entrepreneurial model was still thriving at the national observatory when he joined Gemini and began to collaborate with NOAO staff. Gemini, Mountain argued, needed to adopt a more corporate culture and shed traits characteristic of NOAO's older "four-meter mindset" if it was to survive.[94]

Mountain's view was shared by board members like Corbett and Bless who disliked Randall's management of Gemini. In Bless's view, NOAO had "gotten in the practice of acting like somebody who builds things in his basement" without enough attention to cost and schedule issues. "Telescopes are now extremely complicated systems," Bless said, "which are much more akin to a complicated spacecraft than to building Palomar." Bless saw a mismatch between the traditional telescope-building culture Randall knew and the approach followed by NASA and the aerospace industry.[95] Randall's specialized skills as a telescope engineer mattered more in the first model than in the second. The latter favored attending to cost, schedule, and management; it was less important that Gemini was a telescope rather than a communications satellite.

Mountain and Wolff encouraged closer interaction between Gemini's scientists and engineers as they prepared for their crucial review. Potential critics of the telescopes were also enlisted into the project's advisory structure. AURA elected Frank Low to Gemini's oversight committee and Jim Houck became a member of Gemini's board. In this way, Gemini offered these scientists a genuine stake in the project's success, thus neutralizing their dissent. Gemini's staff also participated in meetings of what became known as the 8-Meter Club, a group of scientists and engineers from Japan and Europe using meniscus mirrors who shared technical advice with Gemini.[96]

Gemini adopted new tools and techniques to help prepare for its upcoming design review. Extensive modeling of the telescopes' enclosures was done at the San Diego Supercomputing Center. Engineers augmented these studies with water tunnel tests to see how air currents would move over, through, and around the telescope to flush out warm air. Another tool Mountain and other Gemini staff relied on were "error budgets." These allowed engineers to account, literally, for how each telescopes' different subsystems affected its total performance. For example, one of the most critical specifications of Gemini was its final image quality. Its primary mirror was to focus 50 percent of infrared light (measured at 2.2 microns) into a small spot only 0.1 arcsecond in size. This meant that when seeing quality in Hawaii or Chile was at its very best, atmospheric conditions, not the telescope, would determine the image quality Gemini produced. Performance gains in one area could be traded off to another so long as the final inaccuracy produced by the telescope stayed within the overall budget. A standard design tool for large telescopes like Keck and Gemini, error budgets and other tools reflected the project's increased focus on comprehensive systems engineering and greater middle management.

Robert Gehrz from the University of Minnesota chaired Gemini's three-day preliminary design review in December 1993. Years earlier, Gehrz had led the committee that chose the design for the ill-fated NNTT project. Unlike the Houck committee, Gehrz's nine-member panel included people like Jacques Beckers and Jerry Nelson who were at the cutting-edge of telescope design. Gehrz's personal notes on the surprisingly uneventful three-day meeting ended with the statement, "There are no technical show stoppers."

A disappointed Angel told a local paper, "It was a committee basically to sign off on a plan. . . . I don't think anyone would call these solutions elegant or inexpensive."[97] Houck was satisfied by Gemini's demonstration of how it could successfully integrate Corning's meniscus mirrors into its design. Years later, Bahcall recognized that "not everything I thought initially about Gemini was right . . . Things I thought were not good ideas initially turned out to be good ideas, especially regarding the choice of the mirror . . . I accepted technical advice which turned out to be wrong."[98]

Gemini's review came at a bittersweet time for Bob Bless. While Gemini's advocates successfully defended the project, hundreds of miles above Tucson, astronauts were repairing the Hubble Space Telescope. Part of the daring mission required removing the High-Speed Photometer, the instrument to which Bless had devoted so many years of his career. When January arrived, however, astronomers had a rejuvenated Hubble, and Gemini, for the first time in the project's history, sailed into a new year with confidence and optimism.

CHAPTER 8

Joining the 8-Meter Club

For many astronomers, views of the sky from the southern hemisphere are especially breathtaking. The Milky Way, often best observed from below the equator, was of particular interest to scientists studying stars and galaxies. Visible within the Milky Way is the Carina Nebula, the largest diffuse nebula in the sky and home to Eta Carinae, a powerful source of infrared radiation about eight thousand light years away. Slightly detached from the Milky Way are the Magellanic Clouds, two irregular galaxies near our own that appear as luminous wisps of ethereal beauty in the night sky.

American astronomers first visited Chile in October 1849. James M. Gilliss arrived in Santiago just as the austral spring was beginning. Gilliss, an American naval officer, was there to lead a scientific expedition financed by the United States government. Its goal was to use parallax measurements to calculate the distance between the earth and the sun.[1] They brought with them the largest telescope yet made in America—a refractor with a lens almost six and half inches in diameter.

Besides his expedition's scientific agenda, Gilliss hoped these experiences would make Chilean interest in astronomy "burn brightly" so that his expedition could "boast that Santiago through our influence established the first national observatory in South America." Bad weather and poor coordination with northern observatories limited the scientific accomplishments of Gillis' trip. Nevertheless, his expedition stirred Chilean interest in astronomy. Three students from the

University of Chile in Santiago assisted the expedition. Before Gilliss's team made the long journey home, the entire observatory was turned over to the Chilean people.

In 1960, telescopes in the southern hemisphere had only a fraction of the light-collecting power of those in the north. Within the next decade, three major new observatories were established by astronomers from the northern hemisphere. With NSF funding, AURA established the Cerro Tololo Inter-American Observatory in 1963 on a 7,000 foot peak of the same name 45 miles southeast of the sleepy seaside town of La Serena.[2]

In the mid-1960s, American astronomers collaborated almost entirely with scientists and staff from the Universidad de Chile, the only Chilean institution with a prominent astronomy program. A few years later, Chile extended privileges to AURA similar to those enjoyed by United Nations staff. AURA established a gated compound for Cerro Tololo's headquarters in La Serena high atop a rocky hill with a stunning view of the Pacific Ocean and the colonial-era town below. The community of Americans and Chileans formed a tightly knit group, and marriages between them were not uncommon. In the small town of La Serena or high atop Cerro Tololo, it was easy for scientists to immerse themselves in their work. One visitor noted that the observatory resembled a frontier fort from an old western movie. "It's a community that does nothing else but astronomy. It has its own culture and when you use it you temporarily join the culture."[3]

Conditions in northern Chile were especially favorable to astronomers. The cold Humboldt Current flows northward along Chile's coast, creating an inversion layer that minimizes atmospheric turbulence. Cerro Tololo is adjacent to the southern part of the Atacama Desert, one of the world's driest places, making it especially good for infrared research because water vapor absorbs infrared radiation. During the austral summer, the night sky is typically crystal clear and familiar stars like Betelgeuse appear bright and steady.

Telescopes sprouted on Cerro Tololo. By November 1967, a small Schmidt telescope for wide-field observing, a 1-meter, and a 1.5-meter telescope had been installed. The observatory director was Victor M. Blanco, a Puerto Rican–born astronomer who later studied at the University of Chicago. The completion of the 1.5-meter telescope marked Cerro Tololo's official inauguration. That same year, the observatory's future appeared even brighter when Chilean President Eduardo Frei and Lyndon B. Johnson jointly announced that the NSF and the Ford Foundation would fund the largest telescope in the southern hemisphere, a 4-meter companion to the one already underway at Kitt Peak.

Although Cerro Tololo was the best equipped observatory in Chile through the 1970s, other institutions began to take advantage of Chile's excellent conditions for astronomy and build telescopes there. The Carnegie Institution of Washington, for example, established an observatory at Las Campanas located about 80 miles from Cerro Tololo. In 1962, the European Southern Observatory purchased the 8,000-foot ridge of La Silla and surrounding land just south of Las Campanas and began the slow process of building several new telescopes. By 1977, European scientists could use a 3.6-meter telescope, its design reminiscent of the 200-inch on Palomar, along with other smaller telescopes.

Unlike AURA, ESO was not an association of universities, but an organization formed by European governments. While AURA found it natural to collaborate with the Universidad de Chile, ESO established its agreement directly with the Chilean government through an international treaty.[4] These differences caused tensions. European astronomers had the rights accorded diplomats because ESO's land was a small piece of Europe in Chile. Many American astronomers at Cerro Tololo compared their status in Chile to that of a guest in a host's house while ESO's scientists displayed a more colonial attitude. Perhaps most offensive was that, while Chilean astronomers routinely received observing time on AURA's telescopes, ESO did not extend a similar invitation to Chilean scientists until the 1990s.

The construction of new observatories attracted many young Chileans to astronomy. Maria Teresa Ruiz was one such person. In the late 1960s, she was a student at the Universidad de Chile. While earning the equivalent of an American bachelor's degree, she took a course in astronomy from Claudio Anguita, director of the university's astronomy program and an influential leader in Chilean academics. Anguita's class intrigued Ruiz enough to encourage her to take an internship at Cerro Tololo. She recalled being picked up at La Serena's small airport by Victor Blanco, who drove her and other students to Cerro Tololo where they learned the basics of observational astronomy. Nights at Cerro Tololo's small telescopes convinced Ruiz to become an astronomer.[5]

In 1971, Ruiz left Chile and started graduate school at Princeton. This was a difficult time to leave Chile. A year earlier, Salvador Allende became Chile's president after a close election. The country was soon embroiled in Allende's socialist reform attempts and right-wing opposition. Allende's election created difficulties for Cerro Tololo and American astronomers working there. Inflation and food shortages appeared and Victor Blanco warned his staff away from black-market profiteering while trying to ensure that Chilean employees at Cerro Tololo were especially well treated.[6]

At Princeton, Ruiz wrote her dissertation under the guidance of Martin Schwarzschild, a theoretician interested in stellar structure, star formation, and galaxy dynamics. She remained concerned about the deteriorating conditions in her country. In September 1973, General Augusto Pinochet came to power in a brutal coup. Ruiz considered returning to Chile, a trip Schwarzschild, himself a refugee from Nazi Germany, urged her not to take.

By 1979, Ruiz felt the domestic situation was secure enough that she could return home, and she accepted a position at the Universidad de Chile. While she was in the United States, the world of astronomy had left Chile behind. Observing time available to Chilean astronomers at Cerro Tololo and Las Campanas went unused and the scientific infrastructure was degraded. Ruiz and other prominent Chilean astronomers spent the next decade rebuilding their country's presence in astronomy. Much of their work was done at the Universidad de Chile although other schools such as the Pontificia Universidad Católica de Chile (Católica) began to develop astronomy programs as well.

In August 1990, NOAO's Sidney Wolff visited Chile on a fact-finding trip accompanied by Jay Gallagher, an astronomer and AURA representative. ESO's new capabilities distressed and dazzled them. ESO had already commissioned its innovative New Technology Telescope (described in Chapter 6), which they found "extremely bold ... There is no telescope like this in the U.S." They also saw the rapid progress ESO was making on its Very Large Telescope project. This prompted Gallagher to note that "within the next decade, no single U.S. observatory will be fully competitive with ESO. Our community needs to understand that the challenge is now on the international level."[7]

At Cerro Tololo, Gallagher and Wolff met with Claudio Anguita who was there for an observing run. Anguita was interested in seeing "a South American component" to the twin 8-meter telescope project AURA was pursuing and the three astronomers discussed the possibility. Anguita, Gallagher told Oertel, "had already begun to work on some possibilities." Oertel's response, penciled in the report's margin, was simply, "Jay: Outstanding!"

The following year, Gemini needed to secure 10 percent of its funding because of Canada's suddenly diminished participation. With Gemini's board and the NSF struggling to keep the project afloat in the face of American opposition, the challenge of finding additional partners fell largely to Oertel and AURA.[8] Oertel's choice of partners was limited. European countries were already committed to ESO and the Very Large Telescope while Japan had its own national project underway. Most American institutions with strong astronomy programs were already involved in their own telescope projects and, as Leo Goldberg had predicted years earlier, were not interested. AURA's long and productive involvement at Cerro Tololo made Chile an attractive potential partner.

Oertel enlisted Claudio Anguita, then the senior spokesperson for Chilean astronomy and well connected to Chilean politicians, as an ally. Anguita often attended AURA meetings, and in 1992, the Universidad de Chile became AURA's first international member. Another valuable contact was Harry G. Barnes, the U.S. ambassador to Chile from 1985 to 1988. Barnes earned the respect of Chileans by helping encourage a 1988 plebiscite in which Chile's voters rejected Pinochet for a democratically elected president. While in Chile, Barnes met the current director of Cerro Tololo, Robert E. Williams. After his term as ambassador ended, Barnes accepted Oertel's invitation to join AURA's board.

Barnes and Oertel marshaled arguments to entice Chile to join Gemini. One inducement was to emphasize how a new telescope project like Gemini would contribute to Chile's economy. Oertel concluded that, for every Chilean dollar spent, nineteen non-Chilean dollars would come in via Cerro Tololo and Gemini. Without a new large telescope, Oertel believed Cerro Tololo would eventually cease being a viable scientific institution and become "outgunned by ESO and Carnegie."[9]

In April 1992, Barnes used these arguments when he met with Edgardo Boenninger—Chile's Minister to the Presidency (akin to a White House chief of staff)—at a posh hotel in Santiago, a rendezvous Anguita helped arrange. The discussion went well and, when their drinks were finished, Boenninger agreed to Chilean participation in Gemini. The actual agreement's details would be handled by AURA with its longtime partner, the Universidad de Chile. The only bad news was that Boenninger would commit Chile to only 5 percent of Gemini—about $9 million. This still left 5 percent of Gemini's funding unaccounted for but, as Oertel told the NSF, "we are tantalizingly close!"[10]

In early 1992, Australian astronomers began to express an interest in Gemini as well.[11] Australian astronomers had a reputation for producing excellent research and building diverse and innovative instrumentation. Australian astronomers, a community of about 250 people, were divided between purchasing a share of Gemini and spending much more to build their own telescope in Australia. Supporters of the latter view opposed buying into telescopes located overseas and said that a small share in Gemini would not meet their need for observing time. Opponents of what some Australian scientists called "emigrant astronomy" carried the day and Gemini ended 1992 still short 5 percent.[12] This news was demoralizing, coming as it did when the mirror controversy was growing in intensity. It also sent Oertel scrambling to meet the deadline set by Congress (and enforced by Dick Malow) for 50 percent foreign participation.

Two other candidates emerged by the end of 1992. Argentina had several national astronomical facilities, including a 2.2-meter telescope that served a community of about 200 astronomers. Brazil, in contrast, had less of a historical tradition in astronomy; but since 1970, its astronomy community had grown to several dozen scientists. For Argentina and Brazil, even a small share in Gemini represented an opportunity to participate in a major international science endeavor as well as a serious commitment of money. Some members of Gemini's board were less optimistic and wondered whether having two extra partners was worth the logistical complications.[13]

Oertel and Malow continued to wrangle over Gemini's looming deadline for setting foreign contributions. Negotiations with Argentina and Brazil were proceeding at a frustrating pace, and Oertel tired of having to brief Malow on his progress. Faced with another deadline, AURA's president gambled and told Malow that, if he really wanted Gemini to succeed, maybe he should get involved. "I think it was a turning point in our relationship," Oertel recalled, "He said nothing but, not long after that, he was in South America."[14]

In February 1993, Malow met with Claudio Anguita to see how they could encourage Chilean politicians to quickly sign the memorandum of understanding that formally indicated Chile's participation in Gemini. Argentina and Brazil, spurred perhaps by Chile's stake in Gemini and Malow's and Oertel's diplomacy, soon showed interest in contributing 2.5 percent each to Gemini. Malow's and Oertel's efforts had paid off. By May 1993, all of these countries formally announced their participation in Gemini.[15]

The successful completion of the partnerships brought an unexpected dividend. In Chile, as he later phrased it, Malow "caught the spirit of Gemini" and began to take a personal interest in the project. A few months later, at a Washington reception to honor Gemini's new partners, Oertel asked whether the congressional staffer might like to join AURA. During 1993, Malow considered his career options. Oertel eventually made a surprise announcement on December 23, 1993: After two decades on Capitol Hill, Malow was going to become AURA's new Special Assistant for International Relations. With the resolution of Gemini's mirror controversy happening only a few weeks earlier, it seemed that Gemini's fortunes had finally changed.

Some scientists who remembered Malow's opposition to projects like the Hubble Space Telescope joked that his new appointment was like the poacher minding the hen house, but Malow became a powerful ally for Gemini and astronomy in general. Oertel recalled, for example, that as soon as he announced Malow's new appointment, difficulties AURA had had with the University of Hawaii in securing a site for Gemini on Mauna Kea disappeared.[16] Malow went on to help secure Congressional exemption from American customs charges for Gemini's various components built overseas. Another bill he championed waived duties for international science projects in general.[17] Perhaps Malow's biggest contribution to Gemini came in 1996 and 1997 when he helped negotiate a series of crises that arose regarding Chile's continued participation in the project.

Chile was poised to emerge as the real winner among Gemini's new partners. Chilean scientists realized that their clear night skies and ample observing time, like copper and oil, were a valuable resource. In addition to their 5 percent share of both Gemini telescopes, Chile would receive an additional 10 percent of observing time at Gemini South in exchange for providing an excellent site. The University of Hawaii negotiated a similar arrangement for Gemini North. As more observatories built telescopes in Chile, its small community of astronomers realized they might soon have something northern astronomers could only dream about—too much telescope time. The availability of so much telescope time could help propel Chile to the major leagues of international astronomy. Before this could happen, however, Chileans needed to resolve internal tensions about how this resource would be divided and what privileges the government should extend to foreign scientists and their institutions.

Chile joined Gemini in 1993 under conditions that later created problems as the project matured. Chile's entry into Gemini was handled largely by a small circle of Chilean politicians who did not formally consult with its astronomy community. This prompted some Chilean scientists to believe Gemini had been thrust upon them. To make matters worse, when a new administration took power in Santiago in 1994, it was less concerned with Chile's commitment to Gemini.

AURA's original negotiations with Chile regarding Gemini were with the Universidad de Chile, not with the Chile's equivalent of the NSF (the Comisión Nacional de Investigación Científica y Tecnológica or CONICYT). Chilean scientists not based at the national university felt disenfranchised, a problem exacerbated by the historically poor relations between astronomers at the Universidad de Chile and the other schools. Many Chilean astronomers believed Gemini should benefit all Chilean scientists "regardless of home institutions."[18] They also resented that Chilean's share of observing time on Cerro Tololo's older telescopes was available only to scientists from the Universidad de Chile with whom AURA had made its original agreement 30 years earlier. Oertel and the Gemini board were sympathetic to these issues. Chile's participation in Gemini was both as a host country and as a paying partner; the latter, Oertel believed, meant that Gemini should be available to all Chilean scientists. On the other hand, any disputes about access to telescopes, Oertel concluded, was a matter for Chileans, not AURA, to resolve.[19]

Disagreements festered among Chilean scientists and their government about access to all the telescopes managed by AURA. Tensions were further heightened by simultaneous disagreements with ESO. Riccardo Giacconi, ESO's director, admitted that Chilean perceptions of ESO were as if the European scientists had come "from outer space . . . not talking to the natives much."[20] Such paternalistic attitudes, formed in the 1960s when Chile's role in astronomical research was limited, no longer matched the reality of Chile's evolving participation in international astronomical research.

Gemini faced its own mounting set of problems in Chile. Supporters of Católica in Chile's legislature blocked laws to free up their country's financial contributions to Gemini. By 1997, Chile was three years behind on its payments (about $3.6 million). Malow, along with Barnes and Van Citters, traveled to Chile several times to negotiate settlement of the outstanding debt to Gemini.[21]

In December 1996, the situation appeared more serious. Malow and Van Citters heard rumors that the Chilean legislature was considering a bill, referred to as the Gemini Law, that would release Chile's contribution to Gemini. One version under consideration, however, would limit all astronomers associated with AURA—this included staff at Gemini as well as Cerro Tololo—to "second-class 'citizen' status without the standing privileges enjoyed by ESO."[22] These problems, unlike the earlier mirror controversy, did not directly affect Gemini's existence, but financial shortfalls caused by Chile's failure to pay its share threatened to delay the project, which was already under tight budgetary constraints. Gemini objected to the proposed new law on principle as well. If passed, scientists, engineers, and other people working for Gemini would have fewer privileges than astronomers from ESO.[23]

By May 1997, Chile still had not met its financial obligations to Gemini. Persistent rumors that the onerous Gemini Law might pass were the last straw for Gemini's board. Robert Gehrz, the Board's chair, advised Chile's science agency that its role in the project was reduced to that of "host country." Chile would still receive 10 percent of time on Gemini South in exchange for providing the site, but Chile was no longer a full partner. The situation embarrassed some of Chile's politicians as their country was defaulting on an international agreement and proposing a law that would surely offend AURA, an organization with which Chile had a history of excellent relations.

Chilean astronomers, by and large, greeted this news with relief, however. Some saw their participation in Gemini as a mixed blessing and criticized their government for not involving them in the original decision. Chilean scientists believed they already had access to as many telescopes as they needed. Despite the wealth of observing time, no Chilean university offered a Ph.D. program in astronomy. Scientists who favored building up Chile's scientific and educational infrastructure worried that their financial obligation to pay Gemini's annual operating costs (which typically amounted to about 10 percent of what each partner paid originally in construction costs) might strangle Chilean astronomy for decades to come. As one prominent Chilean scientist noted, his country spent only about $1 million annually on all astronomical research, and "$700,000 [in annual operations costs] is a lot of money to pay for the privilege of being a partner."[24]

To keep their options open, the Gemini board gave Chile the chance to rejoin the project if it made its financial contribution by September 1 and passed the Gemini Law without the objectionable provisions. Meanwhile, Gemini's board asked the NSF to begin formally seeking other partners to pick up Chile's 5 percent share. Australian astronomers quickly expressed interest as their earlier plans to build another big telescope in Australia had fallen flat.

Negotiations with Chile went down to the wire. On July 31, 1997, embassy representatives from Argentina, Brazil, Canada, Great Britain, and the United States made a joint appeal to the Chilean government on behalf of Gemini. Three weeks later, Chile's president signed the Gemini Law without the objectionable terms. On August 31, with only minutes to spare, the NSF received notice that Chile had transferred its contribution to Gemini's bank account, making Chile a full partner in the twin telescopes once again.[25]

Gemini and the NSF were still interested in adding Australia to the Gemini partnership as well. Just because Chile's participation was now assured was no reason, as Van Citters noted, for the Australians to spend their money on someone else's big telescope.[26] Gemini's partners were also concerned that, while the project's funds might allow two telescopes to be built, there might not be enough money to operate them as cutting-edge research facilities. Astronomers realized that the scientific instruments for new telescopes like Gemini were going to be much more complex and expensive than they had expected. Australia's contribution offered Gemini several advantages: the expertise of a well-established astronomy community, a way to hedge against any budget shortfall, and additional funds with which to build instruments.

As they weighed the benefits of admitting Australia, science managers like Ian Corbett and Wayne Van Citters had to consider broader issues of science policy. Unlike Corbett's British science agency, the NSF's budget was appropriated by Congress on a yearly cycle. This forced Van Citters to balance his long-term policy goals with short-term political considerations. If Gemini were to encounter budget problems, Congress could blast the NSF for not taking Australia's offer. Corbett replied that "Australia, with their five percent, could potentially contribute more . . . than the South American 10 percent." Corbett also noted that the physics and space science communities had realized the benefits of broad international cooperation long ago, something he believed optical astronomers needed to embrace more enthusiastically. Throughout these debates (and indeed all of Gemini's negotiations), Van Citters remained frustrated by his "immersion in an agency whose approach is defensive and tactical, swaying with each political zephyr . . . when we should lay the groundwork for decades ahead."[27]

In November 1996, after considerable indecision on the NSF's part, Gemini offered Australia a formal invitation.[28] Australian astronomers were delighted when their research council formally accepted it and joined Gemini. Their $9 million contribution, with close to $1 million more for annual operations to follow once the telescopes were built, was one of the country's largest investments in any research facility and gave Gemini more power to "enhance the scientific productivity of the telescopes."[29]

The South American countries brought into Gemini an international mélange of scientists, many of whom had relatively little experience working together. With the addition of Australia, Gemini became a far-flung scientific undertaking that spanned two hemispheres, four continents, and over a dozen time zones. Ultimately, the partnership held together, but the final roster of participants in Gemini was not a reflection of some master strategy. Instead, it was an ad-hoc collection of countries drawn together by prevailing political and financial circumstances.

Building "Hyper-Telescopes"

As Gemini and other large telescopes neared completion, astronomers' conception of the telescope as a research tool continued to shift in new and sometimes radical directions. At telescope conferences in the 1960s and 1970s, astronomers and engineers had discussed the telescope as a collection of individual parts—mirrors, gears, domes, drives. Astronomers and engineers at that time did think about how these parts interacted, yet their conception of an observatory as a holistic system was relatively limited and constrained by the computer-modeling tools then available. When new telescopes became operational in the 1990s, astronomers considered how they might use them while responding to prevailing weather conditions, the availability of scientific instruments, and their own research goals to achieve greater efficiency and productivity.

In 1971, at a telescope conference, one scientist expressed his concern that a large telescope might become just a "big computer with a large optical analog-input at its periphery."[30] A quarter-century later, this was precisely how new observatories like Gemini were frequently described—the first link in a data-gathering and analysis system.[31] Gemini would differ from "most current observatories, in which many of the main observatory components are essentially distinct systems." Instead, Gemini staff wrote, "the entire assembly, from the external enclosure to the primary mirror cell down to the instruments will function as an integrated system" with each part contributing to the quality of the light when it arrived the telescope's detector.[32]

Astronomers' traditional concept of an "observatory" came to embrace a panoply of interdependent elements—the telescope's enclosure, instruments, mirror supports and control systems, data archives, and electronic linkages for remote observing. Tying together Gemini's myriad systems were miles of cables and fibers at each telescope. Further complexity was added by the intimate integration of new tools, such as videoconferencing and Internet access, that would allow astronomers to monitor the telescopes and participate in data collection remotely. Even local weather conditions on Mauna Kea and Cerro Pachón constituted part of the "observatory system." To optimize the thermal control of Gemini's thin primary mirror, for example, engineers needed to keep it close to the predicted temperature of the night air. To do so, Gemini staff needed extensive and accurate information on how the weather on the mountains varied seasonally as well as a system for predicting nighttime observing conditions from real-time meteorological data. Ultimately, what astronomers and engineers created was a "hyper-telescope."

Compared to the political and financial obstacles that Gemini and its precursors encountered, the actual building of the Gemini Observatory proceeded smoothly. Until 1994, Gemini consisted primarily of blueprints and technical reports and was the subject of negotiations and compromises between astronomers, engineers, political supporters, and patrons. After Gemini passed its preliminary design review in December 1993, two new telescopes gradually began to take shape on the summits of Mauna Kea and Cerro Pachón.

Gemini's design, meanwhile, continued to evolve in ways that reflected the project's funding constraints and schedule as well as the research that astronomers anticipated they would do. Two of the most noticeable changes occurred before 1995. Early models of Gemini showed two large platforms located to either side of the primary mirror. These locations are the telescope's Nasmyth foci (named after the nineteenth-century Scottish engineer who developed the concept). Light, after being reflected up from the primary mirror to the secondary mirror and back down again, can be directed to a Nasmyth focus by a third mirror located just above the primary. These platforms offer a stable location for heavy instruments that process and record the light collected. Other large telescope designs, such as Keck, Subaru, and the Very Large Telescope, included instruments at their Nasmyth foci. None of these projects also had Gemini's stringent requirements for infrared optimization and high-resolution images.

After months of design analyses, Gemini's engineers and astronomers departed from the conventional approach to the design of large telescopes by eliminating the Nasmyth platforms.[33] This permitted engineers to raise the primary mirror slightly until it was flush with the mirror support structure. Engineers predicted that removing tons of heat-emitting metal around the primary mirror would improve overall image quality while the elevated mirror position would enable better air flow over it.

One scientist who helped develop and refine this vision of how Gemini might outperform other telescopes was Frederick C. Gillett. Born in 1937, Gillett attended graduate school at the University of Minnesota where he studied with Edward P. Ney, an early pioneer in infrared astronomy. Like many people interested in infrared astronomy at that time, he earned a Ph.D. in physics rather than astronomy. After graduation, Gillett collaborated with Frank Low and used Low's new bolometers and dewars to build an innovative spectrometer that measured radiation in the thermal infrared from about 8 to 13 microns. Stray heat pollution from the telescope and the challenge of tweaking detectors to work properly made this wavelength region an especially challenging research area. Gillett spent long nights at the telescope configuring instruments and collecting data. This experience honed his sense of what an infrared-optimized telescope required.

In 1973, Gillett took a position at Kitt Peak where he helped build user-friendly instruments and became one of the key advocates for the development of infrared detector arrays. Gillett also was a key participant in NASA's Infrared Astronomical Satellite, which became one of the landmarks of modern infrared astronomy. After the satellite's launch in 1983, Gillett noticed an anomaly in calibration data coming back from its observations of the white-hot star Vega 25 light-years away from earth in the constellation Lyra. Eventually, Gillett and other scientists concluded that Vega must be surrounded by a massive circumstellar shell of dust grains.[34] This discovery provided the first direct evidence for the presence of matter around a star from which the formation of extrasolar planets might occur. The normally low-key Gillett found himself appearing on television talk shows with luminaries such as Carl Sagan and astronaut Sally Ride. Sagan used the premise of alien signals emanating from Vega as the basis for his 1985 novel *Contact*. NASA, meanwhile, persuaded Gillett to come to Washington and oversee some of its other infrared projects.

Gillett returned to Tucson as one of the world's most respected infrared astronomers and eventually joined Gemini as its associate project scientist. In 1991, he led a committee that defined how to customize an 8-meter telescope for infrared research. He also chaired Bahcall's panel on infrared astronomy for the decadal survey. Gillett's association with Gemini reassured people like Low and Bahcall that the project would stay true to its emphasis on infrared optimization. As Sidney Wolff recalled, when it came to improving and predicting the performance of Gemini, "Fred kept us honest."[35]

Gillett helped oversee a second and much more controversial change to the telescopes' design. Gemini initially planned for the telescopes to have interchangeable top ends that could hold different secondary mirrors. One would be a small convex mirror, about 1 meter in diameter, that was light enough so that electromechanical actuators could nudge it in three dimensions up a few hundred times a second. Astronomers would use this mirror to subtract the sky's thermal background, an adjustment necessary for infrared observing. Another secondary mirror, twice as large and weighing much more, would enable astronomers to image a much wider field of view than they could with the small, infrared secondary.

Members of Gemini's science committee were divided about whether to retain the telescopes' wide-field capability. Canadian astronomers, eager to add to the reputation they established at the Canadian-France-Hawaii Telescope on Mauna Kea for producing exceptionally sharp images, favored the single small secondary mirror. Many in the British community, however, saw wide-field astronomy as a particular strength and wanted both secondaries. Ultimately, the Gemini Science Committee voted to postpone the wide-field secondary mirror in Gemini's initial configuration. Richard Ellis, a British astronomer on Gemini's Science Committee, was disappointed by the news. "Gemini's wide-field capability was going to be different from Keck. I had gone up on a pedestal many times and said, 'This is what we're going to buy. Give us the money.' One of the big pillars I had been selling to the British science community disappeared."[36]

Gemini's choice of a small and relatively lightweight secondary mirror rippled through other areas of the telescopes' design as well. The secondary mirror could be held in place by two criss-crossing steel vanes only 10 millimeters thick. This reduced the scattering of starlight and the emission of undesirable thermal radiation. If a telescope like Keck represented a Harley-Davidson with a big-block engine, Gemini, seen in Figure 10, was analogous to a fast and nimble

Figure 10. Gemini's streamlined final design with instruments at Nasmyth foci removed and infrared-optimized secondary mirror system shown. The entire telescope is over seven stories tall. Courtesy of Gemini Observatory/ NOAO/AURA/NSF.

sport bike. While these design changes reduced Gemini's science capabilities, they also encouraged the project to focus on specific science goals such as excellent image quality and infrared optimization. The project estimated, moreover, that eliminating the Nasmyth platforms and postponing the wide-field option would save as much as $24 million.[37]

Gemini's management strategy evolved with the telescopes' design. During the mirror controversy, the project began to exhibit more of the corporate culture that people like Matt Mountain insisted was necessary for Gemini's success. In mid-1993, the Gemini Board decided not to renew Larry Randall's contract as project manager. By all accounts, Randall had faced an uphill battle as Gemini's manager. His tenure coincided with the project's most contentious period, making it difficult for him to develop a strategic approach to Gemini. Reflecting its continuing emphasis on comprehensive systems engineering, Gemini's board pushed the project to hire more people with experience in this area.

Richard J. Kurz joined Gemini in late 1993 as its project manager. Kurz began his career as a high-energy physicist but, since 1972, had worked for TRW on a variety of NASA and classified projects. Kurz combined his talents with Jacobus (Jim) Oschmann, who the project hired as a systems engineer at the insistence of Gemini's board the previous year. Like Kurz, Oschmann had previously developed and managed laser and optical systems for defense-related projects costing tens of millions of dollars. A large part of his new job was to coordinate the efforts of the different engineering groups and to ensure that the systems they developed worked harmoniously. When Kurz left Gemini in 1998 for a position with ESO, Oschmann took over as project manager.

Neither Kurz nor Oschmann were astronomers or telescope engineers; they were also not fazed by Gemini's cost and scale. Their hiring reflected Gemini's belief that the project was first and foremost a complex technical undertaking that needed experienced managers. That Gemini was a pair of telescopes was seen as somewhat less relevant. "The project realized that they needed a manager at the center who is organizing and watching things," Kurz recalled, "It might just be a distraction if the person were an astronomer."[38]

The biggest change in the twin telescopes' management came in

October 1994 when Matt Mountain replaced Sidney Wolff as Gemini's director. He took this position at a relatively young age. John Huchra, who chaired the search committee that selected Mountain, recalled that the 38-year-old British astronomer brought "a real appreciation of management" that, combined with his infrared experience, was the "winning plate."[39] Mountain's hiring severed the last line of control NOAO had over Gemini. As the observatory's two telescopes were built, the center of Gemini's operations shifted from Tucson to Hilo, Hawaii, where its new headquarters would be located. After becoming director, Mountain noticed that some NOAO staff resented "having their big toy taken away from them," a point made, he said, at social gatherings when they would "poke me in the chest and complain bitterly about it."[40]

These changes pleased the harshest critics of the project's management. By 1995, even Corbett noted that AURA's performance was consistently "good with a tendency toward very good."[41] With Mountain's hiring as Gemini's director, Gillett was appointed project scientist. Together Mountain, Gillett, Oschmann, and Kurz led Gemini from its construction phase toward its dedication. In October 1994, Gemini's advocates participated in two separate groundbreaking ceremonies on Mauna Kea and Cerro Pachón. They also celebrated Corning's successful casting of Gemini's first mirror, an important and symbolic milestone.

Corning made its specialty mirrors at its plant in Canton, New York, near the Adirondacks State Park. It took months for Corning to deliver over fifty tons of its proprietary low-expansion glass to the Canton plant. Round boules of glass, a meter and a half in diameter, were heated in a special furnace to over 1,700° C and fused together. After sawing them into hexagons, technicians placed forty-two of them in a rough circle. The glass was heated a second time to fuse these pieces into a flat monolithic disk, 8.1 meters in diameter but only 30 centimeters thick. After grinding both sides of the blank and etching it with acid to remove surface flaws, the glass was heated a third time in the furnace until it gradually sagged over a mold and took its final meniscus shape. Corning's entire process—from raw glass to a finished mirror blank—took about two years. In October 1995, the company announced the successful completion of Gemini's first blank and finished the second even faster.

After Corning successfully made the mirror blanks, they were carefully shipped overseas in specially designed containers to REOSC, an optics company located in St. Pierre du Perray south of Paris, to be polished. The company was well acquainted with polishing thin meniscus mirrors like Gemini's, having done four for ESO and one for Japan's Subaru project.

In January 1998, REOSC announced that the first Gemini mirror was complete; it finished the second 11 months later. A press release boasted that they were some of the best optics ever made. If the mirror's area were expanded to the size of the United States, for example, any bumps on its surface would be less than a few inches high. REOSC repackaged the 24-ton pieces of glass and they made their way to Hawaii and Chile.

Progress on Gemini's mirrors was perhaps the most closely followed sign of the project's transition from planning to reality. At the same time, major construction was also proceeding on the mountain tops in Chile and Hawaii. Even though Gemini was a tool for advanced astronomical research, much of its construction more closely resembled that of a massive apartment complex. Dozens of steel components had to be designed, worked into shape, and assembled to form the telescope support structures.

By late 1998, most of Gemini North's staff had moved into their new operations center in Hilo. This would serve as the observatory's nerve center and a hub for the communications network tying the two telescopes together. At this time, Gemini's enclosure on Mauna Kea was near completion and the telescope structure was finished. Workers tested its movement with a 23-ton dummy mirror while the real one received its first coat of aluminum (see Figure 11). The observatory's plans for silver-coated mirrors were delayed as technical problems were worked out. Some of the project's engineers and technicians relocated to Chile temporarily where they continued their work on Gemini South. The project's master schedule called for it to be finished about two years after Gemini North.

As the date for Gemini North's first light approached, technicians and engineers struggled to resolve a steady stream of technical problems. Managers like Kurz and Oschmann deployed organizational tools learned while working on aerospace projects. Engineers had to combine different and complex systems to a new degree of precision

Figure 11. Gemini North's 8.1-meter primary mirror after receiving its first aluminum coating. Courtesy of Gemini Observatory/NOAO/AURA/NSF.

to achieve the performance astronomers expected. As Kurz noted, systems engineering and the tools he and other managers used to implement it were "the glue that held the project together."[42]

As they built Gemini, its managers and engineers needed to understand how the telescopes' different systems would interact and find ways to organize this information in a readily available format. The project prepared hundreds of interface control documents. Each detailed how two different telescope components came together. This could be as simple as the pattern of bolts physically connecting two parts or as complex as communication links between two computer systems. For example, when the telescope moved to point at different places in the night sky, gravity deformed the primary mirror in distinct but predictable ways. The mirror's position and shape needed to be monitored and coordinated with the system that pointed the telescope so the actuators supporting the mirror knew what force to apply to the mirror to maintain its shape. These documents were especially important for a project like Gemini whose diverse systems were built in different countries and then transported to Hawaii or Chile to be integrated into the final telescope.

The concept of tracking the behavior of interconnected telescope systems was not new to Gemini. What distinguished it from earlier and smaller telescope projects was the comprehensive and systematic manner in which Gemini's managers codified and organized information. Without adequate control over mechanical, optical, and computer components through systems engineering, the project would court lengthy delays, if not chaos, when technicians combined the observatory's different subsystems. Extensive technical documentation, subcommittee reports, and videoconference sessions all became part of what historian Alfred Chandler referred to as the "visible hand" of middle management.[43]

Emphasis on middle management was something many high-energy physicists such as Berkeley's Luis Alvarez had embraced 30 years earlier. Building bigger bubble chambers and handling their massive data streams required physicists, as Alvarez said, to put on their "engineering hats" and make the sometimes painful transition from crafting instruments to a more corporate style of instrument building.[44] With projects like Gemini, optical astronomers followed a similar path.

Gemini's Science Instruments

Astronomers played pivotal roles in selling the new generation of large telescope projects to their colleagues and funding sources. Once approved and under construction, their scale and complexity often overwhelmed the contributions individual astronomers could make to them. Defining and building the science instruments for telescopes like Gemini was a notable exception to this truism and, in this realm, astronomers maintained a prominent role.

The size and design of the telescope determines how accurately it can track, how much light it can gather, and so on, while the telescope's instruments determine what happens with all the photons it collects. In other words, the instruments define the telescope's ability to do research and make new discoveries. When the 200-inch telescope dominated the astronomical landscape, some of its most important tools were designed by a single astronomer and built in Mount

Wilson's Optical Shop. The scientist himself could transport the instrument to the telescope in the backseat of a car and often make it work in a short period of time. Even in the 1970s and into the 1980s, some of astronomy's most innovative instruments were designed, built, and debugged by small groups of people.

Observing time on telescopes like Gemini was expected to be too expensive and its associated systems too complex to allow users to hitch on their own instruments. Instead, like those on Hubble, Gemini's science instruments were what astronomers called "facility instruments." These were considered an integral part of the overall telescope system and designed to conform with the telescopes' focal length and field of view. Another fundamental design requirement was that the instruments for Gemini be as reliable and straightforward as possible because, once in place, all scientists granted time on Gemini would use them to collect data. Gemini's long-term plans called for eleven facility-class science tools, each costing around $8–10 million and roughly the size and weight of a grand piano. Designing and building them required more than the skills the average astronomer and a few graduate students could muster. Specialists in computer software, lasers, cryogenics, and optics might be necessary, along with an ever-present manager to monitor the budget and schedule.

There were constraints on the instrument program as well. The budget for Gemini's first generation of instruments was only about $17 million, an amount that Mountain and others soon realized was inadequate to build the suite of complex devices the observatory needed. Scientists from all of Gemini's partner countries anticipated helping build the instruments. As a result, Gemini's board was eager to see an equitable distribution of work among their countries' universities and observatories.

Astronomers planned two basic types of instruments for Gemini: imagers and spectrographs. While optimized for infrared observing, astronomers still planned to use Gemini to observe and take spectra in visible light, and about half of the instruments originally planned for Gemini's two telescopes could be used to some degree at optical wavelengths.[45] Gemini's proposed instruments covered a factor of 60 in wavelength—from violet light (about 0.4 microns) through the near-infrared region (1 to 5 microns) to the mid-infrared regime of 5 to 25 microns.

Gemini's instruments took advantage of the tremendous experience engineers had gained in building bigger CCD and infrared arrays. This increased capability was especially profound for infrared viewing. A typical (civilian) infrared array in 1985 had about 900 pixels. A decade later, astronomers routinely designed instruments with infrared arrays boasting a million or more pixels. Improvements in signal quality and sensitivity were just as dramatic, thanks largely to investments NASA had made in detector technologies.[46]

To hold Gemini's massive instruments, each telescope had an instrument support structure at its Cassegrain focus underneath the primary mirror. Resembling a giant plus sign, the structure was large enough to fit several people inside. Gemini's engineers designed it to support up to five different instruments, each weighing a few thousand pounds. Each telescope had an acquisition and guiding unit along with calibration and adaptive optics instruments. Light reflected from the primary mirror up to the secondary mirror and then back down through the hole in the primary mirror. There it met a tertiary mirror at the Cassegrain focus that directed starlight to one of the instruments. By moving this tertiary mirror, astronomers could switch the beam of light rapidly from one instrument to another without losing valuable observing time.

Because Gemini's telescopes were practically identical, astronomers planned that its instruments could be shuttled between Hawaii and Chile for different observing semesters. Gemini's science plan, however, did call for one instrument to be duplicated so both the southern and north hemisphere would have one. This was the Gemini Multi-Object Spectrograph (GMOS).

Since the 1970s, astronomers had looked forward to taking dozens of spectra simultaneously using large telescopes, and all plans for new telescopes included some type of instrument to accomplish this goal. Some astronomers, for example, hoped to use GMOS to extend earlier surveys measuring the redshifts of distant galaxies in order to map the large-scale structure of the universe. Others wanted the capability to study nearby dwarf galaxies in the Local Group (the cluster of over 30 galaxies about three million light years in diameter that contains the Milky Way) to learn more about overall galaxy formation.[47]

GMOS was designed to take spectra using visible light. More specifically, the GMOS for Mauna Kea would be optimized for red light while its twin in Chile would be enhanced for detecting and recording blue light. Taken together, the two GMOS instruments could analyze light from the near ultraviolet through the visible and out to the beginning of the infrared, making them extremely powerful and versatile. Their capability was enhanced further because GMOS could also produce images in addition to spectra.

To take hundreds of simultaneous spectra, engineers and astronomers designed GMOS to use "aperture masks." Their purpose was elegant and simple. After an astronomer had studied star charts and identified targets of interest in the patch of sky he or she was studying, the coordinates would be sent to the observatory. A machine tool would then use a laser to cut thin slits of the right size and position in a carbon-fiber sheet. The slits would match up perfectly with the objects in the sky. Once light was collected by Gemini's primary mirror and relayed to GMOS, it passed through the aperture mask and an assortment of wave front sensors, filters, and other optical devices. Finally, the spectra were recorded by an array of three CCD detectors, each having more than eight million pixels. Masks for several different scientists' observing programs could be stored in a jukebox-style changer and inserted into the instrument as needed.

Building instruments like GMOS required the effort and resources of the same order of magnitude as astronomers needed 25 years earlier to build an entire telescope. GMOS, for example, was designed and assembled as an international collaboration between three different institutions. The Royal Observatory, Edinburgh, and the University of Durham designed the major components for GMOS. Canada's Dominion Astrophysical Observatory made the optical assembly and the instrument's wave front sensors. Once these parts were finished, they were shipped to scientists and engineers at ROE, where they were integrated and tested before being put into several crates and sent off to Hawaii and Chile. At the project's peak, some twenty people were designing and building the two GMOS units and their final cost approached $10 million.

GMOS was one of Gemini's more successful instrument projects. Others did not fare so well. Astronomers hoped to take advantage of another spectrograph on Gemini called the Gemini North Near-Infrared Spectrograph (GNIRS), which was optimized for high-resolution use in the near infrared region of 1 to 5 microns. NOAO successfully competed to build this spectrograph, but its completion was seriously delayed by schedule and technical difficulties.

When work resumed on the spectrograph after a lengthy management review, its cost was estimated at more than $4 million with a delivery date years later than hoped. Many at Gemini, including Mountain, saw NOAO's failure with GNIRS as another sign that the national observatory was still stuck in an outdated "four-meter mindset" that eschewed the more rigorous management they believed necessary to produce the new generation of large telescopes successfully.[48] Additional problems befell another one of Gemini's workhorse instruments, an infrared imager for Gemini North called NIRI. The NSF gave the contract for this instrument to the University of Hawaii as part of the negotiations for access to Mauna Kea. When the imager finally arrived at Gemini North for integration into the telescope, Gemini staff found that it needed months of redesign work, a process Jean-René Roy, the associate director in charge of Gemini North, found "extremely painful."[49]

After the declassification of adaptive optics in 1991, many astronomers and engineers worked at a frantic pace to understand how the technique (especially the use of artificial laser-guide stars developed at places like the Starfire Optical Range) could contribute to their research. Indeed, the number of technical papers and conferences on adaptive optics grew at a rapid rate after 1991.

Despite their increased cost and complexity, spectrographs and cameras were familiar tools for astronomers. The same could not be said for adaptive optics. It was one thing to have access to the formerly classified technology. It was another matter altogether for astronomers to adapt the technology successfully for their own use or, as many astronomers hoped, to build their own systems.

Gradually astronomers began to report technical successes and scientific results using adaptive optics. The Canadian-France-Hawaii Telescope was one facility equipped with adaptive optics that was especially productive. Astronomers using the CFHT were originally concerned about remaining scientifically competitive in the era of the Hubble Space Telescope and larger ground-based telescopes. The better resolution, image quality, and sensitivity provided by adaptive optics represented one solution, and astronomers started development on a prototype system in 1989. By 1994, they were using adaptive optics (christened PUEO after a type of Hawaiian owl with exceptionally good night vision) to make astronomical observations with the CFHT.

Not all astronomers embraced the new technology immediately. Some scientists were skeptical of the gains to be realized with the new technology given its expense and complexity. There was also recognition that the usefulness of adaptive optics was best suited to relatively bright objects and those with bright stars nearby that could serve as natural guide sources. The possibility of creating artificial guide stars with lasers was promising, but success with these systems would require even more time and money. In other words, adaptive optics would not be an all-purpose panacea for astronomers.[50]

Despite concerns about its expense and complexity, adaptive optics became increasingly important to Gemini's design and its suite of science instruments. The technology was a natural addition to a large telescope project that had excellent image quality as a major priority. The relative ease of using adaptive optics to correct wave fronts from infrared sources also contributed to astronomers' eagerness to include the technology. Even observations that did not require the highest-quality images would benefit from adaptive optics, which, because the technology concentrates more light into a smaller area, offered a way to observe with large telescopes more efficiently. In fact, by 1997, astronomers associated with Gemini predicted that up to 85 percent of science programs done with the twin telescopes would depend on adaptive optics.

The study of early phases of star and planet formation appeared to be an especially promising match between Gemini's infrared instruments and adaptive optics.[51] Astronomers' interest in this topic was fed by the 1995 announcement of the first discovery of an extrasolar planet.[52] Two Swiss scientists, Michel Mayor and Didier Queloz, using a spectrograph on a small telescope near Marseilles, measured a slight Doppler shift in 51 Pegasi, a star about 45 light-years from earth. The shifting spectral lines suggested the presence of an orbiting planet with a mass roughly half that of Jupiter. Astronomers had expected such news for years, but the announcement (and its subsequent confirmation by American astronomers Geoff Marcy and Paul Butler) electrified the public and created a media frenzy. Astronomers using spectroscopic techniques found scores of extrasolar planets and brown dwarfs orbiting other stars. The substantial light-gathering capability of the Keck Telescope helped astronomers with access to it take the lead in searching for objects orbiting stars. Aided by adaptive optics, astronomers hoped to use new large telescopes like Gemini to see the faint companions of nearby stars rather than inferring their presence indirectly through spectroscopy.

In 1994, the Herzburg Institute of Astrophysics, located in Victoria, Canada, won the contract to build the adaptive optics system for Gemini North. This was a technological plum for Canadian astronomers and it allowed them to parlay previous experience designing the PUEO system into something bigger and more capable. Mountain recalled that this decision shocked many mainland American astronomers. "There were lots of U.S. groups who were adaptive optics experts. But no one was actually building anything." American astronomers on the mainland, he believed, were still "doing research grants and putting together complex systems." Meanwhile, Canadian and European scientists "were cleaning up on papers using adaptive optics."[53]

Tools like PUEO were designed to be reliable facility-class instruments that the average astronomer could use. This made the Canadians' skills especially valuable to Gemini, which needed to serve a broader community than telescopes like Keck or Subaru. "We were an observatory, not a development lab," Mountain said, "We wanted something where the user could come and get a corrected beam on the detector right away. The physics was understood. This was a systems engineering problem."

Canadian astronomers and engineers designing Gemini's adaptive optics dubbed the system Altair—short for ALTitude conjugate Adaptive optics for the InfraRed and also the name of a bright star in the Summer Triangle. It was initially designed only to use bright stars in the sky as reference sources; but, in 1997, the Gemini board approved funding to upgrade Altair's versatility by using artificial guide stars produced by laser beams. Engineers designed Altair to fit, like Gemini's other instruments, at the telescope's Cassegrain focus. Light collected by the telescope's main mirror was first directed to Altair to correct for atmospheric disturbances. As photons made their way through Altair, they encountered a small mirror that dozens of actuators could deform over several hundred times a second. The corrected beam of light then continued on to one of Gemini's other instruments.

Scientists and managers from all major telescope projects were consistently surprised as they struggled to equip their new facilities with science instruments that were an order of magnitude more expensive and complex than the previous generation of tools. Like all other aspects of building the next generation of large telescopes, the leap in the cost and complexity of instruments suggested the need to embrace new methods. While the active involvement of teams of specialists was a familiar situation to those who had worked on space-astronomy projects, for ground-based astronomers used to building instruments quickly with a few graduate students and taking them to the telescope, this was a new and sometimes unsettling experience.

Throughout history, the scientific community had rewarded people who used instruments to make discoveries, not the people who built them.[54] Judith Cohen was a Caltech astronomer and the lead scientist for one of the Keck Telescope's instruments. She expressed her concerns about the changing status and rewards for what used to be called "gadgeteers" in a provocative letter to the American Astronomical Society. She noted that there were few professional kudos for building new instruments. Moreover, the timescale required to build them meant that graduate students could not fully participate in the experience by helping build an instrument and then collecting data with it. With the cost and complexity of instruments rapidly increasing, she asked, "Who will dream up the grand designs and new schemes that the next generation of instruments will require?"[55]

CHAPTER 9

Point-and-Click Astronomy

For centuries, using a telescope meant going to the observatory at night, often alone, and collecting data. This style, part of the tradition established at observatories like Yerkes, Mount Wilson, Palomar, and Lick, was so widely practiced that astronomers called it "classical observing." Astronomers who used big telescopes in the classical manner described their experiences in colorful and sometimes contradictory language—romantic, creaky, cold, sometimes inefficient, but above all personal.

Astronomers who spent thousands of hours at the telescope collecting data felt a strong emotional connection with this style of research, often describing the experience with masculine and idyllic imagery. According to Alan Dressler, an observational astronomer at the Carnegie Institution of Washington, "A certain degree of cowboy-ish behavior made a difference." "People talked about who was a good observer," Dressler explained, "It was like saying someone was a good roper. Someone who could get that last ten percent out of the telescope. People made their reputations based on how they performed at the telescope."[1]

Ground-based astronomers who observed in the classical style, of course, took risks. If the weather was poor or cloudy, they would collect little data. If an astronomer had an important observing program that called for the best possible seeing conditions and these did not materialize, then he had to fall back on other programs for that night. Classical observing was, in other words, a gamble. There were more serious risks as well. In 1987, Marc Aaronson, a promising young scientist and skilled observer, was fatally injured while using the 4-meter telescope at Kitt Peak.

The advantage of classical observing, many argued, was that control of the telescope rested with the astronomer. If they desired, scientists could point the telescope somewhere new and hope for a discovery based more on serendipity and instinct than extensive preplanning. In fact, Dressler and other well-respected observational astronomers often claimed that the freedom to change one's observing program quickly to match the night's conditions or respond to a sudden inspiration was an important factor in many of their most important discoveries.

In the 1990s, a new sense of what it meant to be an astronomer and use a telescope emerged and became subject to heated discussion. Matt Mountain was aware of his colleagues' nostalgic attachment to the cowboy mythology and he invoked it to advocate new ways of doing astronomy with Gemini. "I do not believe discoveries are made at the telescope riding the weather variations like a cowboy riding a bucking bronco," he claimed, "Usually, discoveries emerge after long nights at a terminal trying to reconcile an awkward data set with preconceived models."[2] The perception that analysis rather than observing made the scientist shifted the focus of scientific discovery away from the telescope itself.

Astronomers' comments and debates about alternative modes of using the telescope were primarily concerned with the largest and most complex telescopes, like Keck and Gemini. Most astronomers assumed they would continue to operate smaller telescopes in the classical fashion. The practice of observing with a small telescope, alone or with a student, was sometimes referred to as the "key under the mat" mode, reflecting its common and relatively uncomplicated nature.

In contrast, one term for the new style of large telescope use—"queue observing"—meant that the scientific information desired by an astronomer was collected by observatory staff and provided in a standard format. Observatory staff determined the order of research projects, taking into consideration time allocation agreements and seeing conditions. Advocates of queue observing stressed that it allowed the telescope to be used more efficiently. Weather and atmospheric conditions on the mountain changed throughout the night so staff could select appropriate research programs for the conditions at the moment. The concept of queue observing was not new. A few scientists had proposed similar schemes in the early 1970s. In addition, some science projects were better suited for queue observing. John Huchra, for example, estimated that at least a third of the data for the Cambridge redshift survey was collected by staff observers. Until the 1990s, however, this was not the typical way optical astronomers did their job.

The possibility of new ways to use telescopes stirred discussion about a scientist's identity as an astronomer. One member of Gemini's science staff remarked at a gathering of scientists and observatory directors, "We have a new definition of an astronomer now. 'Astronomer' doesn't mean that you go to a telescope and push buttons yourself. It means that you deal with the data."[3] Statements such as these prompted concern that the next generation of researchers might be unfamiliar, even alienated, from their most basic research tools. Indeed, as I have discussed, through the 1970s and into the 1980s, many astronomers expressed serious misgivings about relinquishing their control over the collection of images and spectra. Pedagogical issues added additional concerns. Telescopes were used to train new astronomers and people typically became skilled telescope users only after relentless practice. If someone else collected their data, might not the training of future astronomers be disrupted? "I am really worried about the Nintendo mentality in astronomy, fifty or a hundred years from now," one scientist remarked.[4]

Acceptance of new styles of telescope use was gradual, and its origin cannot be traced to any single observatory, technological innovation, or astronomer. Instead, the transformation occurred subtly over more than one generation of scientists. In the 1950s, auto-guiders, for instance, replaced the tracking of objects by hand. "Gadgeteers" developed new electronic devices that allowed scientists to move from the telescope's observing cage to warmer and more comfortable control rooms. In 1968, astronomers at Kitt Peak did the first remote observing with a 5-inch telescope operated from Tucson. As early as 1984, Sandra Faber, a University of California astronomer, proposed that new telescopes like the Keck "be flexible enough to allow quick changeovers to programs that benefit from good seeing." Achieving this, Faber said, would require "substantial changes in operating philosophy."[5]

Despite these changes, most optical astronomers still collected their own data through the 1980s. The basic application for ground-based telescope time at the national observatory was fairly simple. It included a brief description of the astronomer's science program and a list of nights and instruments that were needed. As one astronomer noted, "I can write a proposal for Kitt Peak in an afternoon that's one page long."[6]

For many scientists accustomed to traditional ground-based telescopes, space-based observatories like the Hubble Space Telescope represented a major transformation in their practice of astronomy. Scientific discoveries aside, astronomers saw Hubble affecting their field sociologically and culturally, particularly in the funding arena. By 1995, NASA had surpassed the NSF as the dominant source of federal funding for astronomy. Part of this transformation occurred because NASA coupled observing time on Hubble with funding for data analysis. The effects of Hubble on the working life of astronomers were more subtle. Its location obviously precluded astronomers from using it in person. For many in the science community, applying for time on Hubble was much more complex than requesting time at Kitt Peak or Cerro Tololo's telescopes. NASA expected scientists to be familiar with the space telescope's instruments (each had many different modes of collecting data) and perhaps do initial ground-based observations to complement the precious time they requested with the Hubble Telescope.

The process of requesting observing time started when an astronomer electronically submitted a lengthy Phase I proposal. This was reviewed by staff at the Space Telescope Science Institute (STScI) and a powerful time allocation committee of outside scientists. Fewer than 20 percent of proposals submitted passed muster. The fortunate scientist (or more likely a team of scientists) who passed this hurdle then prepared a Phase II observing program. With technical help and standardized software from STScI, scientists listed each object they wished to observe, its position, the length of exposure, and the observatory's instrument configuration. This information formed the basis of computerized telemetry commands that NASA staff at the Goddard Space Flight Center sent to the telescope. NASA budgeted observing time on Hubble minute by minute around the clock. Scientists' proposals entered an observing queue and were linked to minimize telescope movements and save time. Astronomers typically did not know when their observations were being completed and few were present at STScI when data was collected.

Raw data received from Hubble was processed by STScI staff using standard software packages and stored in data archives. A few days after observations were made, the astronomer received data tapes in the mail. Only at this point did the astronomer re-enter the picture—to make sense of the data, convert the digital bits into images or spectra, and perhaps incorporate them into a publication. Scientists had a year to do this before their right to use the data exclusively expired and the information went into a public archive where all scientists had access to it. Hubble was, after all, a national observatory, and after some disagreements, the astronomy community accepted the public nature of the data it collected.[7] Even in publishing their science results, astronomers using data from Hubble broke with tradition. Their papers averaged three times as many authors as those originating at telescopes on the ground.[8]

Scientists, even those used to the most complex instruments on ground-based telescopes, sometimes complained about the inflexibility of Hubble's queue system and its high degree of automation. Some questioned whether the highly structured operation diminished scientific effectiveness by locking astronomers into tightly specified observing programs rather than letting them change plans on the fly. Looking at the data was still exciting and there was no doubt that Hubble produced some of the most visually stunning science images ever. Yet some astronomers missed the immediacy and emotional connection to data that they collected themselves and compared receiving data tapes in the mail to hearing a baseball game on the radio instead of witnessing firsthand the ballpark's sights and sounds.[9]

In 1990, much of this process—queue observing, complex proposals, public data archives—was new to many optical astronomers but not to all ground-based astronomers. Radio astronomers, especially those using the national radio arrays, were accustomed to receiving data collected by others from queue-based observing runs. Some optical observatories even operated smaller ground-based telescopes in a queue fashion, partly to evaluate the telescope's effectiveness before the completion of bigger facilities like Gemini. Astronomers' reactions were not uniformly positive. Many astronomers who received queue-observing time on the new 3.5-meter WIYN telescope at Kitt Peak, for example, said they still preferred to gamble on good weather and observe in the traditional fashion.[10] These sentiments amused scientists more accustomed to observing with x-ray or radio facilities. British astronomer Richard Ellis remembered "a famous x-ray astronomer who would always say, 'How's it going, Richard? Still wasting money and time zipping around to make observations?'"[11]

Gemini's staff and patrons anticipated that demand for the twin telescopes would be high. Several thousand astronomers from Gemini's partner countries would be competing for observing time. The large community the telescopes would serve required an approach that differed from the classical observing scenario in many ways. First, if astronomers used classical observing techniques, they might encounter bad weather or fail to get useful data because they were unfamiliar with the equipment. Such inefficient use of the telescope would be an issue, given there were only 365 nights a year to divide among all research programs.

Weather and seeing conditions at Mauna Kea and Cerro Pachón introduced additional complexity. The median quality of seeing on Mauna Kea, for example, was about 0.4 arc-seconds. On a typical night, a hypothetically "perfect" telescope would still only produce images with a resolution of 0.4 arc-seconds. The limitation was imposed by atmospheric conditions, not by technology, and seeing conditions varied nightly over a wide range.

Queue scheduling enabled the observatory to link nighttime conditions with appropriate observing proposals. On Mauna Kea, the best nights might produce optimal, but rare, conditions with 0.1 arc-second seeing that Gemini's telescopes were designed to exploit. Not every observing program required such exquisite seeing or cloudless nights, and such programs could be executed under mediocre or even poor conditions. Infrared astronomers, on the other hand, needed nights when atmospheric water vapor, which absorbed some infrared radiation, was lowest. The challenge facing Gemini and other observatories was how to match observing conditions with those programs that truly required the best possible seeing. "Unless we find ways of quickly adapting observations to exploit atmospheric conditions, realizing the scientific potential of these telescopes will rely heavily on luck and the scientific benefits to the Gemini partnership as a whole," Mountain wrote, "will be diminished."[12]

Once construction of Gemini was underway, the scientists and managers associated with the project directed their attention to modeling and optimizing how the telescopes should be used. The premium placed on efficiency and serving the maximum number of scientists would result in some sacrifice of personal interaction and control on the part of astronomers. Gemini's staff incorporated ideas like queue observing and service observing from space-based observatories like Hubble; these approaches would have been anathema to many scientists fifteen years earlier.

Designers also made instruments versatile to increase their efficiency. The instrument clusters underneath the Gemini telescopes were designed so light collected by the primary mirror could be directed to any instrument and switched between these with relative ease. When a particular instrument was out of service, Gemini's staff could switch to a different instrument and execute other astronomers' observing programs. In many ways, the emphasis at Gemini and other observatories on flexibility, streamlined efficiency, and adapting to prevailing weather conditions resembled Japanese "just-in-time" manufacturing practices admired by American business leaders in the 1980s.

Not all major new telescopes followed this path. As an international facility, Gemini differed from the Keck Observatory. Even though Keck resembled Gemini in size and sophistication, astronomers still used the two 10-meter telescopes in the classical model. Astronomers from Caltech or the University of California received blocks of nights several times a year to collect data. They flew to the big island of Hawaii where they monitored and controlled the process from Waimea via Internet and videoconference links, while, at the 14,000 foot summit of Mauna Kea, technicians pointed the telescopes and operated the instruments. While not physically at the telescope, the researcher remained an active participant in the process. Plans for Japan's Subaru telescope and the Carnegie's Magellan telescopes—all serving a smaller user community than Gemini—were based on similar models.

Debates among astronomers and science managers were associated with the adoption of distinctly different ways of using large ground-based telescopes. As part of this dialogue, scientists from several large telescope projects, as well as older established observatories, came to Hawaii in 1995 and again in 1998 to discuss how to optimize observatory operations and their scientific return. Their discussions indicate how much astronomers' thinking had changed since the large-telescope conferences of the 1970s.

In 1974, for example, when astronomers like Geoffrey Burbidge proposed that some science data might be better collected by service observers, they were met with angry and indignant responses from their colleagues. The suggestion that some observatories, especially ones with larger and more complex telescopes, adopt a hands-off policy with regard to astronomers using the telescopes was resisted.[13] Two decades later, astronomers centered their discussions around questions such as: "Who are our customers?" and "Are we delivering a product, a service, or a capability?" Scientists, if only grudgingly, began to accept that, while being an astronomer "used to mean that you go to a telescope and look through an eyepiece, . . . it doesn't mean that anymore."[14] Such statements aroused nowhere near the outcry they would have produced 20 years earlier.

Determining the catalyst for this shift in astronomers' thinking is difficult. The scientists themselves attributed the changes primarily to technological progress—bigger mirrors, more ambitious telescopes, and much greater computing power. Since the 1970s, the amount of information collected by telescopes worldwide had far outpaced gains made in light-collecting power alone. While the total area of telescope mirrors around the world had increased by a factor of 30 since 1975, the number of pixels in a typical CCD detector grew by a factor of 3,000. The introduction and maturation of new communications tools —the Web, the Internet, videoconferencing, and high-bandwidth links connecting mountain peaks in Hawaii and Chile to astronomers in the mainland United States—also motivated astronomers to adopt new patterns of work and consider changes in their nighttime use of telescopes. Moreover, well-publicized science results from the Hubble Space Telescope encouraged scientists to maximize the capability of their new tools.

Not all changes occurred as a result of technological innovations. Additional impetus came from observatory directors and science managers who argued that the cost of building and operating new telescopes demanded greater efficiency. Some in the community recognized that even with smaller (and supposedly less complex) telescopes, the first night astronomers spent observing was often unproductive.

Unless they were intimately familiar with the telescope and its instrumentation, astronomers might need several hours to calibrate equipment before collecting any data. One ESO scientist pointed out that this way of working could not be carried over to more expensive and complex facilities. When the last of the four 8-meter "unit telescopes" of the Very Large Telescope facility was completed in 2000, ESO managers had 1,480 nights of telescope time to allocate annually among European scientists. Each telescope cost about a dollar per second around the clock. "If a staff member's work can save two nights," one European scientist noted, "his or her salary is already accounted for."[15]

In many respects, astronomers' concerns about cost-effective use of their equipment resembled arguments made by particle physicists three decades earlier. At a 1966 meeting at the Stanford Linear Accelerator, Berkeley's Luis Alvarez noted that physicists were part of "a very large business" costing millions of dollar per year. This and the competition for scientific discoveries encouraged physicists to think about increasing the number of interesting "events per dollar" produced by accelerators, recorded in the bubble chamber, and discerned by scanning devices.[16]

Efforts to project an aura of efficiency and managerial competence were accompanied by astronomers' attempts to measure observatory productivity in terms of users served. "Say you are an American astronomer from a Midwest university," Mountain said. "You may get two nights a year on Gemini. What happens if the weather is cloudy? With classical scheduling, that's it for the year."[17] In the United States, these concerns, especially for a nationally available observatory like Gemini, did not originate solely among astronomers but followed larger trends. In 1993, for example, Congress passed the Government Performance and Results Act requiring each federal agency, including the NSF, to devise yardsticks to measure performance and progress. This mandate included the NSF's science centers such as Kitt Peak and Gemini. The statute helped spark debates about how the NSF would evaluate scientific productivity objectively. Was the proper metric the number of times a publication was cited, the number of "customers" requesting time at a telescope, or the number of scientific problems "solved?" This was not just an American trend; similar studies evaluating telescope productivity were done by European astronomers.[18] The need to demonstrate greater efficiency and productivity encouraged scientists to accept models and metaphors from the business world to describe observatory management and telescope operation. Astronomy, like particle physics, was a big business; astronomers described observatories as "data factories" and the researchers who came to them as "customers."

Not all astronomers, even those steeped in the traditions of classical observing, were entirely apprehensive about the cultural shift in optical astronomy. Caltech's Maarten Schmidt, an astronomer in the classical model if there ever was one, saw changes in astronomers' work habits as a positive step. "I think 30 years ago we would have been called backwards by the physicists. Now we are among the most aggressive users of whatever new thing there is."[19] Richard Ellis was another scientist who first used telescopes in the classical fashion. A member of Gemini's board, Ellis believed the observatory was "owned by its community of users" and was eager to "maximize community involvement." Gemini, he wrote, could be operated like Hubble with only a "few people [visiting] Baltimore to get HST data," but he cautioned against adopting such an extreme model. Without regular visits by scientists and a "synergy between the users and telescope staff," observatory morale might suffer and the "buzzing scientific atmosphere" that was an essential part of doing astronomy might be lost.[20]

Astronomers and managers developed plans for using new telescopes like Gemini in this fertile and sometimes confusing milieu. As they struggled to make sense of their community's changes, astronomers turned to physics and the business world to adapt to the present and anticipate the future. Meanwhile they continued debating telescope use in formal panel discussions, conference presentations, and during clouded-out observing runs. Unlike fifty years previously, there was no consensus as to the single best way to operate telescopes. There was, however, a general recognition, even at Keck where the classical model of telescope still prevailed, that a cultural shift in science practice was occurring.

In 1996, Matt Mountain, after reviewing the operations strategies adopted by ESO and Keck, likened the situation to the Goldilocks fairy tale. ESO's completely programmatic and queue-based approach for the Very Large Telescope was "too expensive" while Keck's approach was "too frenetic." By opting to include elements of each, Gemini was "just right."[21] The plan he and Gemini's science committee proposed was that Gemini would initially reserve half of the telescopes' time for a modified version of classical observing. After flying to Hawaii or Chile, astronomers would control and monitor data collection from Gemini's base facilities in Hilo or La Serena.

The remaining nights would be set aside for queue observing done by the observatory staff. The goal was to match the observations of astronomers to the prevailing weather and seeing conditions. Astronomers would receive small blocks of time, creating access for more users and enabling the staff to make sure that the telescopes and instruments needed for that night's work were ready. "From the perspective of an observational astronomer, astronomy may be less fun, but surely, the big returns," Mountain wrote, "may be worth a change in culture."[22]

Dedication

First light with a new telescope is an arbitrary, largely symbolic, and sometimes anticlimactic event. Once achieved, scientists and engineers still expend months of often-frustrating effort to make the telescope's various systems work harmoniously. Gemini was no exception. In late December 1998, light from Jupiter hit the carefully polished surface of Gemini's 8.1-meter primary mirror. The telescope's million-dollar secondary mirror was not yet in place so light from the fifth planet reflected off the main mirror to a sheet of white paper where astronomers could see it. By using images of familiar objects like Jupiter, technicians began the long and delicate process of calibrating the system of computers and actuators that supported the main mirror.

A month later, on January 29, 1999, Gemini's secondary mirror was finally in place. The telescope's observing slit opened and the telescope quietly moved to point first at, appropriately enough, the stars Castor and Pollux in the constellation Gemini. The computer-controlled systems that Gemini's primary mirrors would depend on were not functioning properly yet, so Jim Oschmann and other engineers made manual adjustments until the star they were using to calibrate their systems transformed from an ill-defined elongated image on a computer screen to a relatively sharp point of light. At 13,000 feet closer to sea level, an exhausted Matt Mountain observed their progress on a computer terminal linked via fiber optics to the systems on the summit. "It was hard to get to this point,' Mountain told a reporter, "and there was a realization that the hard work had just begun."[23]

On June 25, 1999, the first of Gemini's two telescopes (Figure 12) was dedicated in a ceremony atop Mauna Kea. To hear remarks by the governor of Hawaii and Prince Andrew, 150 people trekked to the oxygen-depleted summit of Mauna Kea. While guests at the summit spoke, 300 more people viewed the ceremony via live simulcast—a reflection of how the telescope would actually be used—at the telescope's base facility in seaside Hilo, two and half miles lower and 50 miles away.

Astronomers received an indication of Gemini's capabilities when they temporarily obtained a small adaptive optics unit called Hokupa'a (the Hawaiian name for the North Star) and an infrared imager from the University of Hawaii and jury-rigged it to the telescope. They

Figure 12. The completed Gemini North telescope. Notice the open thermal vents and the telescope visible through the dome's partially opened observing slit. Courtesy of Gemini Observatory/NOAO/AURA/NSF.

pointed Gemini North toward the planet Pluto, six billion miles away, and easily resolved its moon, Charon, in orbit. Charon travels around the ninth planet at a distance of only about 9,000 miles so imaging it was proof that both the telescope's mirror support and borrowed adaptive optics systems were working properly. Pictures taken of NGC 6934, a globular cluster in the Milky Way some 50,000 light-years away, demonstrated Gemini's ability to produce high-resolution images. Without adaptive optics, Gemini could still resolve detail of about 0.6 arc-seconds, already better than the best nights at Palomar and enough to see a blurred cluster of stars. After astronomers engaged the adaptive optics system and opened Gemini's air vents to flush out the dome, the resolution improved dramatically to 0.09 arc-seconds. Individual stars in the cluster appeared as pinpoints of light, a feat comparable to resolving the individual headlights of an oncoming car some two thousand miles away.

On January 18, 2002, astronomers and science managers from seven countries arrived in La Serena to dedicate Gemini South and celebrate the completion of the entire observatory. In the morning, a bus caravan left La Serena and headed east through the Elqui Valley toward Cerro Pachón. Many astronomers had experienced the bus trip from La Serena as a prelude to observing runs on nearby Cerro Tololo. One person conspicuously absent was Fred Gillett. The infrared pioneer had died of cancer in April 2001 at age 64. To recognize his role in shaping Gemini's design and leading its science team, the Gemini Board recommended that Gemini's northern telescope be named in honor of this pioneering astronomer.

Guests milled underneath Gemini South's massive enclosure on its high rocky outcropping. The telescope's thermal vents were open and flags from Gemini's seven partner countries flapped in the breeze as classical music and sunshine lifted people's spirits. Gradually the guests seated themselves and waited while Chile's President, Ricardo Lagos, arrived via helicopter for a tour of the facility. Poster-sized images, including those recently produced by Gemini South's Hawaiian twin (Figure 13), added to the atmosphere.

Lagos welcomed Gemini South to the growing family of large telescopes dotting mountaintops throughout Chile. Like the speakers that followed him, Lagos praised Gemini for helping pave the way to future international scientific projects, including even larger telescopes in

Figure 13. Image of the large spiral galaxy NGC 628 (Messier 74) in the constellation Pisces taken during the "first light" observing run by the Gemini Multi-Object Spectrograph on Gemini North. A copy of this instrument is also on the Gemini South. Courtesy of Gemini Observatory/ NOAO/AURA/NSF.

Chile. Lagos's speech was not without humor. "If it is true," he said, "that the Big Bang existed . . . what is still unclear is why the Big Bang happened, and perhaps the person who has more answers is the Archbishop, who will proceed with the blessing of this observatory."

Unlike the dedication of the first Keck Telescope a decade earlier, those who came to celebrate Gemini's completion could not brag that the telescopes were the biggest or the first to enter service. Speakers at the ceremony instead described how Gemini was a transoceanic system for science research, linked together by high-speed data networks and fiber optic cables, that would provide astronomers with full-sky coverage. Many of the speakers also emphasized another special feature of Gemini—that it was built as an international collaboration between seven partners scattered all over the world. More than just geography separated these countries. Building Gemini, they said, required that its advocates transcend these countries' different historical traditions, capabilities, and ambitions in astronomy.

Gemini's dedication was significant and timely for the NSF. Soon after his selection as President, George W. Bush submitted his 2001 budget. Buried in its details was a request for the National Academy of Sciences to consider whether the NSF's astronomy "responsibilities" (in other words, its funding and facilities) should be transferred to NASA to help fulfill Bush's campaign promise "to make Government more results oriented."[24] Stunned by the idea, astronomers were relieved when the Academy concluded that the White House's suggestion was seriously flawed. At the same time, the report criticized how the NSF had handled some aspects of Gemini and it suggested that more "systematic and transparent project management" could have given the NSF greater leadership opportunities.[25]

Nonetheless, for AURA and the NSF, Gemini was a major management success.[26] The observatory was completed without any major scheduling problems and at a cost of about $184 million (an extra $8 million was added to the original budget of $176 million in 1997). Moreover, Gemini's completion bolstered the NSF's claims of competence in undertaking increasingly bigger and more expensive research facilities. At Gemini's dedication, Rita Colwell, the NSF's director, noted that the director of the U.S. Office of Management and Budget had recently praised the NSF as the "best managed agency in the U.S. government."[27]

Few guests at the dedication had performance metrics and Beltway politics in mind, however, when they visited the deck of Gemini's observing floor and saw the distant snow-tipped peaks of the Andes through the telescope's raised thermal vents. The floor of the dome rotated as the telescope pointed and slewed, giving at least one person some minor disorientation and a giddy sense of riding the world's most expensive carousel. The sheer immensity of an 8-meter telescope and the instrument cluster at its Cassegrain focus, impossible to document in press photographs, revealed itself as guests milled about taking pictures of Gemini's multimillion dollar mirror.

That evening, guests attended a dinner replete with celebratory irony. Matt Mountain and William S. Smith, Goetz Oertel's successor as AURA's president (Figure 14), presented a special recognition to Dick Malow, who "had worked tirelessly . . . to ensure that the money kept flowing, enabling us to hold the Gemini Project to budget and schedule." This was achieved, Mountain said, in a nod to Malow's earlier career, "under Dick's close scrutiny." The special recognition of Malow's role—the award cited his "rigorous approach to 'the numbers'"—reaffirmed that the completion of a large project like Gemini demanded not just scientific and technical accomplishments but political acumen as well.

Dedicating a telescope is partly a celebratory, even theatrical, event. In this grand tradition, speakers at Gemini's dedication drew attention

Figure 14. Matt Mountain (left) and Dick Malow at the dedication dinner for the Gemini Observatory in January 2002. William Smith, AURA's president as of February 2000, is at the right. Courtesy of Gemini Observatory/Kirk Puùohau-Pummill.

to the telescope as a symbol. In 1948, when the 200-inch was dedicated, Raymond Fosdick, president of the Rockefeller Foundation, described the new telescope as "the lengthened shadow of man at his best."[28] Fifty-four years later, Colwell struck a similar tone in her speech, one that reflected the values and challenges of the time. International science projects like Gemini, she said, were needed more than ever to help "transcend national boundaries and cultural divides."

Speakers at Gemini's dedication also offered visions of astronomical research that spanned the decades between the reign of telescopes like the 200-inch and the new era of large telescopes astronomers helped bring into existence during the 1990s. Colwell told a metaphorical story about astronomers' work that spoke of an era that was quickly becoming a romantic anachronism. Folklore, she said, held that champagne was discovered by accident when Dom Perignon tasted a bottle of wine that had refermented. Delighted with his discovery, she told how the monk shouted, "Come quickly, I am tasting the stars!" Afterwards, Matt Mountain displayed pictures that reflected a new era of how astronomers would use large telescopes—images of Gemini's control room with its video monitors, computer terminals, and other digital interfaces—and intoned, "Welcome to point-and-click astronomy."

Gemini Nights

In the 1990s, the international astronomy community anticipated doing research with a dozen or more telescopes, each much larger than the venerable 200-inch on Palomar. The new generation of large telescopes sported more glass and cost more money than all other telescopes in the world combined and represented the largest advance in astronomers' light-collecting power ever. It is too soon to evaluate the impact that these new large telescopes—Gemini, Keck, the Very Large Telescope, and so on—has had on astronomy as a science. It is, however, possible to describe what a night at a large telescope was like.

In 1997, as completion of Gemini neared, the observatory staff wrote a description of an "ideal night" at the twin telescopes:

> It is nightfall on Cerro Pachón and the system operator and staff astronomer are . . . discussing with his colleague in La Serena which mix of observations will make the best use of the night's conditions . . . In

the same room, an engineer is trouble-shooting an off-line instrument via videoconference link to Mauna Kea . . . [Later] as the service observer on Cerro Pachón is starting an infrared spectroscopic observation of a high redshift galaxy, it becomes obvious that there is something peculiar about its emission lines . . . the PI [monitoring the process from England] decides to log on from home and look at the extracted spectrum himself. The next observation is listed as classical remote observing and the observers in Tucson and São Paulo have been waiting patiently to connect to the system and observe. In Tucson the PI dials up the Gemini ISDN number and establishes a connection to the site—giving her control of the science operations . . . The same video links allow both the observers in Tucson and São Paulo to interact with the System Operator and the Service Observer on Cerro Pachón. Next month, the two observers are scheduled to use the remaining four hours of their allocation to repeat the same observations on a northern sample, using the Mauna Kea Gemini Telescope.[29]

Gemini's transition to this level of efficiency and convenience took a considerable period of time. Engineers tuned and tested Gemini North for months before handing it over for part-time science operations in mid-2000. For its first semester of science operations, the observatory received over 180 proposals for time, despite only having the most basic instruments mounted under the main mirror. The requests for time represented more than four times the number of observing hours available, a good indication of the high level of interest the astronomy community had in Gemini.

The application for time on Gemini was modeled on the complex process used for the Hubble Space Telescope. An interested astronomer submitted an initial Phase I proposal to his or her country's "National Gemini Office." Each partner country had one. The American office was located at NOAO in Tucson, the last remaining link between the national observatory and Gemini. In addition to a scientific justification, the astronomer's proposal had to include the amount of time needed and the types of objects to be observed. Twice a year, the national office ranked proposals in two lists, one for classical and one for queue observing.

These lists were sent to Gemini, where they were merged, using a set of metrics and algorithms, into initial queue and classical schedules.

Gemini's directors and a second, more powerful international committee created the final schedule. Each partner in Gemini had a representative on the international committee, and Mountain compared the two time allocation committees to the U.S. House of Representatives and Senate, respectively.

Developing the final set of proposal rankings was an inexact process, and it was unlikely that each partner country would receive exactly the amount of time it was due under the partnership. The U.S., for example, had about 117 nights on each telescope. Gemini's goal was to make the time each partner received proportional to its financial contribution over the long run.[30] The international committee first ranked proposals into four groups. To help eliminate favoritism and conflicts between partners, all proposals in each group were later randomly shuffled. The observing programs would then be executed in order of the group's ranking—those in the top group were done first, and so forth.[31]

Meanwhile, successful applicants prepared more detailed Phase II proposals. If the Phase I proposal was comparable to a movie plot in storyboard format, the Phase II proposal was like the detailed instructions given by a film director regarding camera position, lighting, and such. With assistance from Gemini staff scientists, the astronomer used standardized software provided by Gemini to create a detailed set of instructions that told the telescope and its instruments what to do: an exact list of objects to observe and their coordinates, what filters to insert into the instrument, nearby guide stars to use in conjunction with the adaptive optics system, how long to observe each object, and an exact description of how the telescope itself would move. For example, an observer might ask the telescope to move over a square that was an arc-second on each side for 300 seconds, move off to take an image of the sky background for a few more minutes, and subtract this signal from the data before moving to the next object. For people new to the process of using a large telescope like Gemini, preparing the Phase II component could be time-consuming and intimidating. Like dancers with favorite steps, those scientists who became familiar with the process developed sequences of preferred "telescope moves" that they used routinely as part of their nightly work.

An astronomer arriving at the headquarters of the Gemini Observatory in Hilo, perhaps coming to monitor her observing run scheduled for that night's queue, found a single-story building marked by flags from Gemini's seven partner countries. Gemini's headquarters was located on a veritable "Observatory Row" near the buildings for the United Kingdom's Joint Astronomy Center, Japan's base for the Subaru telescope, and a building for the University of Hawaii's Institute of Astronomy. On a clear day, it was possible to see the nearby Pacific Ocean as well as glinting white and silver telescope domes on Mauna Kea.

A person entering the headquarters building immediately encountered Gemini's Operations Center, a large room with extensive windows through which guests could watch the staff and visiting astronomers at work. Filled with rows of computer monitors and workstations, the room was air conditioned amply enough to prompt one visitor to wonder if an effort was being made to recreate the frigid conditions on Mauna Kea's summit. Large television screens provided real-time video links between the telescope's control room at the summit and the base facility. Gemini staff entered and left the control room at all hours of the day as the observatory was operated continuously by three shifts of staff astronomers and technicians. The first shift, from late morning to early evening, was taken by a staff scientist. Using scheduling tools and weather-forecasting models, this scientist made plans for that night's observing run and attended to crucial details such as establishing equilibrium between the temperature of the telescope's mirror and the ambient air.

At around six p.m., the staff scientist passed control of the telescope to the nighttime observing team. For queue-observing runs, there might be just one staff astronomer working at sea level and communicating via phone, e-mail, and video link with the support staff at the summit who actually controlled the telescope and its instruments. Many of the specialized staff at Gemini were system support associates. In classical observing, these people "drove" the telescope for the astronomer. The SSAs took on more specialized roles at Gemini and were not simply telescope grunts but rather professional members of the observatory staff. Because operating a telescope such as Gemini was now a much more complex operation, the nighttime role of the SSAs—Gemini had six available at each telescope—was more demanding. At the beginning of the night's observing run, for example, one SSA adjusted the telescope during a short pointing test and "tuned" the primary mirror to remove any residual optical aberrations and adjust it to the temperature of the night air. The SSA then pointed the telescope at a target and began tuning the wave front sensors that helped keep the mirror and telescope aligned and calibrated throughout the night.

As observing commenced, astronomers at sea level could monitor the process in real time. Data collected by the telescope was instantaneously displayed on computer screens in Hilo and at the summit. Staff astronomers pulled up the Phase II computer files that formed the basis of instructions that went to the telescope, the detectors, and the instruments. These electronic files allowed the staff to communicate with and control the telescope's different systems while the main mirror collected light and converted it into information. Meanwhile, an SSA monitored the status of the telescope and the outside conditions. The SSA alone had the authority to shut down for the night if the wind was too strong or if dangerous snows started on the mountain.

Links between staff at sea level and those at the summit were not without occasional frustrations. In Hilo, it was easy to forget that the summit staff breathed 40 percent less oxygen. This could sometimes cause short-term memory lapses, a situation exacerbated when people at both places were working together on complex tasks like debugging an instrument's software.[32] Once the staff completed the night's observing run, the telescope was handed over to the third shift at sea level while astronomers and operators at the summit drove down to a dormitory at about 9,000 feet.

From about dawn until around noon, the third shift of SSAs monitored the shutdown of the telescope, discussed engineering problems with the technical staff, and began to archive the gigabytes of data collected the previous night. Information collected by the telescope went through what was called a "data pipeline." This used standard software and mathematical algorithms to reduce data for each observing proposal to a standard format. When the data emerged from the pipeline, astronomers could log on to the database from wherever they were based, view the digitized information, and get a rough sense of how successful the night's work had been. Mountain referred to Gemini's system of information management as "data flow," a term in use at other astronomy institutions like STScI and ESO. The data flow, he said, was "an overall description of how you start with an idea and end with a data set in your office." The goal was to link seamlessly all of the different stages of making an observation—having an idea, applying for time, executing an observation, processing the data, and assessing the observatory's success—rather than having these be discrete activities.[33]

Observing with a telescope like Gemini, even for classical-style observing, was a team-oriented process requiring contributions from many different people. The staff included, of course, Gemini's director and two associate directors who oversaw the management and productivity of the two telescopes as a single integrated observatory. Gemini hired new Ph.D. astronomers for fixed short-term appointments in order "to create the new generation of astronomers"—presumably those familiar with using large, modern telescopes—who might "cycle back into the user community."[34] In addition to supporting Gemini's operations, these "Gemini Fellows" were granted observing time for their own research. Gemini also employed senior staff astronomers to support Gemini's operations and do personal research. In all, Gemini required a myriad assortment of engineers, computer programmers, and instrument gurus—some 100 people overall—to function in a productive manner. Like the organization of scientific work established in big factory-like accelerator laboratories, the operation of the new large telescopes was based on close interactions between scientists, engineers, and other technical experts. This entailed a greater amount of task specialization and accompanying hierarchical organization.

Stéphanie Côté, a young Canadian astronomer, was one of the scientists who supported Gemini as it made the transition from engineering to science operations. After getting her Ph.D. from the Australian National University, where she spent considerable time using modest-sized telescopes, she helped operate ESO's 3.5-meter New Technology Telescope in Chile. Like many astronomers of her generation, Côté did not consider herself to be an optical astronomer exclusively. Instead, she freely used observations taken at a variety of wavelengths at ground and space-based telescopes to address specific scientific problems, joining the movement away from what Malcolm Longair called "wavelength chauvinism."

In 2000, Côté was working at the Herzberg Institute of Astrophysics in Victoria where she was part of Canada's Gemini Office. In November, she spent three weeks helping make service observations with Gemini North as part of the commissioning of a new instrument called OSCIR. Built by a team at the University of Florida, astronomers had used it at other telescopes before its builders loaned it to Gemini in exchange for observing time. OSCIR could detect mid-infrared radiation from about 8 to 25 microns and use it to produce images and low-resolution spectra. Côté's task was to help get the instrument working on Gemini and try to collect some useful science data. She spent a considerable amount of her time at the summit of Mauna Kea learning about the OSCIR's quirks from the Florida team and Philip Puxley, a British astronomer who later became the associate director of Gemini South.

The first few nights were "total chaos," according to Côté. The telescope exhibited many problems. Most were minor but, combined with software errors, they made integration of the delicate new instrument with an already complex telescope more difficult. Tempers flared and several nights' worth of observing were lost as astronomers at the summit and Hilo attempted to diagnose problems and determine which were caused by the telescope and which were instrument-related. Finally, the Florida and Gemini staff resolved the problem and began collecting data on the newly installed instrument.

Côté also helped match observers' requests with the weather and sky conditions to create the queue for the night's observing. A typical night started around 4 p.m. when she discussed with technicians and the SSAs via teleconference the previous night's observing run and any new telescope problems found during the day. She then helped plan the next night's observing run before setting off to the summit to arrive before sunset. Eager to get some science data collected, she and the other staff tended to select observing proposals that looked to be fairly easy to do from the highest-ranked group. If the conditions weren't right for the top programs, they moved down the list until they found a suitable program to execute.

The data Gemini staff collected with OSCIR formed the basis for Gemini's first major science result, announced in a press release the following year, as had become the norm at all major observatories. A team of seven researchers, led by Eric Perlman, an astronomer at the University of Maryland, used OSCIR to look at M87, an elliptical galaxy in the constellation Virgo some 50 million light-years away.[35] While Gemini's secondary mirror subtracted the sky background, OSCIR's mid-infrared detectors peered through the dust that obscured optical radiation at the faraway galaxy for seven hours and produced several images. These were the most sensitive mid-infrared images yet taken with a ground-based telescope.

Astronomers had been interested in M87 for several decades. A massive jet of dust and gas appeared to stream away from it for thousands of light-years. Most theorists believed this plume was generated by a massive black hole at the galaxy's center whose signature, a bright optical and infrared emission in the region around the black hole, Perlman and others had imaged with the Hubble Space Telescope in 1998. The jet coming from M87 originates in a disk of superheated gas swirling around this black hole and is propelled and concentrated by intense, twisted magnetic fields. The light that can be seen is produced by electrons twisting along magnetic field lines in the jet, a process known as synchrotron radiation. With OSCIR, Perlman's group saw several bright knots in the synchrotron jet emanating from M87, which they compared with images produced by radio, optical, and x-ray telescopes.

Until the observations that Perlman's team made of M87, astronomers thought all black holes were surrounded by a massive doughnut-shaped cloud of warm gas and dust, which they had seen before. This dusty torus absorbs high-energy radiation from material that is heated before it falls into the black hole and re-emits it at infrared wavelengths, but its small size and location in M87's dusty environment prevented astronomers from imaging it. Contrary to expectations, the observations Perlman's team made did not show the bright thermal emission they expected. Instead, the torus region in M87 was about a thousand times fainter than the jets coming out from it when compared to other well-studied active galactic nuclei. The observations also suggested that, if there were a torus around M87's black hole, it contained far less material than other similar objects.

Perlman's observations earned him (and Gemini) press coverage in major science journals. Just as importantly, the observations showed Gemini's former detractors and future user community that it was capable of producing good science. At the June 2001 meeting of the American Astronomical Society, Gemini had its scientific coming-out party. Astronomers, including Perlman, presented several papers based on data collected at Gemini to an appreciative and interested audience of about two hundred people, many of whom would go on to apply for observing time themselves. For Mountain, discoveries such as Perlman's were proof Gemini was an excellent infrared telescope. They also suggested research areas in which Gemini could outperform more established facilities like Keck and Hubble.

After returning home, Côté reflected on her experience using Gemini. Because Gemini was still being debugged, she found it "still very capricious. But when it was working it was great."[36] Compared to the smaller telescopes she had used before, Gemini was quite different. "The bigger the telescope," she said, "the more removed you are from its workings." With a smaller telescope, Côté would do everything observing at night required. In contrast, Gemini needed several people to do what she called the "direct talking to the telescope." As the astronomer in charge, Côté described her role as comparable to the "manager of a team" who made choices such as what targets to point at or what system calibration to run. "Bonding with the telescope," as she phrased it, still took place. This, however, happened not so much while using the telescope but rather when she went into the dome, saw the stars reflected in the giant mirror, and watched its shadowy form in the dark, its motion controlled by computers. "The process of observing has changed quite radically," Mountain commented. "If you don't invest time in learning the complexity, you find the observing has gone beyond you. It's the machine that does the observing, not you."[37]

Conclusion: Telescopes, Postwar Science, and the Next Big Machine

On Valentine's Day 2002, I visited John Huchra at his office at the Harvard-Smithsonian Center for Astrophysics. Radio stations that morning had announced the death of Waylon Jennings and, as I took off my coat and took out my tape recorder, the astronomer played a song cowritten by the country singer. "The Highwayman" tells of a bold and solitary spirit sailing a starship across an eternal divide. "That," Huchra reflected, "is what astronomy is all about."

Huchra had come a long way since his days as a graduate student at Caltech when Jesse Greenstein flunked him during his Ph.D. qualifier. The embarrassment, Huchra claimed, made him set a goal (reached in the early 1990s) of outpublishing Greenstein.[1] By 2002, over 500 papers featured Huchra's name, making him one of the most cited astronomers in history.

At Caltech, Huchra learned how to use a telescope by starting small, working with faculty mentors and other students, and getting what he called a "feel for the organism." "As an observer, one of the key things that enabled me to do things was that I spent so much time looking at the universe," he said, "through a telescope." By 2002, he probably had spent as much time at a telescope as anyone else in the world. For Huchra and many others, the solitude of nighttime scientific research was part of why they became astronomers in the first place. Macho phrases such as "attacking the night" and "being in the slot" animated Huchra's descriptions of what telescope observing was like. "There is a certain romance in doing astronomy," Huchra noted, "in going to interesting places, and being one person against the darkness while trying to discover things."

To astronomers like Huchra, the advent of new observatories like Gemini and the Very Large Telescope represented the closing of an era and, with it, a loss of the romance they associated with telescopic observing. "One of my biggest fears is that the new generation of telescopes, which is wonderful for astronomy, may be bad for astronomers. In the sense that we are at the stage where they don't want to go to the telescope," Huchra said. "I think it's going to change who becomes an astronomer. I think we'll lose the romantics."[2]

In the introduction to this book, I note that many characteristic features of high-energy physics are abnormal in science. Big Science as exemplified by endeavors like particle physics, the Human Genome Project, and fusion experiments is but one extreme toward which the scientific enterprise can move. If one accepts the traditional metrics (copious funding, big machines, and large teams of researchers) that demarcate what historians and scientists have broadly defined as Big Science, then it is clear that traditional optical astronomy has gravitated steadily in the last 30 years toward this pole.

As a student, Matt Mountain helped write software code for a big physics experiment at CERN. Mountain later found that using a small telescope in the Canary Islands was more to his liking. Mountain recalled that at CERN "you ended up as a small part of a very big team. In 20-20 hindsight, this is now the model I am promoting for astronomy. The irony is rich."[3] Mountain's perspective, as Gemini's lead scientist and then its director, about the path astronomy had taken in the postwar period was not unique. The consensus of most astronomers I interviewed was that they had experienced a metamorphosis of astronomy that began after 1980 and continued accelerating through the 1990s. Scientists freely acknowledged optical astronomy's resemblance to high-energy physics, as evidenced by larger research teams addressing common research problems, an increased reliance on automated and computerized tasks; international collaborations to build bigger and more expensive equipment; and a growing separation between observers, instrumentalists, and theorists. They experienced this transition most noticeably during the period in which they planned, designed, and built the current generation of new telescopes. Many scientists believed that astronomy's growing resemblance to particle physics simply reflected the field's inevitable "natural evolution."[4]

After the dedication of the 200-inch on Palomar, the federal government became American astronomy's major sponsor. Successfully developing support for a new observatory required considerable community activism, something that physicists learned long ago. In the postwar political economy, the decadal reviews prepared by senior astronomers under the guidance of the National Academy of Sciences became powerful tools that helped determine which new facilities would be built. Only with the enlistment of patrons to provide money, advocates to guide the project through difficult times, and shrewd political lobbying to offset any opposition could large telescopes begin to emerge from the realm of blueprints and meeting minutes. Failure to have all of these elements in place could have catastrophic consequences, as witnessed by the demise of the National New Technology Telescope or, more visibly for physicists, the 1993 cancellation of the Superconducting Super Collider. Gemini's completion was as much a triumph of political and managerial proficiency as of scientific and technical prowess.

At observatories like Gemini, nightly use of a large telescope became a coordinated effort involving many people with specialized skills and defined roles. The individual research astronomer was no longer a central or even essential part of the actual data-collection chain. John Huchra predicted that the introduction of more complex instruments and adaptive optics systems would push optical astronomy further toward the pole exemplified by high-energy physics. "On a really big telescope you might end up having five or six people sitting at different consoles, each of whom [is] operating different pieces of the telescope. You'll have a telescope operator whose job is primarily to point the telescope," he said. "You'll have an adaptive optics specialist whose job will be to keep the system running, including perhaps an artificial star laser. You'll have somebody who is a mirror specialist. Then you'll have the astronomer and the data reduction analyst sitting there, all at the same time."[5]

In some cases, astronomers explicitly compared their new observatories to particle accelerators. "The telescope," said Mountain, "is almost like an accelerator these days. We produce high quality, corrected beams of light pointed in the right direction at good instruments and detectors and collect the data."[6] Modern telescopes, of course, are not identical to accelerators. An accelerator creates its own transient phenomena, which scientists analyze. Astronomers are not creating novel experimental conditions at the observatory. Their work, whether done in person or remotely, remains intimately connected to phenomena as local as cloudy night skies.

Teams of interdisciplinary workers became, of course, one of the defining features of postwar large-scale physics. Alan Thorndike of Brookhaven National Laboratory's Cloud Chamber Group remarked in 1967, "The experimenter then is not one person, but a composite. He might be 3, more likely 5 or 8, possibly as many as 10, 20, or more . . . One thing, however, he certainly is not. He is not the traditional image of a cloistered scientist working in isolation at his laboratory bench."[7]

Most publications of contemporary astronomers, unlike particle physicists, are not written by scores of authors whose individual contributions are difficult to determine. The average number of authors on an astronomy paper (including theoretical papers, which tend to have fewer authors) began to increase noticeably after 1970 but, in 2000, still remained just under four.[8] There were noticeable exceptions, however, as astronomers formed large teams to tackle specific scientific problems. In the late 1998, for example, two groups of rival scientists announced that distant supernovae were farther away than expected. This discovery provided evidence that the expansion of the universe was accelerating and was called the breakthrough of the year by *Science* magazine.[9] Over 40 scientists in all shared the honor. With astronomers undertaking more big projects, such extreme examples of multiauthor papers combining the talents of observers, instrument wizards, theoreticians, graduate students, and postdocs promised to become more commonplace.

In the 1960s, particle physicists adopted a host of prepackaged programs to accommodate the increasing amount of data produced by accelerators and recorded by bubble chambers. These data analysis programs, many of which were automated, were widely distributed among laboratories. The programs commandeered the "sanctum sanctorum of the physicist's physics," taking over tasks that previously had defined what it meant to be an experimental physicist.[10] In the early 1980s, similar prepackaged software packages designed for astronomy appeared.

Astronomers at Kitt Peak National Observatory helped lead the way in developing community software that offered standardized tools for data reduction.[11] By the end of the decade, scientists around the world had adopted programs like Kitt Peak's IRAF (Interactive Reduction and Analysis Facility) as part of their basic toolkit, despite comments from some scientists that they represented a form of "creeping socialism."[12]

The distribution of communally shared software tools was part of astronomers' reaction to the increasingly digital nature of the data their telescopes collected. Programs like IRAF reduced data to a common format and helped pave the way for electronic data archives. Data collected by national facilities like the Hubble Space Telescope and Gemini was placed, after an initial proprietary period, in archives that all scientists could access. These repositories not only served as permanent resources for scientists, but they offered the potential of improved observatory management by allowing statistical studies of telescope use.[13]

Enthusiasm and debate about digital data archives was fueled by articles in leading science journals describing what astronomers began to call the National Virtual Observatory. The idea was to pull together the data collected by telescopes at many different wavelengths from around the world (and from space) and deposit it in a single archive that any astronomer, regardless of institutional affiliation, could search. The possibility of bigger survey telescopes was one impetus for the NVO. The Sloan Digital Sky Survey came online in 2000 and released its first data several months later. Meanwhile, Roger Angel and other astronomers proposed an even more ambitious Large-aperture Synoptic Survey Telescope. Using a special optical design, its 8.4-meter mirror would collect over a dozen terabytes of data every night, enabling research on the distribution of dark matter" and perhaps alerting astronomers to the presence of near-earth asteroids. The increased capability and reduced cost of computers, the emergence of the Internet and the Web as legitimate research tools, and the exponentially growing flood of information collected nightly by new telescopes all drove astronomers' interest in the NVO.[14]

NVO advocates gushed about its possibilities and journalists described it with phrases such as "mining the sky." "For all the clever people who don't have access to a big telescope, NVO will allow them to do first-rate observational astronomy," said George Djorgovski of Caltech. Another scientist claimed it could "lead to a true democratization of astronomy."[15] Not everyone was so sanguine. One astronomer predicted that virtual observatories would only help "breed a generation of astronomers who sift through data without knowing about instruments."[16]

In May 2000, the National Academy of Sciences released its fifth decadal survey. Chaired for the first time by two astronomers, Nobel laureate Joseph Taylor of Princeton and Christopher McKee from Berkeley, the committee recommended spending $60 million on the NVO over the next ten years to link "major astronomical data assets into an integrated, but virtual, system."[17] This endorsement gave further impetus to astronomers' plans for some form of an easily accessible national database of astronomical research.

Like scientists' plans to use their telescopes in new ways, discussion about virtual observatories sparked debate over what it meant to be an astronomer. The NVO, one advocate claimed, would make getting data and doing routine observations from one's own telescope "a thing of the past" because "astronomers will instead get their data through data bases." Stories reported in major journals like *Science* and *Nature* fed concerns among astronomers about the future of their field and the entire nature of scientific discovery. Where once being an astronomer meant adhering to a demanding and uncomfortable nighttime observing schedule, science journalists remarked that "astronomers keep the same hours . . . and go through the same motions as everyone else: they come to their office in the morning, sit down at a work station, and surf the net."[18] This was a far cry from the romantic image of the solitary watcher of the cosmos presented a half-century earlier.

Debates among scientists about how to deal with the growing torrent of data from new and bigger machines represented a stage of development in the astronomy community that physicists had passed through long before. In the 1960s, physicists held opposing views about how they should analyze the thousands of events recorded in bubble chambers, and competition between human-centered and fully automatic data readers was contentious. Berkeley's Luis Alvarez placed a premium on people personally interacting with the data on at least some level via computing and measuring devices. At CERN, however, the emphasis was on using automated systems to read bubble chamber photographs, replacing human intervention altogether. As Alvarez noted in 1963, in choosing a path, one must consider both "economic factors, and one's basic philosophy about the interaction of physicists with their own data."[19]

Astronomers' contemporary debates about telescope use, virtual observatories, and data mining are strikingly similar in tone and content to those of physicists forty years earlier. At a 1964 meeting of particle physicists, one scientist commented that mining data bases "frightens me a little because it would mean that in a few years if one wants to do a high energy experiment one would not go to start a new experiment but one would just go to the archives, get a few magnetic tapes, and start to scan the tapes from a new point of view. That would be the experiment." Another physicist commented, "It is important for the physicist to examine his data to look for the unexpected. If the results come only exactly as planned, very little is learned." The same might have easily been said by an astronomer in 1998 arguing in favor of classical observing.[20]

The parallels noted thus far between contemporary ground-based astronomy and a "pathological" example of Big Science like high-energy physics exist mainly within the confines of the observatory. They are connected to issues such as the best way to use telescopes, the autonomy of the individual researcher, control over equipment, how papers are written, and the handling of larger amounts of data. What similarities do we see when we consider the broader history of science in the twentieth century? One of the most salient is scientists' persistent eagerness for what may be called the "Next Big Machine."[21]

In the early twentieth century, George Ellery Hale's efforts to obtain new and ever-larger telescopes for the Mount Wilson Observatory were remarkable for many reasons. Chief among these was that Hale, an inveterate planner and consummate politician, was constantly thinking about the Next Big Machine. In 1908, when the 60-inch telescope on Mount Wilson was completed, Hale had already begun maneuvering for the 100-inch. Once this was built, he lobbied the Rockefeller Foundation for money to build the 200-inch.

In physics, Ernest O. Lawrence followed a similar pattern. In 1931, the cyclotron was a table-top instrument that a scientist could build for less than $1,000. Through his charisma and his association of bigger tools for physics with the possibility of medical cures, Lawrence opened the coffers of the Rockefeller Foundation and the state of California to build new machines at Berkeley's Radiation Laboratory. By 1939, the cyclotrons Lawrence oversaw at the Rad Lab were massive devices costing a thousand times more money than their predecessors.

The efforts of Hale and Lawrence to build bigger instruments suggest more than just a superficial resemblance between astronomy and large-scale physics. This history not only reflects truths about science instruments that remained valid in the immediate postwar period but may also shape the efforts of astronomers who build the Next Big Machine in the twenty-first century. First, scientists desire new and bigger instruments in the belief that they enable new discoveries. Second, the length of time it takes scientists to raise funds for, design, and build new and larger facilities is significant enough that planning for the Next Big Machine often must take place before scientists have exhausted the usefulness of current instruments. Third, new and bigger machines, whether for astronomy, space research, oceanography, or particle physics, may catalyze changes in a research community's sociology and priorities. Perhaps most importantly, achieving the Next Big Machine is often critical not only to scientists' careers but also to the institutional health of observatories and laboratories.

These issues warrant further scrutiny in light of astronomers' recent successes in building the current generation of large telescopes like Keck and Gemini as well as their appetite for even more ambitious telescopes. It is clear that building new and bigger research facilities is seen by many researchers—physicists, astronomers, and others—as part of the continued expansion and health of their discipline. What are the effects, however, of failing to build the Next Big Machine?

The concept of a national system of facilities for ground-based optical astronomy was controversial from the moment Leo Goldberg and others in the 1950s advocated for it. The continued expansion and federal support of Kitt Peak and Cerro Tololo drew the ire and concern of people like Jesse Greenstein and many of his colleagues at institutions with access to privately operated telescopes. The institutional health of the U.S. national observatory was greatly affected by its failure to build the 25-meter Next Generation Telescope and later the 15-meter National New Technology Telescope. The final blow came when the Gemini board formally severed the connection between the twin 8-

meter telescopes and the National Optical Astronomy Observatories in the early 1990s, thereby eliminating any control NOAO might have had over building a bigger telescope facility. As Goetz Oertel noted, "Gemini does not prop up the national observatory the way a national 8-meter project would have done and the way its four meter telescopes did."[22] NOAO's failure to build the Next Big Machine froze it in an era of 4-meter-class telescopes, while disagreement over the observatory's mission fostered confusion and rancor in the science community.

For those scientists who resented the NSF's support of NOAO or believed resources should be spent elsewhere, the separation of Gemini from the national observatory was a positive step. Less obvious is what it was a step toward. From the mid-1970s onward, the astronomy community has been divided by questions about the role that AURA and the national observatory should take (or be given). In 1976, Leo Goldberg proposed a more prominent role for Kitt Peak that entailed operating even bigger national telescopes. Anger from astronomers at places like Caltech helped curtail this ambition. A decade later, AURA's advisors recommended the national observatory pursue a "first among equals" approach. Ultimately, this philosophy led to no new, larger facilities for the national observatory.

Despite their intentions, these efforts did not resolve the entrenched dissension within the astronomy community over enduring issues such the role of the national observatory, the types of facilities the national observatory should operate, and how access to national telescopes should be granted. In the mid-1990s, the debate surfaced once again.

One influential viewpoint was offered by Sandra Faber of the University of California (an institution with considerable access to private facilities) in a widely circulated letter.[23] Faber, who did her dissertation research at Kitt Peak, was struck by differences between the national radio and optical facilities, especially how these were perceived by their constituencies. The national optical observatory had never got "squarely out in front of the technical challenges" posed by astronomers while its user community was "more fragmented, less senior, less committed," circumstances that, according to her, ultimately circumscribed its influence. NOAO, placed in the unenviable position of competing to build cutting-edge facilities with groups that had substantial private resources while still serving those without their own telescopes, confronted a paralyzing tension. One solution, Faber suggested, was for the NSF to "give away all existing NOAO facilities" to university and private consortia, which would presumably operate them more efficiently.

These criticisms damaged morale further at the national observatory and outraged some astronomers, especially those who relied on the national telescopes. As a former president of the American Astronomical Society from a large Midwestern school said, "They [astronomy's "haves"] don't give a flying fuck about the rest of us. They would just as soon take it all."[24] Steve Strom, as chair of Massachusetts's Five College Astronomy Department, formally responded to Faber's critique. Historically, astronomy's "most senior and influential voices" had refused to accept that the national observatory could take a "lead role in developing frontier facilities." Enumerating how the elitism inherent in Faber's proposal would be detrimental to the overall community, Strom opined that American astronomers "must squarely face a choice between two different visions of the future." The optical community could either emulate American radio astronomers and the international optical community and support a healthy national observatory or return to an era when "a privileged few" dominated the field.[25]

Responding to the renewed sense of confusion and crisis (which in the eyes of some had never gone away), the National Academy of Sciences released two reports within five years that addressed NOAO's future.[26] The first report, issued in 1995, concluded that the NSF's support of optical astronomy needed to give highest priority to Gemini, even if this meant closing smaller telescopes at Kitt Peak and Cerro Tololo. NOAO, it concluded, was not nor would be preeminent in all areas of optical astronomy. Instead, it should concentrate only on those areas "where it has the best chances to assert scientific leadership."[27]

The 2000 McKee-Taylor decadal survey reached an even larger audience. Academy reports for fields like astronomy frequently clothed criticisms and recommendations in bland language. In contrast, the McKee-Taylor report surprised many with its candor. Of the three components that comprise America's national observatory system—radio astronomy, solar astronomy, and optical/infrared astronomy—the first two were functioning well. However, the latter, according to the survey, was essentially broken.

While scientists had come to accept and rely on national observatories to do research in radio and solar astronomy, optical astronomy still enjoyed many privately funded telescopes that competed very successfully with the national system. This weakened the national observatory's influence and reduced its potential importance. From the 1970s onward, influential members of the optical astronomy community (unlike radio astronomers) had rejected plans for larger national facilities at Kitt Peak and elsewhere. NOAO's poor health was one unfortunate result of the failure to achieve unity and consensus and build the Next Big Machine.

Such ill effects were not limited to American astronomy. In the United Kingdom, the rejection of plans championed by Alec Boksenberg and the Royal Greenwich Observatory to place the next big British telescope on La Palma had contributed significantly to the decline and eventual closure of that venerable institution in 1998. The results of not building (or not being able to build) the Next Big Machine were not confined to astronomy, either. The history of high-energy physics offers numerous examples in which the failure to construct another more powerful accelerator spelled doom and suggests another, albeit less sanguine, link between the two fields. In the 1980s, for example, Brookhaven National Laboratory was in the midst of developing an innovative and expensive new particle accelerator called ISABELLE. Difficulties with the proton-proton collider's design and subsequent cost and schedule slips led the Department of Energy to kill the program and, ironically, transfer support to the Superconducting Super Collider, itself eventually canceled. The effects on Brookhaven were severe, far-reaching, and caused, among other things, a major shift in the lab's priorities from high-energy to nuclear physics.[28] Instances from other scientific fields—space science, gravity wave experiments, oceanography—all testify that the health of scientific institutions (and even scientific communities) can be hurt when they fail to build the Next Big Machine.

Questions about the Next Big Machine go beyond the declining fortunes of particular institutions. First, not all astronomers believe that a larger telescope equals a better telescope. Recall that in the 1970s, when Goldberg and astronomers at Kitt Peak were considering designs for the Next Generation Telescope, one possibility was a suite of many small telescopes. Supporters of this proposal argued that small telescopes were a way to increase the community's light-collecting area in a democratic and cost-effective way while avoiding the path high-energy physicists had taken in clustering around a few large laboratories.[29] Advocates of this view did not persuade the astronomy community that smaller might be better.

As astronomers won support plans for ever-bigger telescopes in the 1980s, small telescopes continued to rack up remarkable scientific discoveries. Most of the data Huchra and Geller used for the Cambridge redshift survey was collected with relatively small telescopes. Astronomers also detected the first extrasolar planet using only a 1.9-meter telescope. Moreover, statistical studies suggested that the majority of astronomical publications (up to 75 percent) were still based on data collected at ground-based optical telescopes smaller than 4 meters. Helmut Abt, a longtime Kitt Peak scientist and former editor of *The Astrophysical Journal*, concluded that the prevailing tendency of National Academy studies to recommend closing smaller telescopes in favor of supporting and building bigger ones was "unrealistic, biased, and a disservice to astronomy."[30]

Such evidence suggests that there was nothing inevitable or natural about the gravitation of astronomers toward bigger telescopes and other features commonly found in high-energy physics. To believe otherwise is to embrace a form of determinism. Telescopes, after all, are social constructions built by astronomers and engineers. The decision to build bigger is not imposed on the science community. One has to ask whether building bigger telescopes remains a sustainable strategy as they, their instruments, and their operating costs become more expensive.

The international astronomy and telescope engineering communities, in many ways, had surprised themselves by successfully summoning the will and resources to produce the biggest burst in telescope construction ever. In twenty-five years they had effectively increased the amount of aluminum-coated glass trained on the heavens more than twenty-fold. Just as remarkable were the successes of astronomers and engineers in overcoming the technical difficulties posed by their desire to increase the size of their telescopes' light-collecting areas while simultaneously keeping cost and weight at a minimum. By 2000, astronomers had three different and apparently successful technologies around which they could build large telescopes. Jerry Nelson's segmented design was working nightly at the Keck Telescopes. Gemini helped prove critics of lightweight thin meniscus mirrors wrong. And, in January 1997, Roger Angel's Mirror Lab finally cast the first mirror for the Large Binocular Telescope, an 8.4-meter piece of glass that was the world's largest monolithic mirror.

In the late 1990s, astronomers began to discuss the Next Big Machine for ground-based astronomy and their plans gained momentum. The McKee-Taylor survey committee gave substance to rumors and lunchroom talk when it recommended a 30-meter telescope as a major American priority for the next decade. Meanwhile, ESO and others in the European community were seriously considering the construction of OWL, a clever acronym for the Overwhelmingly Large Telescope, an astronomy facility with a light-collecting area 100 meters in diameter.[31]

For telescopes of this size, even Angel's Mirror Lab could not hope to compete. The segmented design used in the Keck Telescopes was the only route to facilities of this scale. Their projected costs were as shocking as astronomers' ambitions—estimates for the Next Big Machine ranged from $700 million for a 30-meter to over $1 billion for OWL. As a result, astronomers could not anticipate having a dozen or more of these behemoths at their disposal, as they do now with 8- to 10-meter telescopes. If construction and operating estimates were accurate, only a few such facilities could be financed worldwide. Simply put, not everyone can afford to build bigger and better telescopes. This raises questions about how the world's astronomers might share observing time at such hypothetical facilities and suggests another instance in which astronomy is gravitating toward the model of high-energy physics.

Or does it? On October 28, 2001, the *New York Times* ran the headline "Intel Founder Gives $600 Million to Caltech." Semiconductor pioneer and Caltech graduate Gordon Moore set a new record in philanthropy when he and his wife announced the largest gift ever to a university. Caltech's president, David Baltimore, predicted the money would strengthen the university's research in existing programs. One of the items reputed to be on Caltech's wish list was a giant new optical telescope to be shared, like Keck, with the University of California.[32] A year later, Caltech and the University of California were funding design studies at several million dollars a year while California astronomers were eagerly exploring how a new privately owned 30-meter telescope (called CELT for California Extremely Large Telescope) might contribute to their astronomical research. By 2002, CELT was the centerpiece of a campaign to raise $1.4 billion for Caltech, and the possibility of a 30-meter telescope excited Richard Ellis enough to lure him from Cambridge University to a new position in Pasadena.

News of Caltech's massive new bequest and subsequent progress fostered a feeling of déjà-vu, a repeat of circumstances in the 1980s that led to Gemini—a large philanthropic gift from a rich donor was enabling California astronomers to surge ahead toward their next large telescope. As Matt Mountain told *Science,* "CELT certainly has the momentum and the attention of the rest of the community."[33] Meanwhile, NOAO nurtured its own plans for a 30-meter-class telescope but remained hampered by funding shortages and a lack of unified support in the general astronomy community. Many believed international collaboration or some form of public-private partnership was probably the path NOAO would have to take to acquire bigger facilities.

Despite all of the similarities described thus far between optical astronomy and high-energy physics, one critical difference remains. Unlike high-energy physics, radio astronomy, space-based research, and many other areas of science, private and state institutions still dominate ground-based optical astronomy in the United States. Of the total light-collecting area American astronomers had at their disposal in 2002, only 22 percent was accessible to the entire science community. Jesse Greenstein's fear that national facilities would soon outstrip the power of independent observatories has not been realized; and, despite all of the profound and sometimes wrenching technological and sociological changes experienced by the astronomy community since 1950, the tradition of privately funded telescopes continues as strong as ever. This tenacious feature, a relic from the decades before World War II, set American astronomy apart from high-energy physics, whose practitioners long ago embraced a system of federally funded national laboratories that both cooperate and compete with one another.[34] In this respect, astronomy at the close of the twentieth century resembled the well-publicized and contentious rivalry between privately funded companies like Celera and the federally funded National Institutes of Health to complete the Human Genome Project.

European and Japanese astronomers, whose resources come only from government sponsorship, have avoided what some see as the "destructive 'balkanization' of US ground-based astronomy."[35] At a public hearing at the National Academy of Sciences in July 2001, Riccardo Giacconi, ESO's former director and future Nobel laureate, warned that American astronomers were in danger of being overtaken by their more unified European counterparts. ESO, he said, built its Very Large Telescope array as a "single machine to do science." Success with the VLT, Giacconi said, poised ESO to build its own Next Big Machine like the Overwhelmingly Large Telescope. A poor precedent was set decades earlier, Giacconi observed, when fortunate astronomers at Carnegie and Caltech had access to the world's largest telescope while astronomers using Kitt Peak were relegated to smaller facilities. In the United States, he concluded, the basic problem was not a lack of technology or funding. It was sociological and originated with the inability of private and national observatories to cooperate.[36]

Looking to the future, the importance of and emphasis on building the Next Big Machine for astronomy poses serious implications for institutions like the national observatory and universities as well as national science policy. With the release of the McKee-Taylor decadal survey, astronomers in the United States began to talk with more vigor about reshaping American ground-based optical astronomy as a "single integrated system." Stiff competition from the more united European astronomy community has been one spur, as were comments from people like Robert Eisenstein, the NSF's Director of Mathematical and Physical Sciences. In December 2000, prominent astronomers from public and private observatories gathered for a semiannual meeting at the National Academy of Sciences. Eisenstein warned them that the "days of entitlement for astronomy were over." The astronomy community, he said, desperately needed to speak with a single unified voice—as the high-energy physicists had done decades earlier, he noted—to ensure funding for the Next Big Machine.[37]

What remained to be seen was how the integration of the private and public observatories might be achieved. What incentive was there for scientists from historically independent institutions like Caltech or the Carnegie Observatories to combine forces and funding with government entities like NOAO? Carnegie's Alan Dressler offered one hopeful view at the Academy meeting. He compared the current landscape of astronomy to a biological ecosystem in which each telescope had its own niche that was part of an integrated whole. Embracing this analogy, he suggested, could help heal the wounds caused by "decades of non-productive, even destructive competition between national and university/independent observatories."[38]

Such language—ecosystem, niche, single integrated system—suggested that astronomers' conceptualization of the telescope had continued to evolve. Fifty years previously, the telescope has been understood as a system of separate parts, including the mirror, truss, and dome. A half-century later, astronomers thought of their principal research tool as a "hyper-telescope." The observatory had become a complex assemblage of intimately interconnected and interdependent optical, mechanical, and computer systems, with the telescope simply the most visible node of an astronomical data-collecting network. The emergence of the "hyper-telescope" began to give rise to optimistic talk of an entire network of observatories, all cooperatively linked yet each maintaining some degree of autonomy, which together would enable the best science research.

While concerns held by scientists about the equitable distribution of America's astronomical resources have not entirely disappeared, encouraging signs of reconciliation have begun to appear among astronomers and science managers. For example, private observatories like Keck have opened their doors partway to visiting astronomers in a reciprocal exchange for additional NSF support and a public-private partnership featuring the University of Arizona and NOAO has been seriously considering a giant telescope to scan the entire night sky weekly. Even more dramatic was the formal agreement in May 2003 between representatives of AURA, Caltech, and the University of California that recognized that cooperation among them was necessary to realize a 30-meter mega-telescope for U.S. scientists.

In 1978, shortly before retiring from Caltech, Jesse Greenstein said, "I think wanting too many telescopes is an American disease." This craving, he said, had forced scientists to choose between "democratic choice" and "elitist, snobbish concentration of effort."[39] Astronomers and engineers built the Gemini Observatory in the space created by the tension between these two traditions after scientists like Greenstein and Goldberg fought raucous and divisive battles over how their community should organize itself, divide its resources, and design new telescopes. As American astronomy entered the new millennium, these competing voices appeared to have taken the first steps toward harmony within observatory domes around the world.

GIANT TELESCOPES

SOURCES

ABBREVIATIONS

NOTES

ACKNOWLEDGMENTS

INDEX

GIANT TELESCOPES

Telescope and/or institution	Location	Mirror size in meters (inches)	Date completed
Hale Telescope	Palomar Mountain, California	5 meters (200″)	1948
Shane Telescope, Lick Observatory	Mount Hamilton, California	3 meters (120″)	1959
Mayall Telescope	Kitt Peak, Arizona	4 meters (158″)	1973
Blanco Telescope	Cerro Tololo, Chile	4 meters (158″)	1974
Anglo-Australian Telescope	Mt. Stromlo, Australia	3.9 meters (153″)	1975
European Southern Observatory	La Silla, Chile	3.6 meters (141″)	1976
Bolshoi Azimuthal Telescope	Nizhny Arkhyz, Russia	6 meters (236″)	1976
Canada-France-Hawaii Telescope	Mauna Kea, Hawaii	3.6 meters (141″)	1979
United Kingdom Infrared Telescope	Mauna Kea, Hawaii	3.8 meters (150″)	1979
Multiple Mirror Telescope	Mt. Hopkins, Arizona	4.5 meters (177″)	1979
William Herschel Telescope	La Palma, Canary Islands	4.2 meters (165″)	1987
Keck I	Mauna Kea, Hawaii	10 meters (393″)	1993
ARC	Apache Point, New Mexico	3.5 meters (138″)	1994

Telescope and/or institution	Location	Mirror size in meters (inches)	Date completed
Starfire Optical Range	Albuquerque, New Mexico	3.5 meters (138")	c. 1994
WIYN	Kitt Peak, Arizona	3.5 meters (138")	1994
Keck II	Mauna Kea, Hawaii	10 meters (393")	1996
Hobby-Eberly Telescope	Mt. Fowlkes, Texas	9.2 meters (363")	1997
Very Large Telescope	Cerro Paranal, Chile	4 × 8.2 meters (322" each)	1998–2000
Subaru Telescope	Mauna Kea, Hawaii	8.3 meters (326")	1999
Gemini North	Mauna Kea, Hawaii	8.1 meters (319")	1999
Multiple Mirror Telescope (with new mirror)	Mt. Hopkins, Arizona	6.5 meters (255")	2000
Walter Baade Telescope	Las Campanas, Chile	6.5 meters (255")	2000
Landon Clay Telescope	Las Campanas, Chile	6.5 meters (255")	2002
Gemini South	Cerro Pachon, Chile	8.1 meters (319")	2002
Large Binocular Telescope	Mount Graham, Arizona	2 × 8.4 meters (330" each)	Est. 2003

Sources

The contributions to science from Gemini and the other large telescopes discussed in this book are just beginning. As the lifetime of an observatory is typically measured in decades, future historians will have the opportunity to evaluate the impact of the current generation of large telescopes on the practice and intellectual content of astronomy. Nonetheless, researching and writing this book gave me access to a wide range of resources that may not be available to the historian looking back at this period with an extra fifty years of perspective. This brief essay describes the sources I used, how they were collected, and how they are referred to in the Notes.

The history of Gemini and its precursors spans two eras of communication technologies, a fact that poses considerable challenges for the historian. Jesse Greenstein's and Leo Goldberg's early correspondences with each other and their colleagues was handwritten and later typed. By their careers' end, notes and correspondence were often sent via electronic mail. The written evidence I used in writing this book was incredibly varied. It included personal correspondence preserved at various archives, papers from the working files of different institutions and agencies, and correspondence and the personal papers of astronomers who generously allowed me to peruse them and copy those of interest. I also made use of formal government and project reports, Congressional hearings, as well as published scientific and technical papers.

Archival collections of individual scientists' papers used included: at the California Institute of Technology's Archives, the papers of Jesse L. Greenstein and Bruce Rule; the papers of Leo Goldberg at the Harvard University Archives; and the papers of Fred Whipple at the Smithsonian Institution Archives.

Peter Strittmatter graciously gave me supervised access to the working papers of the Steward Observatory and the Steward Observatory Mirror Laboratory at the University of Arizona. An even more valuable source was the collection pertaining to Gemini and other telescope projects at the National Science Foundation's Division of Astronomical Society. Wayne Van Citters was instrumental in helping me secure access to this. Other than proprietary materials or those dealing with personnel issues, I was given free rein to examine this rich collection. Some materials came from Van Citters's own working files, while others belonged to the Astronomy Division's working files. At some point, it is hoped, these government records will be sent to the National Records and Archives Administration for preservation. In the meantime, copies of documents referred to in the Notes are in the author's possession.

Papers, technical reports, grant proposals, and oral history transcripts from several institutions were also a valuable resource. These included materials from the Center for History of Physics at the American Institute of Physics; the library at the National Optical Astronomy Observatories; the Space History collection at the National Air and Space Museum; and the library at the Gemini Observatory in Hilo, Hawaii.

I was not able to review the private records of the Association of Universities for Research in Astronomy. This lack of access, however, did not present an insurmountable obstacle. Duplicate copies of many reports and meeting minutes existed in the records of either the NSF or individuals. Frank K. Edmondson also supplied a few important AURA-related documents from his personal files.

Finally, several astronomers and engineers granted me access to their personal files. These included: Roger Angel at the University of Arizona, Larry Barr at NOAO, Richard Ellis at Cambridge University, Robert Gehrz at the University of Minnesota, and Goetz Oertel in Washington, D.C. These papers not only helped me construct a chronology of events but also afforded me personal perspectives on events and personalities in the science community.

Oral history interviews were an invaluable tool for understanding events. They offered also an opportunity to record the personal experiences and impressions of scientists, engineers, and science managers. I conducted over 120 hours of tape-recorded interviews between 1998 and 2002. I also made use of the excellent oral histories found at the Center for History of Physics, the Caltech archives, and the National Air and Space Museum. Except where indicated in the Notes, all interviews used were conducted by the author. Most of these interviews were conducted for this book. However, several were general oral histories that are now part of the formal collection at the Niels Bohr Library at the Center for History of Physics in College Park,

Maryland. These interview sessions are indicated in the following list with an asterisk. While not all interviewees listed are quoted in the book, each provided rich and valuable background information and personal recollections of particular events and issues. In all cases, interviewees gave their consent to be interviewed as part of my research. I made an effort to talk to a wide variety of people, not just the elite of the science world. I often interviewed the same person more than once as a way of checking facts and resolving conflicting stories. My use of any interview was tempered, of course, with the knowledge that personal memories are selective and people's recollections as well as the significance of their memories change over time.

Interviews

Helmut Abt (1/20/99 and 10/28/99*; Tucson, Ariz.)
Roger Angel (9/17/98, 9/25/98, 11/5/98, 11/10/98, 12/9/98, and 3/19/01*; Tucson, Ariz.)
John Bahcall (12/2/99; Princeton, N.J.)
Lawrence Barr (11/17/98, 12/2/98, and 1/13/99; Tucson, Ariz.)
Jacques Beckers (6/1/99; Chicago, Il.)
Robert Bless (7/28/00; Madison, Wisc.)
Todd Boroson (2/16/99 and 1/27/00; Tucson, Ariz.)
Geoffrey Burbidge (3/17/99; San Diego, Calif.)
Fred Chaffee (10/27/00; Waimea, Hawaii)
Ian Corbett (4/6/00; Washington, D.C.)
Stéphanie Côté (12/19/00; by telephone)
David Crampton (11/7/00; Victoria, British Columbia)
Michael Cusanovich (11/2/98; Tucson, Ariz.)
Roger Davies (6/19/00; Durham, England)
Warren Davison (2/24/99; Tucson, Ariz.)
Alan Dressler (11/15/99 in Washington, D.C. and 12/21/00 by telephone)
Michael Edmunds (6/20/2000; Cardiff, Wales)
Richard Ellis (5/1/00; Pasadena, Calif.)
Richard Ellis and Crystal Martin (9/23/00; Palomar, Calif.)
Sandra Faber (8/1/02*; Santa Cruz, Calif.)
Alex Filippenko (4/7/01; by telephone)
Craig Foltz (12/22/98; Tucson, Ariz.)
Robert Fugate (11/27/00*; Albuquerque, N.M.)
Robert Gehrz (11/23/98 and 5/25/99; Minneapolis, Minn.)
Larry Goble (10/29/98; Tucson, Ariz.)
Donald Hall (10/22/00; Hilo, Hawaii)
George Herbig (10/30/00; Honolulu, Hawaii)
John Hill (9/23/98, 10/9/98, and 1/11/99; Tucson, Ariz.)

James Houck (4/21/02*; Ithaca, N.Y.)
John Huchra (2/14/02* and 6/25/02*; Cambridge, Mass.)
John Jefferies (2/26/99; Tucson, Ariz.)
Karen Kenagy (11/4/98; Tucson, Ariz.)
Robert Kraft (8/1/02*; Santa Cruz, Calif.)
Richard Kurz (4/6/00; Tucson, Ariz.)
Simon Lilly (10/6/99; Washington, D.C.)
Malcolm Longair (6/15/00; Cambridge, England)
Frank Low (11/30/98, 2/23/99, and 5/4/00*; Tucson, Ariz.)
Richard Malow (9/24/99 and 10/13/99; Washington, D.C.)
Steve Maran (7/25/00; Washington, D.C.)
Buddy Martin (9/18/98; Tucson, Ariz.)
Christopher McKee (7/29/02*; Berkeley, Calif.)
Donald Morton (11/6/00; Victoria, British Columbia)
Matt Mountain (10/24/00 in Hilo, Hawaii and 11/21/00 by telephone)
Goetz Oertel (9/24/99, 11/29/99, 3/22/00, 1/21/01*, and 4/9/01*; Washington, D.C.)
J. Beverly Oke (11/7/00; Victoria, British Columbia)
Jim Oschmann (4/17/02; Tucson, Ariz.)
Patrick Osmer (6/2/00; Columbus, Ohio)
Lawrence Randall (7/21/00; Old Forge, N.Y.)
Marcia Rieke (2/22/99; Tucson, Ariz.)
Jean-René Roy (11/100 in Hilo, Hawaii and 11/28/01 and 5/7/02 by telephone)
Marie Teresa Ruiz (12/8/01; Washington, D.C.)
Maarten Schmidt (9/21/00; Pasadena, Calif.)
Malcolm Smith (1/21/02*; La Serena, Chile)
Peter Strittmatter (9/10/98, 10/14/98, 11/16/98, 12/22/98, and 1/15/99; Tucson, Ariz.)
Stephen Strom (1/19/99 and 2/19/99; Tucson, Ariz.)
Virginia Trimble (9/27/00; College Park, Md.)
Wayne Van Citters (10/6/99 and 7/24/00; Washington, D.C.)
Gordon Walker (11/7/00; Victoria, British Columbia)
Russell Warner (11/12/98; Tucson, Ariz.)
Daniel Watson (11/1/98 and 11/8/98; Tucson, Ariz.)
Neville Woolf (9/21/98, 10/27/98, and 1/20/99; Tucson, Ariz.)
Sidney Wolff (11/1098, 12/17/98, 2/22/99, and 10/28/99*; Tucson, Ariz.)

Events and Trips to Observatories

Since 1998, when I began the research for this book, I have also been fortunate to attend several notable events that aided my research immeasurably—the casting of two large mirrors at the Mirror Lab, public sessions sponsored by the National Academy of Sciences and the NSF, meetings of the American Astronomical Society, large-telescope conferences, and Gemini's dedication in Chile in early 2002 all gave me insights into the political and technical aspects of building large telescopes.

I also spent time at various observatories and telescopes. With the assistance and patience of several scientists—Richard Ellis, Robert Fugate, Matt Mountain, and Malcolm Smith, to name a few—I was able to visit Kitt Peak, Gemini North in Hawaii, as well as Gemini South and Cerro Tololo in Chile, the 200-inch on Palomar, and Starfire Optical Range in New Mexico. Some of this time was actually on the mountain at night; other sessions were spent "observing the observers" in control rooms lower down the mountain. My goal was simple—to develop a firsthand sense of what being an astronomer was like. When I began my research, my understanding of what astronomers actually did was naïve. I still pictured them as the romantic and solitary watchers of the cosmos described in the early parts of this book. This view has since been tempered with the reality I observed and, to a small degree, experienced—computer-bleared eyes, sugary midnight snacks, and the sometimes ecstatic but frequently tedious exploration of the cosmos with a "beautiful and cantankerous" telescope.

Abbreviations

Ap. J.	*The Astrophysical Journal*
AST/NSF	Division of Astronomical Science, National Science Foundation, Ballston, Va.
BHR/CITA	Papers of Bruce Rule, Caltech Archives
CHP/AIP	Center for History of Physics, The American Institute of Physics, College Park, Md.
CITA	Other materials from the Caltech archives, other than the specific collections noted here
FLW/SI	Papers of Fred Whipple, Smithsonian Institution Archives
Gemini	Library at the Gemini Observatory, Hilo, Hawaii
GKO	Personal files of Goetz K. Oertel, Washington, D.C.
JLG/CITA	Papers of Jesse Greenstein, Caltech Archives
JRA	Personal files of J. Roger Angel, Tucson, Ariz.
LDB	Personal files of Larry D. Barr, Tucson, Ariz.
LG/HUA	Papers of Leo Goldberg, Harvard University Archives
NASM	Space History collection at the National Air and Space Museum, Washington, D.C.
NOAO Library	Library at the National Optical Astronomy Observatories, Tucson, Ariz.
PASP	*Publications of the Astronomical Society of the Pacific*
RDG	Personal files of Robert D. Gehrz, Minneapolis, Minn.
RSE	Personal files of Richard S. Ellis, Cambridge, England
SO/UA	Steward Observatory, Tucson, Ariz.
SOML/UA	Steward Observatory Mirror Laboratory, Tucson, Ariz.
S&T	*Sky & Telescope*

Notes

Introduction: Beautiful and Cantankerous Instruments

1. I have adopted an astronomers' convention in this book by referring to the size of telescopes in English units or in metric units, depending on the age of the instrument. For example, astronomers call the Palomar telescope the "200-inch"—equivalent to about 5 meters—whereas telescopes built after 1970 are usually referred to in metric units, such as Kitt Peak's 4-meter (158-inch) telescope in Arizona. These numbers describe the diameter of the telescope's main mirror and not the size of the telescope itself.
2. Meinel interviewed by David DeVorkin in "Multiple Mirror Telescope," session 6, May 11, 1989 (Smithsonian Videohistory Program, Smithsonian Institution Archives, RU 9542), p. 42. Additional information comes from personal communications between Meinel and the author.
3. NOAO Press Release (NOAO 99–08), June 25, 1999, AST/NSF Files.
4. The literature on the postwar developments of high-energy physics is too vast to give more than a few examples. See, for example, Peter Galison, *Image and Logic: A Material Culture of Microphysics* (Chicago: University of Chicago Press, 1997); Daniel J. Kevles, *The Physicists: A History of a Scientific Community in America* (New York: Vintage, 1987). Also see Robert Seidel, "Accelerating Science: The Postwar Transformation of the Radiation Laboratory," *Historical Studies in the Physical Sciences* 13 (1983):375–400; and Seidel, "A Home for Big Science: The Atomic Energy Commission's Laboratory System," ibid. 16 (1986):135–175.

5. Whitford interview, July 15, 1977, by David DeVorkin, p. 51, CHP/AIP Collection.
6. Allan Sandage, preface to *The Hubble Atlas of Galaxies* (Washington: Carnegie, 1961).
7. Robert Smith, "Engines of Discovery: Scientific Instruments and the History of Astronomy and Planetary Science in the United States in the 20th Century," *Journal for the History of Astronomy* 28 (Feb. 1997):49–77.

1. Leo and Jesse's Changing World

1. Jesse Greenstein, "An Astronomical Life," *Annual Reviews of Astronomy and Astrophysics* 22 (1984):1. Background material on Greenstein and Goldberg was drawn from a number of sources, including extensive oral history interviews. Spencer Weart's interviews with Greenstein in 1977 and 1978 and with Goldberg in May 1977 were especially helpful. Transcripts from these sessions are in the CHP/AIP collections. Quotes are referred to directly by page number hereafter.
2. Lawrence Aller, "Leo Goldberg," *National Academy of Science's Biographical Memoir* 72 (1994):115–135.
3. Greenstein to Goldberg, Aug. 5, 1941, and Jan. 30 (year not included), "G 1933–46" folder, Box 7, HUG FP 83.10, LG/HUA.
4. Undated letter (probably mid-1980s), Goldberg to his children, Box 10, HUG FP 83.8, LG/HUA.
5. Greenstein to Goldberg, Dec. 10, 1941, "G 1933–46" folder, Box 7, HUG FP 83.10, LG/HUA.
6. David DeVorkin, "The Maintenance of a Scientific Institution: Otto Struve, the Yerkes Observatory, and Its Optical Bureau during the Second World War," *Minerva* 18, 4 (1981):595–623.
7. David DeVorkin and Paul Routly, "The Modern Society: Changes in Demographics," in *The American Astronomical Society's First Century*, ed. D. DeVorkin (Washington: American Astronomical Society, 1999), 122–136.
8. Greenstein interview, July 21, 1977, p. 71.
9. Daniel J. Kevles, *The Physicists: A History of a Scientific Community in America* (New York: Vintage, 1987), 278–279, 372.
10. Goldberg interview, May 16, 1978, pp. 43–47.
11. Ibid., p. 50.
12. Ibid., pp. 47–53. Correspondence between Goldberg and Hulbert that notes this is in HUG FP 83.10, LG/HUA.
13. Owen Gingerich, "The Summer of 1953: A Watershed for Astrophysics," *Physics Today* 47, 12 (1994):34–40.

14. David DeVorkin, *Science with a Vengeance: How the Military Created the US Space Sciences after World War II* (New York: Springer Verlag, 1992), 204–241.
15. Ibid., 207.
16. Greenstein to Goldberg, July 7, 1960, "Greenstein" folder, Box 7, HUG FP 83.10, LG/HUA.
17. Donald Osterbrock, "The Appointment of a Physicist as Director of the Astronomical Center of the World," *Journal for the History of Astronomy* 23, 3 (1992):155–165.
18. Bush to Walter Adams, Oct. 23, 1944, Carnegie Institution of Washington Archives; Bowen to Henry Norris Russell, Oct. 26, 1945, Henry Huntington Library, San Marino, California. From Osterbrock, "Appointment of a Physicist."
19. Greenstein interview, July 21, 1977, pp. 87–89.
20. Helmut A. Abt interview, Oct. 28, 1999, p. 8, CHP/AIP Collection.
21. Greenstein, "An Astronomical Life," p. 18; also see Greenstein's correspondence with scientists such as Martin Schwarzschild, Box 27, JLG/CITA.
22. Goldberg folder, Box 12, JLG/CITA.
23. Nancy Bolton, "Press Pilgrimage to Palomar," *S&T* 10, 1 (1948):59–60. "What Do Astronomers Expect of the 200-Inch Telescope?" Ibid., 61.
24. *Dedication of the Palomar Observatory and the Hale Telescope: June 3, 1948* (Pasadena: California Institute of Technology, 1948).
25. Greenstein interview, July 21, 1977, p. 133.
26. Nathaniel Carleton quoted in interview by David DeVorkin, 1989, in "Multiple Mirror Telescope," session 2 (Smithsonian Videohistory Program, Smithsonian Institution Archives, RU 9542), p. 3.
27. Greenstein interview, July 21, 1977, p. 92.
28. The time allocation process at Palomar and the different research divisions are discussed in a number of oral history interviews in the CHP/AIP collections, including the Greenstein interviews; Ira S. Bowen interview, Aug. 26, 1969, by Charles Weiner; and Horace W. Babcock interview, July 25, 1977, by Spencer Weart.
29. Annual reports from the Mount Wilson and Palomar Observatories list visiting astronomers, their affiliations, the telescope they used, and the type of research they were doing.
30. Bowen interview, Aug. 26, 1969, p. 60.
31. On early instrumentation for the 200-inch, see Ira Bowen, "The Spectrographic Equipment of the 200-Inch Hale Telescope," *Ap. J.* 116, 1 (1952):1–7.
32. Maarten Schmidt interview, Sept. 21, 2000.

33. Quote is from a virtual tour of the 200-inch featuring Greenstein, at <www.astro.caltech.edu/palomarpublic/prime/pf1.html>.
34. J. Beverly Oke interview, Nov. 7, 2000.
35. <www.astro.caltech.edu/palomarpublic/prime/pf7.html>; Greenstein to Virginia Trimble, Mar. 31, 1971, "Trimble" folder, Box 39, JLG/CITA.
36. Donald Osterbrock, "Rudolph Leo Bernhard Minkowski," *National Academy of Science's Biographical Memoir* 54 (1983):286.
37. Donald Osterbrock, personal communication, Jan. 30, 2001.
38. Greenstein interview, Mar. 16, 1982, by Rachel Prud'homme, CHP/AIP Collections, p. 39.
39. Russell to Bowen, Nov. 3, 1945, quoted in David DeVorkin, "Electronics in Astronomy: Early Applications of the Photoelectric Cell and Photomultiplier for Studies of Point-Source Celestial Phenomena," *Proceedings of the IEEE* 73, 7 (1985):1216.
40. DeVorkin, "Electronics in Astronomy." Also see William Baum, "Counting Photons—One by One," *S&T* 17, 5 (May 1955):264–267; 17, 6 (June 1955):330–334.
41. Oke interview, Nov. 7, 2000.
42. H. W. Babcock, B. H. Rule, and J. S. Fassero, "An Improved Automatic Guider," *PASP* 68, 402 (1956):256–258.
43. Halton C. Arp, "Reduction of Photoelectric Observations by an Electronic Computer," *Ap. J.* 129 (1959):507–513.
44. Peter Galison, "Bubble Chambers and the Experimental Workplace," in *Observation, Experiment, and Hypothesis in Modern Physical Science*, ed. Peter Achinstein and Owen Hannaway (Cambridge, Mass.: MIT Press, 1985), 309–373, quotation on p. 342.
45. Ibid., 347.
46. Greenstein to Robert Fleischer, Mar. 25, 1974, "AURA 1974" folder, Box 95, JLG/CITA.
47. Baum, "Counting Photons," *S&T* 17, 6.
48. Greenstein interview, July 21, 1977, pp. 147–148.
49. Ibid., p. 147; Greenstein to Thomas Gieryn, Oct. 9, 1979, "Greenstein Working Files," CHP/AIP Collection.
50. Goldberg to Hulbert, May 20, 1957, "Judge Henry Hulbert" folder, Box 9, HUG FP 83.10, LG/HUA.
51. Frank K. Edmondson, "AURA and KPNO: The Evolution of an Idea, 1952–58," *Journal for the History of Astronomy* 22, 1 (1991):72.
52. John Lankford, *American Astronomy: Community, Careers, and Power, 1859–1940* (Chicago: University of Chicago Press, 1997), 199.
53. Charles A. Federer, "Optical Astronomy's Instrumental Needs," *S&T* 25, 2 (1963):67.

54. Ronald Doel, *Solar System Astronomy in America: Communities, Patronage, and Interdisciplinary Science, 1920–1960* (Cambridge, England: Cambridge University Press, 1996), esp. ch. 6; Joseph Tatarewicz, "Federal Funding and Planetary Astronomy, 1950–75: A Case Study," *Social Studies of Science* 16, 1 (1986):79–103.
55. David DeVorkin, "Who Speaks for Astronomy? How Astronomers Responded to Government Funding after World War II," *Historical Studies in the Physical and Biological Sciences* 31, 1 (2000):55–92. Lankford, *American Astronomy*, ch. 7, describes the considerable power of observatory directors.
56. Kevles, *Physicists*, 355.
57. On ONR funding of astronomy, see DeVorkin, "Who Speaks for Astronomy," and Leo Goldberg, "The Founding of Kitt Peak," *S&T* 45, 3 (1983):228–232.
58. The formal name of the committee was NRC Committee on Astronomy, Advisory to ONR. Members included Ira Bowen (Palomar), C. D. Shane (Lick), Lyman Spitzer (Princeton), and Albert Whitford (Wisconsin).
59. C. D. Shane, "Preliminary Proposals for Support of Astronomy by the National Science Foundation," *Astronomical Journal* 55 (Apr. 1950):86.
60. A. F. Spilhaus, *University Research Potential: A Survey of the Resources for Scientific and Engineering Research in American Colleges and Universities* (Washington: Engineering Research Council, 1951).
61. Goldberg to Greenstein, Apr. 30, 1952, "NSF" folder, HUG FP 83.21, LG/HUA.
62. DeVorkin, "Who Speaks for Astronomy," 78–81. Also Greenstein interview, July 21, 1977, pp. 153–154.
63. Goldberg, "Founding of Kitt Peak," 229; Goldberg to Joel Stebbins, Aug. 18, 1953, "S" folder, Box 22, HUG FP 83.10, LG/HUA.
64. John B. Irwin, *Proceedings of the NSF Astronomical Photoelectric Conference* (Bloomington: University of Indiana Press, 1955), 107–108.
65. J. Merton England, *A Patron for Pure Science: The National Science Foundation's Formative Years, 1945–57* (Washington: National Science Foundation, 1983), ch. 13. On NSF funding for physical versus biological sciences in the 1950s, see Kevles, *Physicists*, and Toby Appel, *Shaping Biology: The National Science Foundation and American Biological Research, 1945–1975* (Baltimore: Johns Hopkins University Press, 2000).
66. Goldberg, "Founding of Kitt Peak," 230; England, *Patron for Pure Science*, 396.
67. Meinel, personal communication, Nov. 9, 1998.
68. Abt interview, Oct. 28, 1999, pp. 24–26.

69. On the process of choosing Kitt Peak, see Edmondson, *AURA and Its US National Observatories*, chs. 5–6.
70. A. E. Whitford, "The Plan for a New American Observatory," *PASP* 68, 401 (1956):116–117.
71. England, *Patron for Pure Science*, 291; Edmondson, *AURA and Its US National Observatories*, ch. 8.
72. Goldberg, "Founding of Kitt Peak," 232.
73. Robert Seidel, "A Home for Big Science: The Atomic Energy Commission's Laboratory System," *Historical Studies in the Physical Sciences* 16 (1986):135–175; England, *Patron for Pure Science*, 292–310.
74. Edmondson, "AURA and KPNO," 77.
75. Described inimitably in Daniel S. Greenberg, *The Politics of Pure Science* (Chicago: University of Chicago Press, 1967), ch. 9.
76. Allan Needell, "Lloyd Berkner, Merle Tuve, and the Federal Role in Radio Astronomy," *Osiris* 3 (1987):261–288.
77. Goldberg to Greenstein, Aug. 13 and Feb. 26, 1968, "Greenstein" folder, HUG FP 83.10, LG/HUA. Letters between Greenstein and George Kistiakowski, 1968, Box 19, JLG/CITA.
78. Nathan Reingold, "Vannevar Bush's New Deal for Research," *Historical Studies in the Physical and Biological Sciences* 17, 2 (1987):307. Also see Daniel J. Kevles, "The National Science Foundation and the Debate over Postwar Research Policy, 1942–45: A Political Interpretation of Science—The Endless Frontier," *Isis* 68, 1 (1977):5–26.
79. DeVorkin and Routly, "Modern Society," 128.
80. Quoted in Otto Struve, "The General Needs of Astronomy," *PASP* 67, 397 (1955):218.
81. Kevles, *Physicists*, 373–374; David Kaiser, "The Postwar Suburbanization of American Physics," manuscript (2001).
82. Goldberg to Struve, Dec. 8, 1954, "Sugargrove Advisory Committee" folder, Box 23, HUG FP 83.10, LG/HUA.
83. Struve, "General Needs of Astronomy," 218.
84. Robert W. Seidel, "Accelerators and National Security: The Evolution of Science Policy for High-Energy Physics, 1947–1967," *History and Technology* 11, 4 (1994):361–391.
85. Goldberg to Green, July 11, 1962, "Origins of the Whitford Committee" folder, Box 5, HUG FP 83.18, LG/HUA. Also Whitford interview, July 15, 1977, by David DeVorkin, pp. 95–96, CHP/AIP collections.
86. Whitford to "Dear Colleague," Feb. 15, 1963, Whitford folder, Box 43, JLG/CITA; A. E. Whitford (chair), *Ground Based Astronomy: A Ten Year Program* (Washington: National Academy of Sciences, 1964), foreword; quote from Whitford interview, July 15, 1977, p. 105.

87. Greenstein to Robert Green, Mar. 16, 1963, "Whitford Report" folder, Box 3, HUG FP 83.18, LG/HUA.
88. Greenstein to Goldberg, Mar. 26, 1963, "Leo Goldberg (1960–76)" folder, Box 12, JLG/CITA.
89. Goldberg to Sandage, Oct. 7, 1963; Sandage to Goldberg, Oct. 11, 1963, "Whitford Committee" folder, Box 5, HUG FP 83.18, LG/HUA.
90. B. T. Lynds, personal communication, July 6, 2001.
91. Lawrence A. Aller, "Space-Age Astronomy and Whale-Oil Lamps," *Science* 139 (Jan. 4, 1963):21–22.
92. Goldberg to Sandage, Oct. 24, 1963, "Whitford Committee" folder, Box 5, HUG FP 83.18, LG/HUA.
93. Burbidge to Greenstein, Apr. 22, 1963, p. 6, "Geoffrey Burbidge" folder, Box 4, JLG/CITA.
94. Greenstein to Burbidge, Apr. 26, 1963, ibid.
95. Whitford interview, July 15, 1977, pp. 97–99.
96. Whitford, *Ground Based Astronomy*, 44–46.
97. Goldberg comment, "Situation of the NGT," p. 3, "NGT Scientific Advisory Committee" folder, Box 31, HUG FP 83.12, LG/HUA.
98. George Herbig to Whitford, May 16, 1963, "Whitford Report" folder, Box 5, HUG FP 83.18, LG/HUA.
99. Goldberg interview, May 17, 1978, pp. 138–139; Joseph Tatarewicz, *Space Technology and Planetary Astronomy* (Bloomington: Indiana University Press, 1990), 101–115.
100. Greenstein to Homer Newell, Feb. 24, 1969, "G-H" folder, Box 2, HUG FP 83.19, LG/HUA.

2. Tradition and Balance

1. George W. Ritchey, "Astronomical Photography with High Magnification," *Journal of the Royal Astronomical Society of Canada* 22, 9 (1928):376.
2. George W. Ritchey, *The Development of Astro-Photography and the Great Telescopes of the Future* (Paris: Societe Astronomique de France, 1929).
3. Biographical material on Meinel comes from personal communications in 1998 and 1999. Also, David DeVorkin interviews Meinel in "Multiple Mirror Telescope," session 6, May 11, 1989 (Smithsonian Videohistory Program, Smithsonian Institution Archives, RU 9542), pp. 36–56.
4. Aden Meinel, "An Overview of the Technological Possibilities of Future Telescopes," in *Optical Telescopes of the Future*, 13–26.
5. Meinel to Gerard Kuiper, Apr. 2, 1953, Gerard P. Kuiper papers, University of Arizona Special Collections, Tucson. These papers were un-

processed at the time of writing; I wish to thank David DeVorkin for providing me with a copy of this letter.

6. Report by Gerard P. Kuiper, Mar. 26, 1953, ibid.
7. Helmut A. Abt interview, Oct. 28, 1999.
8. On Meinel's relations with McMath and AURA see Edmondson, *AURA and Its US National Observatories*, 29–45.
9. Meinel, personal communication, Nov. 11, 1998.
10. Aden B. Meinel, "MIAMI—Technical Report #10," "National Astronomical Observatory" folder, Box 24, BHR/CITA.
11. Abt interview, Oct. 28, 1999.
12. "Multiple Mirror Telescope," session 6, p. 45.
13. A. E. Whitford (chair), *Ground Based Astronomy: A Ten Year Program* (Washington: National Academy of Sciences, 1964), 42–43.
14. Ibid., 46–47.
15. Meinel, personal communication, Nov. 12, 1998.
16. Roger Angel interview, Sept. 27, 1998.
17. Ira S. Bowen, "Telescopes," *Astronomical Journal* 69, 10 (1964):823.
18. I. S. Bowen, "Problems in Future Telescope Design," *PASP* 73, 431 (1961):124.
19. M. L. Humason, N. U. Mayall, and A. R. Sandage, "Redshifts and Magnitudes of Extragalactic Nebulae," *Ap. J.* 61, 3 (1956):97–162.
20. H. L. Helfer, G. Wallerstein, and J. L. Greenstein, "Abundances in Some Population II K Giants," *Ap. J.* 129, 5 (1959):700–719.
21. Jesse L. Greenstein and Maarten Schmidt, "The Quasi-Stellar Radio Sources 3C48 and 3C273," *Ap. J.* 140, 1 (1964):1–34; historical accounts of the discovery of quasars include Dennis Overbye, *Lonely Hearts of the Cosmos: The Story of the Scientific Quest for the Secrets of the Universe*. (Boston: Little, Brown, 1991), and Richard Preston, "Beacons in Time: Maarten Schmidt and the Discovery of Quasars," *Mercury* (1988):2–11.
22. This was later published as I. S. Bowen, "Telescopes," *Astronomical Journal* 69, 10 (1964):816–825.
23. See, for example, papers in W. A. Hiltner, ed., *Astronomical Techniques*, vol. 2 (Chicago: University of Chicago Press, 1962).
24. R. N. Wilson, *Reflecting Telescope Optics*, vol. I (New York: Springer Verlag, 1996).
25. Bowen, "Telescopes," 825.
26. "Proposal for a 200-inch Telescope in the Southern Hemisphere," 14, "200-inch Telescope" folder, Box 38, JLG/CITA.
27. Bagrat K. Ioannisiani, "The Soviet 6-meter Altazimuth Reflector," *S&T* 54 (Nov. 1977):356–362.

28. A transcript of the meeting was published as D. L. Crawford, ed., *The Construction of Large Telescopes* (New York: Academic Press, 1966).
29. "Bruce Rule" folder, Box 32, JLG/CITA.
30. "Original ARPA Contract Ends," *Optical Sciences Center Newsletter* 1, 3 (1967), NOAO Library.
31. A. B. Meinel et al., "The Optical Sciences Center: Its History, Organization, and Relation to Government and Industry," *Applied Optics* 10, 2 (1971):243–247.
32. Frank J. Low interview, Jan. 25, 2000, CHP/AIP Collection.
33. Frank J. Low, "Low-Temperature Germanium Bolometer," *Journal of the Optical Society of America* 51, 11 (1961):1300–1304.
34. "Multiple Mirror Telescope," session 6, p. 50.
35. "Synthetic Aperture Optics Summer Study" folder, Box 23, HUG FP 83.10, LG/HUA.
36. Aden B. Meinel, "Aperture Synthesis Using Independent Telescopes," *Applied Optics* 9, 11 (1970):2501–2504.
37. Fred L. Whipple, "Mount Hopkins and the MMT," in *The MMT and the Future of Ground-Based Astronomy*, ed. Trevor Weekes (Cambridge: Smithsonian Astrophysical Observatory, 1979), 3–7.
38. Larry Mertz, "Design for a Giant Telescope," in *Optical Instruments and Techniques 1969: Proceedings of the Conference Held at the University of Reading during 14th–19th July, 1969*, ed. J. Home Dickson (Reading: Oriel Press, 1970).
39. "Multiple Mirror Telescope," session 6, pp. 53–54.
40. Memo from Frank Low to "Interested Colleagues," June 4, 1970, Smithsonian Astrophysical Observatory; "A Large Astronomical Telescope at Low Cost," SAO Publication 010–98A, Nov. 1970, MMT records, Box 24, FLW/SI.
41. Jesse Greenstein, ed., *Astronomy and Astrophysics for the 1970s: Reports of the Panels* (Washington: National Academy Press, 1973), 46.
42. Excerpts from "Smithsonian Institution, Budget Justification for the Fiscal Year 1972," submitted Jan. 1971, MMT records, Box 24, FLW/SI.
43. Weymann to Goldwater, June 15, 1971, MMT records, Box 24, FLW/SI.
44. "Multiple Mirror Telescope," session 2, p. 25.
45. On the design of the MMT, see, for example, Aden Meinel et al., "A Large Multiple-Mirror Telescope (MMT) Project," in *Instrumentation in Astronomy*, eds. Lewis Larmore and Robert Poindexter, vol. 28 (Bellingham, WA: SPIE, 1972); Nathaniel P. Carleton and William F. Hoffmann, "The Multiple Mirror Telescope," *Physics Today* 31, 9 (1978):30–37.

46. W. P. Goring et al., "Performance of the MMT: MMT Computer Systems," in *Advanced Technology Optical Telescopes*, eds. Geoffrey Burbidge and Larry Barr, vol. 332 (Bellingham, WA: SPIE, 1982), 24.
47. Nathaniel P. Carleton and Thomas E. Hoffman, "The MMT Observatory on Mount Hopkins," *S&T* 52, 1 (1976):21.
48. "Multiple Mirror Telescope," session 2, p. 4; Thomas Hoffman, "The Multiple Mirror Telescope: An Engineering History, Description, and Status," in *Optical Telescopes of the Future*, 185–207.
49. Ray Weymann to Robert Leighton, Dec. 27, 1973, Box 43, JLG/CITA; "MMT Project Review 1975–1976," Box 106, JLG/CITA.
50. Frank L. Low interview, Nov. 30, 1998.
51. Peter Strittmatter, "Multiple Mirror Telescopes," in *Optical Telescopes of the Future*, 180.
52. Carleton and Hoffmann, "Multiple Mirror Telescope," 100–101.
53. Neville J. Woolf, in *The MMT and the Future of Ground-Based Astronomy*, viii. Strittmatter, "Multiple Mirror Telescopes," 181.
54. G. H. Herbig, "Daytime and Other Unanticipated Uses of Telescopes," in *ESO/CERN Conference on Large Telescope Design*, ed. Richard West (Geneva: ESO/CERN, 1971), 478.
55. Arthur Hoag, "'Stellar' Research at the Kitt Peak National Observatory," in *Conference on Research Programmes for the New Large Telescopes*, ed. A. Reiz (Geneva: ESO/SRC/CERN, 1974), 47.
56. *Annual Report of the Director: Mount Wilson and Palomar Observatories, 1968–69.*
57. Edwin W. Dennison, "Electronic Optical Astronomy: Philosophy and Practice," *Science* 174, 4006 (1971):240–241; *1967–68 Annual Report of the Mount Wilson and Palomar Observatories.*
58. Dennison, "Electronic Optical Astronomy," 240
59. J. B. Oke, "A Multi-Channel Photoelectric Spectrometer," *PASP* 81, 478 (1969):11–22; J. Beverly Oke interview, Nov. 7, 2000, and Oke, personal communication, June 14, 2001.
60. J. B. Oke, "Design of Cassegrain Cages," in *Conference on Large Telescope Design*, 357–359.
61. *Conference on Large Telescope Design*, 371.
62. Jacques F. Vallee and J. Allen Hynek, "An Automatic Question-Answering System for Stellar Astronomy," *PASP* 78, 464 (1966):315–323.
63. Jesse Greenstein to Arthur Hoag, Aug. 16, 1968, "Arthur Hoag" folder, Box 15, JLG/CITA.
64. Larry Barr interview, Nov. 11, 1998.
65. Larry Randall interview, July 21, 2000.
66. *Conference on Large Telescope Design*, 349.

67. D. L. Crawford, "AURA's Two 150-Inch Telescopes," ibid., 23–36.
68. Herbig, "Unanticipated Uses of Telescopes," 478.
69. Jesse L. Greenstein, "Large Telescope Astronomy," in *Research Programmes for the New Large Telescopes*, 13, 19.
70. Geoffrey R. Burbidge, "How to Use Large Optical Telescopes to the Fullest Extent," ibid., 363–370.
71. Ibid., 364.
72. Ibid., 368.
73. "Panel Discussion," ibid., 392.
74. Ibid., 394.
75. Ibid., 393.
76. Ibid., 394.
77. Greenstein, ed., *Astronomy and Astrophysics for the 1970s*, table 9.32.
78. CARSO's failed plan and its repercussions are described, in somewhat sanitized form, in chs. 16 and 19 in Edmondson, *AURA and Its US National Observatories*.
79. Ibid., 189.
80. Greenstein interview, May 19, 1978, p. 251; also see memo on his conversation with Schwarzschild and Goldberg, May 1, 1968, "DuBridge" folder, Box 3, JLG/CITA.
81. Goldberg to Babcock, Apr. 5, 1966, "AURA History—CTIO Funding" folder, Box 7, HUG FP 83.41, LG/HUA
82. Goldberg to Bundy, May 5, 1966, "AURA History Post 1958" folder, Box 7, HUG FP 83.41, LG/HUA.
83. Bundy to Stratton, May 25, 1966, "AURA History—CTIO Funding" folder, LG/HUA.
84. Letters between Goldberg and DuBridge, June 13 and June 22, 1966, ibid.
85. Memo from Greenstein to L. A. DuBridge and R. F. Bacher, May 1, 1968, "DuBridge" folder, Box 3, JLG/CITA.
86. Babcock to Frank Edmondson, Dec. 10, 1976, quoted in Edmondson, *AURA and Its US National Observatories*, 324. The identity of Babcock as the letter's author was withheld originally by Edmondson. Subsequent personal communication from Edmondson, June 20, 2002, clarified this point.
87. Greenstein interview, May 19, 1978, pp. 236, 252.
88. Catherine Westfall, "Fermilab: Founding the First US Truly National Laboratory," in *The Development of the Laboratory: Essays on the Place of Experiment in Industrial Civilization*, ed. Frank James (New York: American Institute of Physics, 1989), 208. On the debate over SLAC, see Stuart Leslie, *The Cold War and American Science: The Military-Industrial Complex*

89. Greenstein to William T. Golden, Nov. 12, 1971, "William Golden" folder, Box 12, JLG/CITA.
90. Letters between Mayall and Greenstein, April 17 and April 20, 1967, "AURA History—CTIO Funding" folder, LG/HUA.
91. "Distribution of Effort," "Albert Whitford" folder, Box 43, JLG/CITA.
92. Greenstein interview, May 19, 1978, p. 227.
93. Memo from Greenstein to Steering Committee Members, Aug. 27, 1969, "NAS Astronomy Survey Committee (Greenstein)" folder, Box 3, HUG FP 83.18, LG/HUA.
94. Greenstein to Abt, Sept. 24, 1970, "Abt" folder, Box 96, JLG/CITA.
95. Greenstein interview, May 19, 1978, p. 235.
96. Ibid.
97. Quoted in Gloria Lubkin, "The Decision to Build the Very Large Array" (New York: American Institute of Physics, 1975), 22, in CHP/AIP institutional files.
98. Greenstein to Harvey Brooks, July 23, 1971, cited by Lubkin, ibid.
99. Jesse Greenstein, ed., *Astronomy and Astrophysics for the 1970s; Report of the Survey Committee,* vol. 1 (Washington: National Academy of Sciences, 1972), 8.
100. Abt to Greenstein, July 7, 1971, "Abt" folder, Box 96, JLG/CITA; memo from Mayall to Abt, July 12, 1971, "NAS Astronomy Survey Committee" folder, Box 3, HUG FP 83.18, LG/HUA.
101. Talk to NSB, June 21, 1973, "NSF Science Board Meeting" folder, Box 100, JLG/CITA.

3. Visions of Grandeur

1. Goldberg, 1982 supplement to May 1977 oral history interview, CHP/AIP.
2. Greenstein to Goldberg, Mar. 2, 1971, "Leo Goldberg" folder, Box 12, JLG/CITA.
3. Memo, Bev Oke and Jim Gunn to Greenstein, Nov. 5, 1976, "1976 AURA Mission" folder, Box 106, JLG/CITA; memo, John Danziger to Goldberg, June 30, 1970, "AURA Board Correspondence" folder, Box 1, HUG FP 83.16, LG/HUA.
4. Memo, Roger Lynds to Art Hoag, June 12, 1971, "1970 January AURA Annual Meeting" folder, Box 2, HUG FP 83.16, LG/HUA.
5. Greenstein to A. E. Whitford, June 18, 1970, "Whitford" folder, Box 43, JLG/CITA.

6. Stephen Strom to Goldberg, Mar. 22, 1971, "Correspondence prior to 9/1/71" folder, Box 3, HUG FP 83.6, LG/HUA.
7. Goldberg, 1982 supplement interview, pp. 155–158.
8. Quoted in Goldberg to Bill Howard, Sept. 8, 1982, "AURA History: 4-meter CTIO Funding" folder, Box 7, HUG FP 83.41, LG/HUA.
9. Goldberg to Greenstein, Sept. 13, 1972, Box 7, HUG FP 83.8, LG/HUA. Goldberg to Richard Miller, Aug. 25, 1972, "AURA 1972" folder, Box 95, JLG/CITA.
10. Goldberg, 1982 supplement interview, p. 157; Arthur Hoag, "'Stellar' Research at the Kitt Peak National Observatory," in *Conference on Research Programmes for the New Large Telescopes*, ed. A. Reiz (Geneva: ESO/SRC/CERN, 1974), 43–52.
11. Goldberg to Dubridge, Mar. 11, 1957, "Leo Goldberg" folder, Box 12, JLG/CITA.
12. Larry Barr interview, Nov. 18, 1998.
13. Larry Barr interview, Jan. 13, 1999. Memo, Goldberg to Gil Lee, May 2, 1973, "AURA 1973–74" folder, Box 95, JLG/CITA.
14. Memo, Goldberg to William Bartley, Aug. 22, 1975, Box 9, HUG FP 83.8, LG/HUA.
15. Memo, Goldberg to KPNO staff, Aug. 22, 1975, ibid.
16. On the telescope's design, see a two-volume proposal to the NSF, "The PALANTIR: A Concept for a 25 Meter Telescope" (Tucson, Jan. 1977), NOAO Library.
17. Ibid., vol. 2, section 1, 5–6; section 2, 27–29.
18. After inflation, $104 million in 1977 dollars; ibid., vol. 1, section 1.
19. Astronomers measure the resolving power and field of view of telescopes in degrees, arc-minutes, and arc-seconds. An arc-second is 1/3600 of a degree and is about equivalent to the angle covered by a dime seen from a mile away.
20. Memo, Goldberg to Hall, Nov. 24, 1976, Box 10, HUG FP 83.8.
21. See, for example, H. L. Johnson and W. L. Richards, "The Optimum Size of Infrared Photometric Telescopes," *Ap. J.* 160 (May 1970):L111; J. B. Oke, "On the Optimum Size of Optical Telescopes," *Ap. J.* 162 (Oct. 1970):L77–L78.
22. "NGT Report #5: A Concept for a 25 meter Telescope—The Singles Array" (Tucson: Kitt Peak National Observatory, 1978), NOAO Library.
23. M. J. Disney, "Optical Arrays," *Monthly Notices of the Royal Astronomical Society* 160 (1972):213–232.
24. Michael Disney interview, May 1, 1984, by Robert W. Smith, NASM Space History Collection.

25. Ibid.
26. Barr interview, Jan. 13, 1999.
27. Goldberg to Vera Rubin, Apr. 29, 1977, Box 10, HUG FP 83.8, LG/HUA.
28. Greenstein to William T. Golden, Dec. 17, 1972, "William Golden" folder, Box 12, JLG/CITA.
29. Greenstein to Goldberg, Apr. 24, 1972, "AURA 1972" folder, Box 95, JLG/CITA.
30. Memo, Goldberg to Helmut Abt, Nov. 13, 1972, Box 7, HUG FP 83.8, LG/HUA.
31. Revised CARSO Proposal for a Southern 200-inch Telescope from 1966, p. 25, "200-inch Telescope" folder, Box 38, JLG/CITA.
32. General memo from Graham Berry, Nov. 1974, and memo, C. J. Pings to Greenstein, Dec. 5, 1974, "KPNO Public Relations" folder, Box 19, JLG/CITA.
33. Goldberg to Gil Lee, Aug. 2, 1973, Box 9, HUG FP 83.8, LG/HUA.
34. Memo, Greenstein to R. F. Christy, Feb. 11, 1974, "AURA 1973–1974" folder, Box 95, JLG/CITA.
35. Greenstein to William Golden, Mar. 18, 1974, "AURA 1974" folder, Box 95, JLG/CITA.
36. Ibid.
37. Greenstein interview, May 19, 1978, by Spencer Weart, CHP/AIP Collection.
38. John Teem interview, Sept. 26, 1984, by Robert W. Smith, NASM Space History Collection.
39. Greenstein to William Golden, Oct. 9, 1974, "AURA 1974" folder, Box 95, JLG/CITA.
40. Described in Edmondson, *AURA and Its US National Observatories*, 237–239.
41. Memo, Goldberg to KPNO staff, Aug. 11, 1975, Box 9, HUG FP 83.8., LG/HUA.
42. On the debate over management of the Hubble Space Telescope, see Robert W. Smith, *The Space Telescope: A Study of NASA, Science, Technology, and Politics* (Cambridge, England: Cambridge University Press, 1989), ch. 6.
43. Goldberg, 1982 supplement interview, p. 163.
44. Diary note by Greenstein, Nov. 26, 1975, "AURA 1975" folder, Box 95, JLG/CITA.
45. Victor Blanco to Al Hiltner, Sept. 2, 1976, "Search Committee KPNO" folder, Box 2, HUG FP 83.6, LG/HUA.
46. Diary note by Greenstein, Nov. 26, 1975.

47. Goldberg to Greenstein, Apr. 16, 1976, "Search Committee KPNO" folder, Box 2, HUG FP 83.6, LG/HUA.
48. "The Future for AURA," "AURA 1976" folder, Box 108, JLG/CITA.
49. Addendum, Aug. 1976, "AURA Log," Box 7, HUG FP 83.41, LG/HUA. Both Goldberg and Greenstein kept logs of the conversations they had with each other and with other astronomers about issues surrounding Goldberg's retirement and salary as well as larger policy issues.
50. Randall interview, July 21, 2000, and memo, Randall to Goldberg, Sept. 14, 1976, "AURA History" folder, Box 7, HUG FP 83.41, LG/HUA.
51. Memo, Steve Strom to Beverly Lynds, Aug. 30, 1976, ibid.
52. Memo, Golden to Goldberg, May 14, 1976, "AURA Organization Committee" folder, Box 108, JLG/CITA. Golden to Greenstein, Oct. 26, 1976, "AURA 1976" folder, Box 108, JLG/CITA.
53. "Resumé of Conversations and Actions," Feb. 1976, "AURA History 1975–1977" folder, Box 108; Allan Sandage to Greenstein, undated, and George Preston to Greenstein, Nov. 4, 1976, "1976 AURA Mission" folder, Box 106, JLG/CITA.
54. "Draft of AURA Mission Statement," Sept. 1976, "1976 AURA Mission" folder.
55. Bev Oke and Jim Gunn to Greenstein, Nov. 5, 1976, ibid.
56. "Draft of AURA Mission Statement," Dec. 1976 (emphasis added), ibid.
57. Personal memo by Goldberg, Aug. 23, 1977, Box 10, LG/HUA.
58. Jerry E. Nelson interview, June 2, 1992, by Timothy Moy, p. 14, CITA.
59. Donald E. Osterbrock, John R. Gustafson, and W. J. Shiloh Unruh, *Eye on the Sky: Lick Observatory's First Century* (Berkeley: University of California Press, 1988), 264–265.
60. Nelson interview, June 2, 1992, p. 11.
61. Ibid., p. 17.
62. Ibid., p. 19.
63. "NGT report #3: A Concept for a 25 meter Telescope—The Steerable Dish" (Tucson: Kitt Peak National Observatory, 1977), NOAO Library.
64. Wampler's original idea was for 7-meter mirror. He later scaled it up to 10 meters; E. J. Wampler, "The University of California 10 Meter Telescope Project" in *Symposium on Recent Advances in Observational Astronomy*, eds. Harold Johnson and Christine Allen (Ensenada, Mexico, 1981), 147–151. The 200-inch mirror's thickness, for comparison, averages 22 inches.
65. David S. Saxon interview, Jan. 29, 1997, by Shirley K. Cohen, CITA.

66. Jerry Nelson, "The Proposed University of California 10-Meter Telescope," in *Optical Telescopes of the Future,* 133–141.
67. Ivan R. King, "Space and Ground-Based Optical Astronomy," ibid., 27–34.
68. Greenstein presented two talks: "Astronomical Implications of Future Very Large Telescopes" and "Conference Resume," both in *Optical Telescopes of the Future,* 525–536.
69. Barr interview, Nov. 18, 1998.
70. Geoffrey Burbidge interview, Mar. 17, 1999.
71. Reports of the Users' Committee appear in the *KPNO Newsletter,* for example, Feb. 1980 and Apr. 1981.
72. *KPNO Newsletter* (June 1981): 8.
73. Jacob Lubliner and Jerry Nelson, "Stressed Mirror Polishing 1: A Technique for Producing Nonaxisymmetric Mirrors"; Jerry Nelson et al., "Stressed Mirror Polishing 2: Fabrication of an Off-Axis Paraboloid," *Applied Optics* 19, 4 (1980):2332–2340, 2341–2351.
74. See Paul Ciotti, "Mr. Keck's Bequest," *Los Angeles Times Magazine,* May 24, 1987, 17–20; and the Keck Telescope oral histories at CITA. For a different view, see Harland Epps interview, June 27, 1988, by David DeVorkin, NASM Space History Collection.
75. William C. Keel, "Galaxies through a Red Giant," *S&T* (1992):626–632.
76. Fred H. Chaffee interview, Oct. 27, 2000.
77. Ray Weymann et al., "Multiple-Mirror Telescope Observations of the Twin QSOs 0957+561 A,B," *Ap. J.* 233 (1979):L43–46; Frederic H. Chaffee, Jr., "The Discovery of a Gravitational Lens," *Scientific American* (Nov. 1980):70–78.
78. Barr interview, Jan. 13, 1999.
79. Curtis Peebles, *Guardians* (Novato, Calif.: Presidio, 1987), 118.
80. Robert Smith and Joseph Tatarewicz, "Replacing a Technology: The Large Space Telescope and CCDs," *Proceedings of the IEEE* 73, 7 (1985):1221–1234.
81. James Janesick and Morley Blouke, "Sky on a Chip: The Fabulous CCD," *S&T* (1987):238–242.
82. Leo Goldberg, "Complementarity between Space and Ground-Based Developments: An Overview," in *Optical and Infrared Telescopes for the 1990s,* ed. Adelaide Hewitt (Tucson: KPNO, 1980), 131; "Panel Discussion: Space vs. Ground—Competition or Collaboration," ibid., 1151–1172.
83. "Planning and Reality: Money, Politics, and Science," in *Optical and Infrared Telescopes for the 1990s,* 1192.

84. Lawrence Randall, "A Broad Picture of Various Proposals for Large Telescopes," ibid., 1173.
85. "Planning and Reality: Money, Politics, and Science," ibid., 1191.
86. "Panel Discussion," ibid., 1165.
87. "Planning and Reality," ibid., 1181.

4. Paper Telescopes

1. M. J. Disney, "Post-Detector Arrays," in *Optical and Infrared Telescopes for the 1990s*, ed. Adelaide Hewitt (Tucson: KPNO, 1980), 447.
2. Sandy Faber, "Large Optical Telescopes: New Views into Space and Time," *Annals of the New York Academy of Sciences* 422 (1984):171.
3. Ibid., 172.
4. Robert P. Kirshner et al., "The Big Blank—Void in Space," *Scientific American* 246, 2 (1982):75.
5. George Field (chair), *Astronomy and Astrophysics for the 1980s: Report of the Astronomy Survey Committee* (Washington: National Research Council, 1982).
6. Larry Barr interview, Nov. 18, 1998. There was an earlier effort, for example, in 1978 for an innovative but short-lived Advanced Technology Telescope, which had an alt-az mount and a very thin primary mirror. The NSF supported the project by adding $100,000 to the NGT budget. In June 1979, Kitt Peak asked the NSF for $2 million to build the Advanced Technology Telescope, which an NSF subcommittee endorsed; shortly afterward, the NSF abandoned plans for it. "A Proposal for an Advanced Technology Telescope—the 2-meter ATT" (Tucson, June 1979) and "A Review of the KPNO 2-Meter ATT Project by the NSF's Optical/Infrared Subcommittee" (Washington, Sept. 1979), RDG personal papers.
7. Harlan Smith to Joseph Wampler, July 3, 1979, copied to Geoffrey Burbidge and John Teem, "UV-OIR panel 1980" folder, Box 106, JLG/CITA.
8. *KPNO Newsletter* 8 (Apr. 1980): 1.
9. Barr interview, Nov. 18, 1998.
10. *KPNO Newsletter* 17 (Dec. 1981): 3.
11. Hughes discusses this idea in a number of places, for example, *American Genesis: A Century of Invention and Technological Enthusiasm* (New York: Penguin, 1989), 71–72, and "The Evolution of Large Technological Systems," in *The Social Construction of Technological Systems*, eds. Weibe Bijker, Thomas Hughes, and Thomas Pinch (Cambridge, Mass.: MIT Press, 1987), 73–76.

12. Edward Nather quoted in Harlan Smith and Thomas Barnes, eds., *Report of the Optical Conference on the 7.6 meter Telescope* (Austin: University of Texas, McDonald Observatory, 1982), 43.
13. T. G. Barnes, "University of Texas 7.6-meter Telescope Project," in *Advanced Technology Optical Telescopes*, ed. Geoffrey Burbidge and Larry Barr (Bellingham, WA: SPIE, 1982), 102–108.
14. Roger Angel to George Field, July 13, 1979, SO/UA Working Files.
15. Neville Woolf and Roger Angel, "MT-2," in *Optical and Infrared Telescopes for the 1990s*, 1062–1150.
16. Peter Strittmatter interview, Sept. 29, 1998.
17. Memo to Kitt Peak staff, Dec. 9, 1977, "AURA Miscellaneous 10/1/77" folder, Box 1, HUG FP 83.16, LU/HUA.
18. Strittmatter interview, Oct. 14, 1998.
19. Ibid.
20. Background material on Angel comes from several interviews I did with him, including one on Mar. 19, 2001.
21. Roger Angel interview, Mar. 19, 2001.
22. Ibid.
23. John M. Hill interview, Sept. 24, 1998.
24. Leif J. Robinson, "Roger Angel: The Glass Giant of Arizona," *S&T* (July 1985):10–11.
25. Robert Kirshner quoted in William Sweet, "Keck Foundation Offers Caltech $70 million for 10-m Telescope," *Physics Today* 38, 2 (1985):74.
26. Susan Teska to Floyd Lance, Aug. 13, 1981, SO/UA Files.
27. Roger Angel and John Hill, "Honeycomb Mirrors of Borosilicate Glass," in *Scientific Importance of High Angular Resolution at Infrared and Optical Wavelengths* (Garching, Germany: ESO, 1981), 61–65.
28. This was not a new concept. In 1909, a professor of experimental physics at Johns Hopkins University used a rotating dish of mercury to obtain reasonably good images of the moon.
29. Mirror Lab Oven Pilot Logbook, vol. 1, SOML/UA Files. Logbook entries go back to Dec. 1981.
30. J. R. P. Angel, "Very Large Ground-Based Telescopes for Optical and IR Astronomy," *Nature* 295 (Feb. 25, 1982):651–657; "Large New-Technology Optical Telescopes Proposed," *Physics Today* 34, 8 (1981):17–20.
31. Dan Watson interview, Nov. 8, 1998.
32. Dennis Mammana, "The Incredible Spinning Oven," *S&T* (July 1985):7–9.
33. Roger Angel to Harlan Smith, May 20, 1982, and various letters in 1983 and 1984 between Kiichi Kodaira and Angel, SO/UA Files.
34. Sweet, "Keck Foundation Offers Caltech $70 million," 73.

35. Leif Robinson, "Geoffrey Burbidge: KPNO's Man at the Top," *S&T* (Sept. 1983):196.
36. Geoffrey Burbidge interview, Mar. 17, 1999.
37. W. C. Lewis and W. D. Shirkey, "Mirror Blank Manufacturing for the Emerging Market," in *Advanced Technology Optical Telescopes*, 307–309.
38. "Notes on the NNTT Design Workshop: Technology Development Program Report #3" (Tucson: Kitt Peak National Observatory, Nov. 1982).
39. *KPNO Newsletter* 22 (Aug. 1982):1.
40. The two competing designs are summarized in L. D. Barr et al., "The 15 Meter National New Technology Telescope: The Two Design Concepts," in *Advanced Technology Optical Telescopes II*, eds. B. Mack and L. Barr, vol. 444 (Bellingham, Wa.: SPIE, 1983), 37–46.
41. Garth Illingworth to Jesse Greenstein, June 4, 1980, copied to all members of UV/O/IR panel, "UVIOR Panel, NAS Astronomy Survey Committee (1980)" folder, Box 106, JLG/CITA.
42. Christopher A. Leidich and R. Bruce Pittman, eds., *Large Deployable Reflector Science and Technology Workshop* (Moffett Field, Calif.: Ames Research Center, 1982).
43. Joseph Wampler to Garth Illingworth, June 25, 1980, "UVIOR Panel, NAS Astronomy Survey Committee (1980)" folder, Box 106, JLG/CITA.
44. Field (chair), *Astronomy and Astrophysics for the 1980s: Report of the Astronomy Survey Committee*, 137.
45. This was originally the Subcommittee on Housing and Urban Development-Independent Agencies. The name was later changed to VA, HUD, and Independent Agencies. Richard Munson, *The Cardinals of Capitol Hill: The Men and Women Who Control Government Spending* (New York: Grove Press, 1993).
46. "Hearings before a Subcommittee of the Committee on Appropriations, Subcommittee on VA, HUD, and Independent Agencies," 97th Cong., Mar. 24, 1981, 107.
47. Memo, Goldberg to Donald Hall, Sept. 2, 1981, and memo, Donald Hall to KPNO staff, Dec. 9, 1981, HUG FP 83.10, LG/HUA.
48. Goldberg to Stuart Rice, July 13, 1982, and Rice to Goldberg, July 26, 1982, "AURA" folder, Box 7, HUG FP 83.41, LG/HUA.
49. Memo, Geoffrey Burbidge to the Scientific Advisory Committee, Mar. 9, 1983, RDG personal papers.
50. "Mar. 28, 1983 notes from SAC meeting," RDG personal papers.
51. Robert Gehrz interview, May 25, 1999.

52. Roger Davies interview, June 19, 2000.
53. Memo entitled "Science with the NNTT," Nov. 9, 1983, LDB personal papers.
54. Robert Gehrz interview, Nov. 23, 1998.
55. "Reasons for Selecting the Segmented Design for the National 15-meter Telescope," unpublished paper by Jerry Nelson and Terry Mast with running commentary by Roger Angel and Neville Woolf, June 1984, LDB personal papers.
56. Malcolm Savedoff to Geoffrey Burbidge, June 20, 1984, RDG personal papers.
57. *KPNO Newsletter* 32 (May 1984): 5.
58. Bruno Latour, *Science in Action* (Cambridge, Mass.: Harvard University Press, 1987), 253.
59. William Sweet, "Plans for New Technology Telescope Come into Sharper Focus," *Physics Today* 38, 1 (1985):92.
60. M. Mitchell Waldrop, "Mauna Kea: Coming of Age," *Science* 214, 4 (1981):1110–1114.
61. Sidney Wolff interview, Oct. 28, 1999, CHP/AIP Collection; *KPNO Newsletter* 32 (May 1984):1–2.
62. Jacques Beckers interview, June 1, 1999.
63. John T. Jefferies interview, Feb. 26, 1999.
64. Paul Ciotti, "Mr. Keck's Bequest," *Los Angeles Times Magazine* (May 24, 1987): 17–20.
65. Larry Barr interview, Dec. 2, 1998.
66. "July 13, 1984 Minutes of July SAC Meeting," RDG personal papers.
67. Gehrz interview, May 25, 1999.
68. "July 14, 1984 Minutes of July SAC Meeting," RDG personal papers.
69. Gehrz interview, May 25, 1999.
70. Barr interview, Dec. 2, 1998.
71. Memo, Burbidge to SAC members, Aug. 6, 1984, RDG personal papers.

5. Growing Pains

1. "Making Large Glass Honeycomb Mirrors," proposal to the NSF by NOAO and the University of Arizona (Tucson, Nov. 1987), SOML/UA Files.
2. Scott Barker, "Seeing Stars: The Vatican Comes to Mt. Graham," *Tucson Lifestyle* (Sept. 1993):105–108.
3. Robert Kirshner (chair), "Report of the Subcommittee on Large O/IR Telescopes" (Washington: NSF, 1986), AST/NSF Files.

4. Memo from NSF, Sept. 23, 1985, AST/NSF Files.
5. "UA Large Telescope Project," undated (probably 1984), SO/UA Files.
6. Memo from NSF, Sept. 23, 1985, AST/NSF Files.
7. This is far short of the 0.05 arc-second resolution of the Hubble Space Telescope, orbiting above the earth's atmosphere, and one of the reasons astronomers impatiently anticipated its launch.
8. Conrad Istock and Robert Hoffmann, *Storm over a Mountain Island: Conservation Biology and the Mt. Graham Affair* (Tucson: University of Arizona Press, 1995).
9. *NOAO Newsletter* 10 (June 1987):26–27.
10. Roger Angel interview, Oct. 8, 1998.
11. Ibid.
12. Maarten Schmidt interview, Jan. 21, 1992, and Wal Sargent interview, July 9, 1991, both by Timothy Moy, CITA.
13. See Paul Ciotti, "Mr. Keck's Bequest," *Los Angeles Times Magazine* (May 24, 1987):17–20, and a series of interviews on the history of Keck Telescope held at CITA.
14. William Sweet, "Keck Foundation Offers Caltech $70 million for 10-m Telescope," *Physics Today* 38, 2 (1985):71–75.
15. Jesse L. Greenstein, "All Hail the Keck Ten-Meter Telescope Project," *Physics Today* 38, 2 (1985):136.
16. Gerry Smith interview, Oct. 10, 1992, by Judith Goodstein, p. 28, CITA.
17. Jerry Nelson interview, June 2, 1992, by Timothy Moy, p. 62, CITA.
18. Ibid. Also Nelson interview, Apr. 12, 1996, dealing with scientific collaborations, CHP/AIP Collection.
19. David A. Allen, "Infrared Astronomy: An Assessment," *Quarterly Journal of the Royal Astronomical Society* 18 (1977):189.
20. H. W. Babcock, "The Possibility of Compensating Astronomical Seeing," *PASP* 65, 386 (1953):229–236.
21. Eva C. Freeman, *MIT Lincoln Laboratory: Technology in the National Interest* (Cambridge, Mass.: MIT Press, 1994).
22. John W. Hardy, "Adaptive Optics," *Scientific American* 270, 6 (1994):60–65.
23. Robert Q. Fugate interview, Nov. 27, 2000, CHP/AIP Collection.
24. Graham Collins, "Making Stars to See Stars: DOD Adaptive Optics Work Is Declassified," *Physics Today* 46, 2 (1992):18.
25. Numerous books describe the technology and politics of SDI; of particular interest for its broad coverage is Frances Fitzgerald, *Way Out There in the Blue: Reagan, Star Wars, and the End of the Cold War* (New York: Simon and Schuster, 2000).

26. Eliot Marshall, "Researchers Sift the Ashes of SDI," *Science* (1994):620–623.
27. Robert Fugate, personal communication, Oct. 9, 2001.
28. "3.5-meter Mirror Project Review," Nov. 20–21, 1991, SOML/UA Files.
29. Simon P. Worden, "Why Astronomers Should Love SDI," *S&T* 74, 4 (1987):340; also, responses to Worden's editorial in the Feb. and Mar. 1988 issues of *S&T* from Jerry Nelson and Timothy Ferris. On scientists' reaction to SDI in general, see Daniel S. Greenberg, *Science, Money, and Politics: Political Triumph and Ethical Erosion* (Chicago: University of Chicago Press, 2001), ch. 18.
30. "NOAO FY 1987 Program Plan," Apr. 1987, p. 26, AST/NSF Files.
31. Bob Davis, "Quiet Clout: How a House Staffer Wields Great Power over Policy Decisions," *Wall Street Journal* (June 30, 1989).
32. Richard Munson, *The Cardinals of Capitol Hill: The Men and Women Who Control Government Spending* (New York: Grove Press, 1993), 29. Richard Malow interview, Oct. 13, 1999.
33. Robert W. Smith, *The Space Telescope: A Study of NASA, Science, Technology, and Politics* (Cambridge, Mass.: Cambridge University Press, 1989), 173–174.
34. Malow interview, Oct. 13, 1999.
35. Irwin Goodwin, "VLBA: A Congressman's Victory over NSF Project," *Physics Today* 37, 10 (1984):56.
36. Irwin Goodwin, "Erich Bloch: On Changing Times and Angry Scientists at NSF," *Physics Today* 41, 8 (1988):47–52.
37. Greenberg, *Science, Money, and Politics*, 110.
38. Maarten Schmidt interview, Sept. 29, 2000. Mr. Bloch did not respond to requests for an interview.
39. Walter Sullivan, "New Design Makes Giant Telescopes Possible," *New York Times* (Feb. 12, 1985): C1.
40. "Hearings before a Subcommittee of the Committee on Appropriations, Subcommittee on VA, HUD, and Independent Agencies," 99th Cong., Mar. 26, 1985, 193–195.
41. Goetz Oertel and Richard Malow interview, Sept. 24, 1999.
42. "Hearings before a Subcommittee of the Committee on Appropriations, Subcommittee on VA, HUD, and Independent Agencies," 98th Cong., Mar. 14, 1984, 125.
43. George B. Field, "Three Years after the Field Report, How Is Astronomy Faring?" *Physics Today* 38, 4 (1985):144.
44. Sweet, "Keck Foundation Offers Caltech $70 Million," 74–75.
45. Jacques Beckers interview, June 1, 1999.

46. Ibid.
47. John Jefferies interview, Feb. 26, 1999. Memo, Wayne Van Citters to Sidney Wolff, Oct. 26, 1987, AST/NSF Files.
48. Kurt Riegel to John Jefferies, Dec. 4, 1985, AST/NSF Files.
49. Walter Sullivan, "Large Computer-Guided Telescopes Readied," *New York Times* (Aug. 5, 1986): C1.
50. Valérie De Lapparent, Margaret Geller, and John Huchra, "A Slice of the Universe," *Ap. J.* 302 (1986):L1–L5.
51. D. Lynden-Bell et al., "Spectroscopy and Photometry of Elliptical Galaxies: Galaxy Streaming toward the New Supergalactic Center," *Ap. J.* 326 (1988):19–49.
52. Goetz Oertel interview, Sept. 24, 1999. Also, Oertel interviews, Apr. 23, 1999; Nov. 29, 1999; and Apr. 9, 2001 (the last at CHP/AIP).
53. Sidney Wolff interview, Dec. 18, 1998.
54. Oertel interview, Sept. 24, 1999.
55. Memo, Goetz Oertel to AURA Board Retreat Participants, July 21, 1987, "Agenda and Issues for Board Retreat," SO/UA Files.
56. Steve Strom interview, Jan. 19, 1999. Frank Low interview, Nov. 30, 1998.
57. Steve Strom (chair), "Report of the Future Directions for NOAO Committee," AURA, Sept. 15, 1987, NSF-AST files.
58. Frank Low interview, Feb. 23, 1999.
59. Ibid.
60. Strom (chair), "Report of the Future Directions."
61. *NOAO Newsletter* 12 (Dec. 1, 1987):1.
62. Roger Davies interview, June 19, 2000.
63. "Report of Subcommittee on NOAO," prepared by the Advisory Committee for Astronomical Sciences (Gerald Tape, chair), June 1988, AST/NSF Files.
64. Greenstein to Lawrence Aller, June 4, 1990, "Aller" folder, Box 1, JLG/CITA.

6. Astropolitics

1. John Irvine and Ben Martin, "Assessing Basic Research: The Case of the Isaac Newton Telescope," *Social Studies of Science* 13 (1983):50.
2. Brian Salter and Ted Tapper, "The Application of Science and Scientific Autonomy in Great Britain: A Case Study of the Science and Engineering Research Council," *Minerva* 31, 1 (1993):38–55.
3. Malcolm Longair interview, June 15, 2000.
4. Matt Mountain interview, Oct. 24, 2000.

5. Michael Edmunds interview, June 20, 2000; Roger Davies interview, June 19, 2000.
6. Edmunds interview, June 20, 2000.
7. David H. Smith, "Alec Boksenberg: King of the Castle," *S&T* 67 (Apr. 1984):312–315.
8. Longair interview, June 15, 2000.
9. Mountain interview, Oct. 24, 2000.
10. Richard Ellis interview, May 1, 2000.
11. M. G. Edmunds and R. S. Ellis, "A UK Strategy for Future Ground-Based Telescopes," Oct. 23, 1986, RSE personal papers; David S. Smith, "British Plan for the Year 2000," *S&T* 73, 5 (1987):497.
12. John Jefferies to Richard Ellis, Mar. 12, 1987; Ellis to Jefferies, May 11, 1987; Sidney Wolff to Ellis, May 28, 1987, RSE personal papers.
13. M. G. Edmunds, "United Kingdom Plans for Very Large Telescopes," in *Very Large Telescopes and Their Instrumentation*, ed. M. H. Ulrich (Garching, Germany: ESO, 1988), 193–197.
14. Ian Corbett interview, Apr. 6, 2000.
15. Goetz Oertel to Richard Ellis, Sept. 3, 1987, RSE personal papers.
16. David Smith to Richard Ellis, Mar. 8, 1988, RSE personal papers. "Astronomers Set Their Sights on Bigger, Better Telescopes," *Times Higher Education Supplement* (July 1, 1988).
17. Richard M. West, "Europe's Astronomy Machine," *S&T* 75 (May 1988):471–481.
18. *ESO Messenger* 50 (Dec. 1987):1.
19. "Hearings before the House Subcommittee on Science, Research, and Technology," 101st Cong., Mar. 14, 1989, 437.
20. Daniel Fischer, "A Telescope for Tomorrow," *S&T* 78 (Sept. 1989):249–252.
21. "Fine Images from New Technology Telescope," *Nature* 343, 6260 (1990):685.
22. Sidney Wolff interview, Dec. 18, 1998.
23. "Hearings on Science, Research, and Technology," Mar. 14, 1989, 375–391.
24. Alan MacRobert, "U.S Astronomy in Crisis," *S&T* 75 (May 1988):469–470.
25. "Hearings on Science, Research, and Technology," Mar. 14, 1989, 453.
26. "Columbus on the Shoals," *S&T* 83 (Apr. 1992):370.
27. Letters from Oertel and Wolff, Apr. 9, 1988, and Aug. 15, 1988, RSE personal papers; "Budget Ax Falls," *S&T* 75 (Sept. 1988):239.
28. "The NOAO 8-m Telescopes" (Washington: Association of Universities for Research in Astronomy, 1989), NOAO Library.

29. Ibid., vol. 1, 17.
30. Ibid., 3.
31. Ibid., 22.
32. Ibid., ix.
33. Ibid., 20.
34. "Panel Discussion," in *Conference on Research Programmes for the New Large Telescopes,* ed. A. Reiz (Geneva: ESO/SRC/CERN, 1974), 394.
35. "Report to Accompany H.R. 5158," VA/HUD/IND Agencies Appropriation Bill, 101st Cong., Sept. 26, 1990, 153.
36. "Hearings before a Subcommittee of the Committee on Appropriations, Subcommittee on VA, HUD, and Independent Agencies," 98th Cong., Mar. 14, 1984, 97.
37. "Hearings before a Subcommittee of the Committee on Appropriations, Subcommittee on VA, HUD, and Independent Agencies," 100th Cong., Mar. 3, 1987, 21.
38. Bloch to Tor Hagfors, Aug. 25, 1988, AST/NSF Files.
39. "Minutes of Joint Working Group," Oct. 17–18, 1988, p. 1, AST/NSF Files.
40. Richard Malow interview, Sept. 24, 1999.
41. Report, Sidney van den Bergh to James Hesser, Sept. 19, 1988, AST/NSF Files.
42. Simon Lilly interview, Oct. 6, 1999.
43. *NOAO Newsletter 22* (June 1990):2.
44. Goetz Oertel interview, Sept. 24, 1999.
45. Ann Gibbons, "Astronomers Want New Optical Telescopes, but . . ." *Science* 248 (May 18, 1990):806–807.
46. E-mail (with author deleted), May 7, 1991, circulated at the NSF, NOAO, and among Canadian astronomers, AST/NSF Files.
47. Gibbons, "Astronomers Want New Optical Telescopes," 807.
48. "SERC Funds 8 Metre Telescope Design Study," Jan. 16, 1989, SERC press release (PN 03:89), RSE personal papers.
49. "Spoilt for Choice," *Economist* (July 14, 1990).
50. "Summary for LT Options by Alec Boksenberg," Mar. 2 1990, RSE personal papers.
51. T. Lee to Richard Ellis, Jan. 31, 1990, RSE personal papers.
52. Van Citters to A. E. Hughes, June 11, 1990, AST/NSF Files.
53. Erich Bloch to William Mitchell, July 12 and Aug. 10, 1990, AST/NSF Files.
54. Richard Ellis interview, May 1, 2000.
55. Marcia Bartusiak, *Einstein's Unfinished Symphony: Listening to the Sounds of Space Time* (Washington: Joseph Henry Press, 2000).

56. "Hearings before a Subcommittee of the Committee on Appropriations, Subcommittee on VA, HUD, and Independent Agencies," 101st Cong., Feb. 27, 1990, 72.
57. Ibid., 77.
58. Ibid., 115.
59. Dick Malow interview, Sept. 24, 1999.
60. Corbett interview, Apr. 6, 2000.
61. Paul Hanle, "Astronomers, Congress, and the Large Space Telescope," S&T 69 (Apr. 1985):300–305.
62. John Bahcall interview, Dec. 2, 1999; Irwin Goodwin, "Academy Group Observes Astronomy in Cloudy Period of Budget Cuts," *Physics Today* 42 (Dec. 1989):45–46.
63. Bahcall interview, Dec. 2, 1999.
64. John Bahcall (chair), *Astronomy and Astrophysics Panel Reports: Working Papers* (Washington: National Academy of Sciences Press, 1991), section 14, 2–9.
65. See papers by Helmut Abt, including "Some Trends in American Astronomical Publications," *PASP* 93, 553 (1981):269–272; "The Growth of Multi-wavelength Astrophysics," *PASP*, vol. 105 (1993):437–439; "Trends toward Internationalization in Astronomical Literature," *PASP* 102, 649 (1990):368–372.
66. The NSF and NASA agreed in March 1959 that ground-based astronomy was in the NSF's bailiwick. Over time, this separation blurred somewhat as NASA funded ground-based work that strongly supported its space program. See also *Federal Funding of Astronomical Research* (Washington: National Academy of Sciences Press, 2000).
67. Between 1988 and 1997, 42 percent of the most-cited astronomical papers—a good metric of the research astronomers consider to be most important—were based on information collected from ground-based optical telescopes. Helmut Abt, "The Most Frequently Cited Astronomical Papers Published During the Past Decade," *Bulletin of the American Astronomical Society* 32, 3 (2000):937–941.
68. Wayne Van Citters interview, July 24, 2000.
69. Robert Q. Fugate interview, Nov. 27, 2000, CHP/AIP Collection. See also Malcolm Browne, "Anti-Missile Technology Delights Astronomers," *New York Times* (Aug. 6, 1991): B1.
70. John Bahcall (chair), *The Decade of Discovery in Astronomy and Astrophysics* (Washington: National Academy Press of Sciences, 1991), 82–3.
71. R. Cargill Hall, "Missile Defense Alarm: The Genesis of Space-Based Infrared Early Warning," NRO Office, July 1988. My thanks to Dr. Hall for providing a copy of this formerly classified paper.

72. NASA engineers began trying to declassify some of the military's technology in the late 1970s; see "Infrared Astronomy: Scientific/Military Thrusts and Instrumentation," in Nancy Boggess and Howard Stears, eds., *Infrared Astronomy: Scientific/Military Thrusts and Instrumentation*, vol. 280 (Bellingham, Wash.: SPIE, 1981); William E. Burrows, *Deep Black: Space Espionage and National Security* (New York: Random House, 1986).
73. David Edge, "Mosaic Array Cameras in Infrared Astronomy," in *Invisible Connections: Instruments, Institutions, and Science*, eds. Robert Bud and Susan Cozzens (Bellingham, Wash.: SPIE Optical Engineering Press, 1992), 160; Ian Gatley, D. L. DePoy, and A. M. Fowler, "Astronomical Imaging with Infrared Array Detectors," *Science* 242 (Dec. 8, 1988):1264–1270.
74. Bahcall (chair), *Decade of Discovery*, 75.
75. Bahcall interview, Dec. 2, 1999.
76. Frank Low interview, Feb. 23, 1999.
77. Frank Low interview, Nov. 30, 1998.
78. Quotes from Bahcall interview, Dec. 2, 1999.
79. John Bahcall interview, Aug. 26, 1992, by Robert W. Smith, pp. 34–36, NASM Space History Collection.
80. Robert Smith, "The Biggest Kind of Big Science: Astronomers and the Space Telescope," in *Big Science: The Growth of Large Scale Research*, eds. P. Galison and B. Hevly (Stanford: Stanford University Press, 1992), 184–211.
81. Greenstein to Bahcall, Apr. 11, 1991, "Bahcall" folder, Box 2, JLG/CITA.

7. Smoke and Mirrors

1. Robert W. Smith, *The Space Telescope: A Study of NASA, Science, Technology, and Politics* (Cambridge, England: Cambridge University Press, 1989), 400–401. Eric Chaisson, *The Hubble Wars* (Cambridge, Mass.: Harvard University Press, 1994), gives a personal account of the HST program.
2. Smith, *Space Telescope*, 414; Leif J. Robinson, "Hubble's Troubles: Reflections from the Editor." *S&T* 80 (Oct. 1990):340–341.
3. Bertram Schwarzschild, "Hubble's Primary Mirror Has the Wrong Shape," *Physics Today* 43, 8 (1990):17.
4. Bertram Schwarzschild, "Hubble Investigation Board Finds Out What Went Wrong," *Physics Today* 43, 11 (1990):19–21; Angel quoted in Chaisson, *Hubble Wars*, 227.

5. "The Hubble Space Telescope Optical Systems Failure Report" (NASA, Nov. 1990), esp. 9-2, 9-4.
6. Daniel Kevles, "Big Science and Big Politics in the United States: Reflections on the Death of the SSC and the Life of the Human Genome Project," *Historical Studies in the Physical and Biological Sciences* 27, 2 (1997):269–297.
7. Jerry Nelson interview, June 2, 1992, by Timothy Moy, CITA.
8. E-mail, Richard Ellis to Large Telescope Panel and SERC, July 5, 1990, RSE personal papers.
9. Robert C. Bless interview, Nov. 3, 1983, by Robert Smith, NASM Space History Collection; and Bless interview, July 28, 2000.
10. Bless interview, July 28, 2000.
11. Goetz Oertel to Jim Houck, Jan. 14, 1993, AST/NSF Files.
12. E-mail, Corbett to Van Citters, July 1, 1992; Corbett to Bernard Burke, Mar. 4, 1993, AST/NSF Files.
13. Goetz Oertel interview, Mar. 22, 2000.
14. Oertel to the AURA Board, Nov. 5, 1991, AST/NSF Files.
15. Ian Corbett interview, Apr. 6, 2000.
16. Richard Malow interview, Sept. 24, 1999.
17. "Hearings before a Subcommittee of the Committee on Appropriations, Subcommittee on VA, HUD, and Independent Agencies," 102nd Cong., Mar. 5, 1991, 93.
18. Ibid., 101.
19. Peter Aldhous and David Lindley, "Canada Gets Cold Feet," *Nature* 351 (June 27, 1991):680.
20. David Crampton and Gordon Walker interviews, Nov. 7, 2000 (Canadian astronomers involved in Canadian negotiations). Also Donald Morton to "Dear Colleague," July 11, 1991, AST/NSF Files.
21. Bahcall interview, Dec. 2, 1999.
22. Bahcall to Sidney Wolff, Dec. 4, 1991, AST/NSF Files.
23. Bahcall to Survey Members, Dec. 4, 1991, AST/NSF Files.
24. Hall to Wayne Van Citters, Feb. 4, 1992; confidential memo from Corbett, Mar. 9, 1992, RSE personal papers.
25. Sen. Daniel Inouye to Walter Massey, Aug. 3, 1992, AST/NSF Files.
26. Malcolm Longair interview, June 15, 2000.
27. Internal memo from Corbett, Jan. 10, 1992, RSE personal papers; Corbett to Van Citters, Dec. 24, 1992, AST/NSF Files (emphasis in original).
28. Oertel to John Bahcall, Nov. 27, 1991, AST/NSF Files.
29. Oertel to Malow, Feb. 14, 1992, AST/NSF Files.
30. Corbett's confidential "Note of Record: Interim Gemini Board," early May 1992, RSE personal papers.

31. Ibid.; "Minutes of May 1992 Gemini Board Meeting," AST/NSF Files; Larry Randall interview, July 21, 2000.
32. Corbett to Oertel, Dec. 11, 1992, RSE personal papers.
33. Leif J. Robinson, "Let's Have the Truth about Gemini," *AAS Newsletter* 65 (June 1993):2.
34. "Minutes of Interim Gemini Board of Directors Meeting," Oct. 10–11, 1991, AST/NSF Files.
35. Wayne Van Citters and Julie Lutz to Michael Cusanovich, Oct. 28, 1991, AST/NSF Files.
36. Steve Hinman, personal communication, Nov. 3, 1998.
37. Leif Robinson, "Spinning a Giant Success," *S&T* (July 1992):26–31.
38. Anthony J. Goffigan to Roger Angel, Apr. 1, 1992; Angel to Goffigan, Apr. 8, 1992, SO/UA Files.
39. "Proposal for Fabrication of Telescope Primary Mirror Blanks for the Gemini Project," Option 3 and Option 6, SOML/UA Files.
40. Direct information on Corning's bid remains proprietary. Indirect information came from Mirror Lab correspondence (for example, "What Happened with Gemini?" undated internal Steward Observatory memo from fall 1992), several oral history interviews, and Leif Robinson and Jack Murray, "The Gemini Project: Twins in Trouble?" *S&T* (May 1993):26–32.
41. ESO was thought to have paid around $8 million and Japan about $10 million for single mirror blanks.
42. W. Lewis and W. Shirkey, "Mirror Blank Manufacturing for the Emerging Market," in *International Conference on Advanced Optical Telescopes*, eds. Geoffrey Burbidge and Larry Barr, vol. 332 (Tucson: SPIE, 1982), 307–309.
43. Margaret B. W. Graham and Alex T. Shuldiner, *Corning and the Craft of Innovation* (New York: Oxford University Press, 2001), esp. 391–449. Also informal interview with Corning manager Robert Jones, July 20, 2000.
44. "Gemini 8-M Mirror Solicitation, Source Selection Advisory Committee Report to Larry Randall," Aug. 1992, p. 6, AST/NSF Files.
45. B. E. Powell to Larry Daggert, Feb. 6, 1989, SO/UA Files.
46. For example, if a representative of AURA wanted to visit the Mirror Lab, Gemini would be billed by the hour while the person was on the premises. "Final Evaluation Report for Request for Proposal #800842: Fabrication of the Gemini Project Primary Mirror Blanks," Aug. 4, 1992, AST/NSF Files.
47. "Gemini 8-M Mirror Solicitation, Source Selection Advisory Committee Report to Larry Randall," Aug. 1992, p. 6, AST/NSF Files.

48. Charles Beichman to Wayne Van Citters, Apr. 20, 1992, and e-mail, Jim Houck to Van Citters, Apr. 22, 1992, AST/NSF Files.
49. Matt Mountain interview, Oct. 24, 2000.
50. "What Happened with Gemini?" internal draft document, Fall 1992, SO/UA Files; Larry Barr interview, Dec. 3, 1998.
51. Roger Angel interview, Nov. 5, 1998.
52. Harland Epps quoted in Robinson and Murray, "Gemini Project," 27.
53. John McGraw to Bob Bless, Oct. 13, 1992, SO/UA Files.
54. Longair interview, June 15, 2000.
55. Memo, Stephen Hinman to Thomas Thompson, Feb. 14, 1992, SOML/UA Files.
56. Steve Hinman, personal communication, Nov. 3, 1998.
57. NASA did contribute some funding for the second Keck and a fraction of observing time was subsequently made available to the community. The lion's share of time on the two Keck telescopes, however, still went to Caltech and UC astronomers.
58. Jay Gallagher to Van Citters, Nov. 5, 1992, AST/NSF Files.
59. Ibid.
60. Draft letter, Roger Angel to Walter Massey, Fall 1992, SO/UA Files.
61. John McGraw to Bob Bless, Oct. 13, 1992, SO/UA Files.
62. E-mail, Van Citters to Sidney Wolff, Sept. 24, 1992, AST/NSF Files.
63. E-mail, Marcia Rieke to John Bahcall, Nov. 11, 1992, AST/NSF Files.
64. An Apr. 22, 1992 e-mail from James R. Houck to Van Citters complained, for example, that Gemini had diminished science potential because it "promises to be all things to all people"; AST/NSF Files.
65. Roger Davies and Keith Raybould, "Technical Description of the U.K. Large Telescope," in *Advanced Optical Telescopes IV*, vol. 1236 (Bellingham, WA: SPIE, 1990), 26–40; Roger Davies interview, June 19, 2000; Simon Lilly interview, Oct. 6, 1999.
66. E-mail, Bless to Van Citters, Nov. 6, 1992, AST/NSF Files.
67. "Minutes of Gemini Informational Meeting," Nov. 8, 1992; Peter Strittmatter to R. C. Bless and Peter S. Conti, Nov. 12, 1992, AST/NSF Files.
68. "Notes on Gemini Board Meeting, 9–11 Nov. 1992," RSE personal papers.
69. Longair interview, June 15, 2000.
70. Mountain interview, Oct. 24, 2000.
71. Augustus Oemler to Van Citters, Nov. 12, 1992; William van Altena to Sen. Barbara Mikulski, Nov. 13, 1992; Bruce Carney to John Bahcall, Nov. 13, 1992, all in AST/NSF Files.
72. E-mail, Wolff to Van Citters, Nov. 19, 1992, AST/NSF Files. Other ac-

counts of the meeting include memo from Peter Strittmatter, Nov. 30, 1992, SO/UA Files, and memo, Aaron Asrael to Van Citters, Jan. 25, 1993, AST/NSF Files.
73. Richard Malow interview, Oct. 13, 1999.
74. Barbara Mikulski and Bob Traxler to Walter Massey, Dec. 9, 1992, AST/NSF Files.
75. John Huchra interview, Feb. 14, 2002, CHP/AIP Collection.
76. Chaisson, *Hubble Wars,* 201–202; Gunn to Van Citters, Dec. 28, 1992, AST/NSF Files.
77. Roberta Humphreys to Van Citters, Jan. 14, 1993, AST/NSF Files.
78. Oertel to Houck, Jan. 14, 1993, AST/NSF Files.
79. Draft letter, Sandra Faber to Committee on Astronomy and Astrophysics, Apr. 19, 1993, AST/NSF Files.
80. John Huchra, personal communication, Feb. 14, 2002.
81. "Findings of the Gemini Review Committee," Feb. 13, 1993, AST/NSF Files.
82. Mountain interview, Oct. 24, 2000; Randall interview, July 21, 2000.
83. Quoted in Robinson and Murray, "Gemini Project," 28.
84. Kapper to William Harris, Feb. 19, 1993, AST/NSF Files. Kapper, Letter to the Editor, *S&T* (Sept. 1993): 7.
85. John Bahcall to Bob Bless, Feb. 11, 1993, AST/NSF Files.
86. "Potential Impacts of Cancellation of Mirror Blank Contract with Corning," Feb. 9, 1993, internal NSF memo, AST/NSF Files.
87. *Gemini Project Newsletter* 5 (June 1993): 3.
88. William Harris to Kevin Kelly, Feb. 24, 1993, AST/NSF Files.
89. E-mail, Houck to his committee, Mar. 3, 1993; William Harris to Barbara Mikulski, Mar. 1, 1993, AST/NSF Files.
90. E-mail, Bless to Van Citters, Apr. 9, 1993, AST/NSF Files; Robinson and Murray, "Gemini Project," 32.
91. See "Gemini Project Controversy," *AAS Newsletter* 64 (Mar. 1993): 2–3; Faye Flam, "Mirror, Mirror, Which Is the Fairest?" *Science* 260 (Apr. 23, 1993): 483–484.
92. Mountain interview, Oct. 24, 2000.
93. Gerry Smith interview, Oct. 10, 1992, by Judith Goodstein, p. 34, CITA.
94. Mountain interview, Oct. 24, 2000.
95. Bless interview, July 28, 2000.
96. For example, "Mirror Technology Workshop," Gemini Project Report RPT-O-G0050, Gemini Library, Hilo, HI.
97. Jim Erickson, "UA Mirrors Rejected for Telescopes," *Arizona Daily Star* (Dec. 13, 1993): 1B.
98. Bahcall interview, Dec. 2, 1999.

8. Joining the 8-Meter Club

1. For Gilliss's trip and quotations from his notes, see Wendell F. Huffman, "The United States Naval Astronomical Expedition (1849–52) for the Solar Parallax," *Journal for the History of Astronomy* 22, 3 (1991):208–220.
2. Background on CTIO is drawn from several sources, including Victor M. Blanco, "Telescopes, Red Stars, and Chilean Skies," *Annual Reviews of Astronomy and Astrophysics* 39 (2001):1–18, and Blanco's unpublished account of early CTIO history from Goetz Oertel's personal files.
3. Goetz Oertel interview, Nov. 29, 1999.
4. Adriaan Blaauw, *ESO's Early History: The European Southern History from Concept to Reality* (Munich: ESO, 1991).
5. Maria Teresa Ruiz interview, Dec. 8, 2001.
6. Malcolm Smith interview, Jan. 21, 2002.
7. Memo from Gallagher to Goetz Oertel, Aug. 27, 1990, GKO personal papers.
8. Richard Malow interview, Sept. 24, 1999.
9. Goetz Oertel interview, Sept. 24, 1999.
10. Harry Barnes to Goetz Oertel, Apr. 27, 1992, GKO personal papers. Memo from Oertel to Fred Berenthal, May 26, 1992, AST/NSF Files.
11. "Minutes of May 1992 Gemini Board Meeting," AST/NSF Files.
12. A. W. Rodgers to Bob Bless, Feb. 19, 1992, AST/NSF Files. E-mail from group of Australian astronomers to A. R. Hyland, July 1992; M. H. Brennan to Walter Massey, Nov. 16, 1992, RSE personal papers.
13. "Minutes of the Feb. 1993 Gemini Board Meeting," AST/NSF Files.
14. Goetz Oertel interview, Apr. 9, 2001.
15. *Gemini Newsletter* 5 (June 1993):1.
16. Oertel interview, Apr. 9, 2001.
17. "The Gemini Telescope Project Act of 1996" was first introduced by Hawaii's Sen. Inouye. Robert Gehrz to Goetz Oertel, Nov. 26, 1996, AST/NSF Files.
18. For example, Leopoldo Infante and Nikolaus Vogt to Goetz Oertel, Sept. 2, 1992, AST/NSF Files.
19. Draft letter from Oertel, Sept. 19, 1993, AST/NSF Files.
20. Quoted in Denis Cioffi, "ESO Ships Steel to Chile Despite Labor and Land Conflicts," *Physics Today* 47, 12 (1994):55.
21. Telegram from U.S. Embassy in Santiago to the NSF, Apr. 4, 1997, AST/NSF Files.
22. E-mail from Van Citters to Donald Morton and Ian Corbett, Dec. 9, 1996, AST/NSF Files.

23. "Chile Contribution to the Gemini 8-Meter Telescope Program," NSF report, June 1997, AST/NSF Files.
24. Reaction to Chile's membership: "Brief on Chilean Default," NSF internal memo, June 12, 1997, AST/NSF Files. Jeffrey Mervis, "Gemini Woos Australia to Replace Chile," *Science* 277 (Aug. 8, 1997):758–759.
25. E-mail from Van Citters, Aug. 29, 1997, AST/NSF Files. Toni Feder, "Chile Rejoins Gemini as a Full-Fledged Member," *Physics Today* 50, 10 (1997):94–95.
26. Feder, "Chile Rejoins," 95.
27. E-mail exchanges between Corbett and Van Citters, Sept. 17 and 18, 1997, AST/NSF Files. I should note that Corbett reminded Van Citters his views were personal and not representative of British science policy.
28. Draft e-mail (never sent) from Van Citters to Hugh van Horn, Oct. 7, 1997; memo from Van Citters et al. to Neal Lane, Nov. 21, 1997, AST/NSF Files.
29. Peter Pockley, "Seventh Heaven for Australia's Optical Astronomers," *Nature* 391 (Feb. 19, 1998):724.
30. Page 371 of *Large Telescope Design*.
31. Patrick Osmer and Todd Boroson, "Report on October Workshop on O/IR Ground-Based System," presented at a Dec. 6, 2000, meeting of the Committee on Astronomy and Astrophysics at the National Academy of Science. Meeting notes in author's collection.
32. Douglas Simon, David Robertson, and Matt Mountain, "Gemini Telescopes' Instrumentation Program," in *Infrared Detectors and Instrumentation for Astronomy*, ed. Albert Fowler (Bellingham, Wash.: SPIE, 1995), 296–307.
33. Sidney Wolff to the Gemini Board, May 17, 1993, AST/NSF Files.
34. H. H. Aumann et al., "Discovery of a Shell Around Alpha Lyrae," *Astrophysical Journal* 278 (Mar. 1, 1984):23–27.
35. Frank Low, "Frederick Gillett (1937–2001)," *Nature* 411 (June 21, 2001); "Passing of Fred Gillett, Infrared Astronomy Pioneer," NOAO Press Release 01–08, Apr. 26, 2001.
36. Richard Ellis interview, May 1, 2000.
37. "Notes on Instrument Development Program in Minutes of November 1996 Gemini Board Meeting," AST/NSF Files.
38. Richard Kurz interview, Apr. 6, 2000; Jim Oschmann interview, Apr. 17, 2002.
39. John Huchra, personal communication, Feb. 14, 2002.
40. Matt Mountain interview, Nov. 21, 2001.
41. Corbett to Van Citters, Mar. 9, 1995, RSE personal papers.

42. Richard Kurz interview, Apr. 6, 2000.
43. Jim Oschmann interview, Apr. 17, 2002. Alfred Chandler, *The Visible Hand: The Managerial Revolution in American Business* (Cambridge, Mass.: Harvard University Press, 1977).
44. See Peter Galison, *Image and Logic: A Material Culture of Microphysics* (Chicago: University of Chicago Press, 1997), esp. 405–406.
45. Douglas Simons, Fred Gillet, and Richard McGonegal, "Gemini Instrumentation Program Overview," in *Optical Telescopes of Today and Tomorrow: Following in the Footsteps of Tycho Brahe*, ed. Arne Ardeberg, vol. 2871 (Bellingham, Wash.: SPIE, 1996).
46. Fred C. Gillett, "Infrared Arrays for Astronomy," in *Infrared Detectors and Instrumentation for Astronomy*, 2–7.
47. Roger Davies et al., "GMOS: The GEMINI Multiple Object Spectrometer," in *Optical Telescopes of Today and Tomorrow*, 1100.
48. Matt Mountain interviews, Oct. 24, 2000, and Nov. 21, 2001.
49. Jean-René Roy interview, Nov. 28, 2001.
50. Stephen T. Ridgway, "Scientific Programs in Adaptive Optics: An Overview and Commentary," in *Adaptive Optical System Technologies*, eds. Domenico Bonaccini and Robert K. Tyson (Bellingham, Wash.: SPIE, 1998), 438–446.
51. F. Gillet et al., "Future Gemini Instrumentation: Report of the First Gemini Instrumentation Workshop" (Tucson: Gemini 8-M Telescopes Project, 1997).
52. Michel Mayor and Didier Queloz, "A Jupiter-Mass Companion to a Solar-Type Star," *Nature* 378 (Nov. 23, 1995):355.
53. Mountain interview, Oct. 24, 2000.
54. Steven Shapin, "The Invisible Technician," *American Scientist* 77 (1989):554–563.
55. Judith Cohen, "Letter: How to Ensure That No New Instruments Are Built for Ground-Based Telescopes," *AAS Newsletter* (1998):2. In my interviews with astronomers, I typically asked about the status of instrument builders in their community; almost all echoed Cohen's concerns.

9. Point-and-Click Astronomy

1. Alan Dressler interview, Nov. 15, 1999.
2. Quote in Todd Boroson, John Davies, and Ian Robson, eds., *New Observing Modes for the Next Century* (San Francisco, Calif.: Astronomical Society of the Pacific, 1996), 240.
3. Ibid., 249.

4. Ibid., 256.
5. Sandy Faber, "Large Optical Telescopes: New Views into Space and Time," *Annals of the New York Academy of Sciences* 422 (1984):173.
6. Stephen Cole, "Astronomy on the Edge: Using the Hubble Space Telescope," *S&T* 84, 4 (1992):391.
7. Ibid., 390–392.
8. Helmut Abt and Sarah Stevens-Rayburn, "Publication Statistics for Recent Papers from the Hubble Space Telescope," *Bulletin of the American Astronomical Society* 34 (2001):935–937.
9. Jim Houck interview, Apr. 24, 2002. His comments were affirmed by many other astronomers.
10. T. Boroson et al., "The WIYN Queue: Theory Meets Reality," in *Observatory Operations to Optimize Scientific Return*, ed. Peter Quinn (Bellingham, Wash.: SPIE, 1998), 41–49.
11. Richard Ellis interview, May 1, 2000.
12. "Project Scientist's Outlook," *Gemini Newsletter* 9 (Oct. 1994).
13. See, for example, "Panel Discussion," in *Conference on Research Programmes for the New Large Telescopes*, ed. A. Reiz (Geneva: ESO/SRC/CERN, 1974), 392.
14. "Panel Discussion," in Boroson, Davies, and Robson, eds., *New Observing Modes for the Next Century*, 245, 249.
15. ESO astronomer Dietrich Baade, ibid., 40.
16. Peter Galison, *Image and Logic: A Material Culture of Microphysics* (Chicago: University of Chicago Press, 1997), 403–404.
17. Matt Mountain interview, Oct. 24, 2000.
18. C. R. Benn and S. F. Sanchez, "Scientific Impact of Large Telescopes," *PASP* 113, 781 (2000):385–396.
19. Maarten Schmidt interview, Sept. 21, 2000.
20. E-mail from Ellis, Aug. 4, 1997, RSE personal papers.
21. "Minutes of June 1996 Gemini Board Meeting," AST/NSF Files.
22. Matt Mountain, in Boroson, Davies, and Robson, eds., *New Observing Modes for the Next Century*, 244.
23. From p. 111 of "A New Eye Opens on the Cosmos."
24. Executive Office of the President, *A Blueprint for New Beginnings: A Responsible Budget for America's Priorities* (Washington: GPO, 2001), 161.
25. Norman R. Augustine, chair, *U.S. Astronomy and Astrophysics: Managing an Integrated Program* (Washington: National Academy Press, 2001), 38–39.
26. "AURA Comments on NSF OIG Report: Large Infrastructure Project Management," undated (Fall 2001); copy provided by Matt Mountain and in author's personal papers. Goetz Oertel interview, Nov. 29, 1999.

27. Rita Colwell, "Comments at Gemini Dedication," Jan. 18, 2002. Available at <www.aura-astronomy.org/nv/colwell.htm>.
28. *Dedication of the Palomar Observatory and the Hale Telescope: June 3, 1948* (Pasadena: California Institute of Technology, 1948).
29. "Gemini Operations Plan—version 3.1," Oct. 10, 1995, RSE personal papers.
30. P. J. Puxley et al., "Gemini Observatory Science Operations Plan," in *Observatory Operations to Optimize Scientific Return,* ed. Peter J. Quinn (Bellingham, Wash., 1998), 63–72.
31. Jean-René Roy interview, May 7, 2002.
32. Stéphanie Côté interview, Dec. 19, 2000.
33. Matt Mountain interview, May 7, 2002.
34. Puxley et al., "Gemini Observatory Science Operations Plan."
35. "Gemini Observation Deepens Mystery of Local Active Galaxy," Gemini Observatory press release, Oct. 29, 2001, <www.gemini.edu/project/announcements/press/2001-3.html>. Also Eric Perlman et al., "Deep 10 Micron Imaging of M87," *Ap. J.* 561, 1 (2001):L51–L54.
36. Côté interview, Dec. 19, 2000.
37. Matt Mountain interview, Oct. 21, 2001.

Conclusion: Telescopes, Postwar Science, and the Next Big Machine

1. Jesse L. Greenstein passed away in Arcadia, California, on October 21, 2002, at age 93 as this book was undergoing final revisions; one obituary noted that he was considered the "father figure" for an entire generation of astronomers at Caltech.
2. John Huchra interview, Feb. 14, 2002, CHP/AIP Collection.
3. Matt Mountain interview, Oct. 24, 2000.
4. A viewpoint explicitly stated in *Federal Funding of Astronomical Research* (Washington: National Academy of Sciences Press, 2000), 12.
5. Huchra interview, Feb. 14, 2002.
6. Matt Mountain interview, Oct. 25, 2000.
7. From the frontpiece of Peter Galison, *Image and Logic: A Material Culture of Microphysics* (Chicago: University of Chicago Press, 1997).
8. Helmut A. Abt, "Astronomical Publication in the Near Future," *Publications of the Astronomical Society of the Pacific* 112, 777 (2000):1417–1420.
9. James Glanz, "Cosmic Motion Revealed," *Science* 282, 5397 (1998):2156–57.
10. Galison, *Image and Logic,* 392.
11. *KPNO Newsletter* 16 (Aug. 1981): 6.
12. Noted in Steve Strom to Goetz Oertel, Aug. 12, 1993, AST/NSF Files.

13. *Gemini Newsletter* 17 (Dec. 1998).
14. Geoff Brumfiel, "The Heavens at Your Fingertips," *Nature* 420, 6913 (2002):262–264.
15. Ron Cowen, "Mining the Sky: Taking Some Big Bytes of the Universe," *Science News* 159, 8 (2001):124–125. Govert Schilling, "The Virtual Observatory Moves Closer to Reality," *Science* 289 (July 14, 2000):238–239.
16. Tim Chapman, "Astronomy Enters Age of Virtual Reality," *Physics World* 15, 3 (2002):13. Fred Hapgood, "Astronomy and the Internet," *Beam Line* (1997):49–51. Toni Feder, "Astronomers Envision Linking World Data Archives," *Physics Today* 55, 2 (2002):21.
17. Christopher McKee and Joseph Taylor (chairs), *Astronomy and Astrophysics in the New Millennium* (Washington: National Academy Press, 2000), 132.
18. Hapgood, "Astronomy and the Internet," 49.
19. Galison, *Image and Logic,* 399.
20. Ibid., 398–399.
21. Several papers on the topic "Building the Next Big Machine" were given at the 2001 Annual Meeting of the History of Science Society in a session organized by Robert W. Smith and the author. My thanks to Dr. Smith for his insightful views on this issue.
22. Goetz Oertel interview, Sept. 24, 1999.
23. Sandra Faber to Committee on Astronomy and Astrophysics, Apr. 19, 1993, AST/NSF Files.
24. Robert Gehrz interview, Nov. 23, 1998.
25. Steve Strom to Goetz Oertel, Aug. 12, 1993, AST/NSF Files.
26. Richard McCray (chair), *A Strategy for Ground-based Optical and Infrared Astronomy* (Washington: National Academy Press, 1995); McKee and Taylor (chairs), *Astronomy and Astrophysics in the New Millennium.*
27. McCray (chair), *Strategy,* 22.
28. Michael Riordan, "The Demise of the Superconducting Super Collider," *Physics in Perspective* 2, 4 (2000):411–425. Robert Crease, personal communication, Apr. 23, 2002. A similar case exists for the Cambridge Electron Accelerator, as described in Elizabeth Paris, "A Laboratory's Life: Consequences of *Not* Being Allowed to Build the Next Machine," paper presented at the Annual History of Science Society Conference, Denver, 2001.
29. Michael Disney interview, May 1, 1984, by Robert W. Smith, NASM Space History Collection.
30. Helmut A. Abt, "The Productivity of Ground-Based Optical Telescopes of Various Apertures," presented at the 199th meeting of the Ameri-

can Astronomical Society, Washington, Jan. 2002. I am grateful to Dr. Abt for a copy of his paper.
31. Govert Schilling, "Telescope Builders Think Big—Really Big," *Science* 284 (June 18, 1999):1913–1915.
32. Jocelyn Kaiser, "Caltech Lands Record-Breaking $600 Million," *Science* 294 (Nov. 8, 2001):979.
33. Robert Irion, "California Astronomers Eye 30-Meter Scope," *Science* 298, 5596 (2002):1151–1153.
34. Peter J. Westwick, *The National Labs: Science in an American System, 1947–1974* (Cambridge, Mass.: Harvard University Press, 2002).
35. Matt Mountain to Claude Canizares, July 23, 2001, AST/NSF Files.
36. Riccardo Giacconi, "Reflections of a Former ESO Director General," presentation to the NAS, June 13, 2001, copy in author's papers.
37. Meeting of NAS Committee on Astronomy and Astrophysics, Washington, Dec. 6, 2000; Robert Eisenstein, personal communication, Jan. 3, 2001; notes for both in author's papers.
38. Alan Dressler, "Report to the CAA,; December 6, 2000;" copy in author's papers. Cooperative interaction between public and private observatories was recommended, in fact, by the 2000 decadal survey through the Telescope System Instrumentation Program, the goal of which was, in part, to "bring the national and private observatories together as a coherent research system." *Astronomy and Astrophysics in the New Millennium*, p. 40.
39. Jesse L. Greenstein; May 19, 1978 interview with Spencer Weart, p. 243-244; CHP/AIP Collection.

Acknowledgments

Writing this book has been more fun and rewarding than I should admit. In the process, I have accumulated a number of debts, personal and professional, that deserve recognition.

Two research grants from the National Science Foundation provided funding while the Friends for the Center for History of Physics and the Maurice A. Biot fund at California Institute of Technology's Archives contributed travel grants. Librarians and staff at the United States Naval Observatory, the National Optical Astronomy Observatories, and the archives at Caltech, Harvard University, and the Smithsonian Institution were especially accommodating. I was also fortunate to spend time with colleagues at the George Washington University's Center for History of Recent Science. Roc Reimer and Joel Parriott helped me gain access to meetings at the National Academy of Sciences. This book was written while I was at the Center for History of Physics at the American Institute of Physics; I benefited enormously from my interaction with Spencer Weart and the other staff as well as from easy access to the excellent collections of the Niels Bohr Library.

In the course of researching the postwar history of telescope construction and use, I interviewed dozens of astronomers, observatory directors, science managers, and telescope engineers. Some people, however, were especially helpful; special thanks go to Richard Ellis, Robert Fugate, Robert Kirshner, Matt Mountain, Goetz Oertel, Don Osterbrock, Leif Robinson, Malcolm Smith, and Peter Strittmatter. Wayne Van Citters deserves special recognition for allowing me to examine essential documentary materials in the working files of the NSF's Division of Astronomical Sciences.

Several friends and colleagues—notably, David DeVorkin, Randy Papadopoulos, Robert W. Smith, Olivia Walling, and Spencer Weart—made valuable suggestions for improving this book; Richard Malow provided valuable last-minute insights into Washington politics; and Michael Fisher, Sara Davis, Elizabeth Collins, and the staff at Harvard University Press helped make publishing enjoyable. I did not take all advice offered. Nonetheless, I appreciate the suggestions from those noted and others I have neglected to name. Responsibility for whatever errors remain is my own.

Finally, I wish to thank Olivia Walling for ringing my bell. *Per Amore, Ad Astra.*

Index

Aaronson, Marc, 266
Abt, Helmut A., 21, 40, 55, 83
Active optics, for meniscus mirror, 181
Adams, Walter S., 35, 50
Adaptive optics, 154–159, 261–262; for infrared wavelengths, 155, 159; for meniscus mirror, 181; declassification of, 197–198; for Canada-France-Hawaii Telescope, 261–262; for Gemini 8-Meter Telescopes Project, 262–264
Ad Hoc Panel on Astronomical Facilities, 38
Advanced Development Program, 139–140, 159
Advanced Research Projects Agency, 61
Advisory Panel for a National Astronomical Observatory, 39–40
Advisory Panel for Astronomy, 44
Allende, Salvador, 240
Aller, Lawrence, 47
Altair (ALTitude conjugate Adaptive optics for the InfraRed), 263
Altitude-azimuth mount, 59
Alvarez, Luis W., 32, 104, 295–296
American Astronomical Society, 16
Anderson, John, 51
Angel, James Roger Prior, 120–124, 205, 206. *See also* Steward Observatory Mirror Lab
Anglo-Australian Telescope, 173–174
Anguita, Claudio, 239, 240–241

Anti-Semitism, 17
Apache Point telescope, 149
Aperture synthesis, 64
Argentina, in Gemini 8-Meter Telescopes Project, 242
Arp, Halton C., 32
Artificial star, 156–158
Associated Universities, Inc., 40
Association of Universities for Research in Astronomy, Inc. (AURA): development of, 39–41; membership in, 41, 94, 139; Goldberg appointment at, 49; Carnegie Southern Observatory and, 77–78, 79; board membership of, 94; conflicts of interest of, 94–101; Greenstein appointment at, 96–97; Hubble Space Telescope management by, 98; Goldberg on, 98–99, 100, 101; Greenstein on, 100; mission statement by, 100–101; Oertel appointment at, 166–167; goals survey by, 183; 8-meter telescope proposal by, 183–186; international collaboration and, 189–190; Universidad de Chile in, 241; Malow appointment at, 243. *See also* Gemini 8-Meter Telescopes Project
Astronomer, definition of, 267
Astronomy Missions Board, 49
Astrophysical Journal, The, 21
Astrophysical Research Consortium, 149

Atkins, Chester, 192–193
AURA. *See* Association of Universities for Research in Astronomy, Inc.
Australia, in Gemini 8-Meter Telescopes Project, 242, 246–247

Baade, Walter, 1, 18, 26, 29, 74
Babcock, Horace W., 76–77, 154–155, 156
Background noise, 63
Bahcall, John N., 195–197, 204, 211–212, 223, 228, 235
Barnes, Harry G., 241
Barr, Lawrence, 88–89, 91, 93, 116, 163
Baum, William A., 31
Baustian, William, 52, 89; Kitt Peak appointment of, 60
Beckers, Jacques M., 139–140, 159, 163, 169
Big Science, 4, 5–6, 8–9, 291–293
Blaauw, Adriaan, 75
Blanco, Victor M., 238, 240
Bless, Robert C., 206–208, 213, 231–232, 234
Bloch, Erich, 161–162, 182, 187–188, 190
Boehlert, Sherwood, 229
Boksenberg, Alec, 174, 190–191, 300
Boland, Edward P., 132–133, 160, 161, 162
Bolshoi Teleskop Azimutal'ny, 59, 108–109
Borosilicate mirrors, 122–125
Bowen, Ira S., 19–20, 38, 52–53, 58–59
Boyce, Peter, 233
Brahe, Tycho, 4–5
Branscomb, Lewis M., 187
Brazil, in Gemini 8-Meter Telescopes Project, 242
Brookhaven National Laboratory, 300
Bundy, McGeorge, 77–78
Burbidge, Geoffrey R., 47, 110; on data collection, 74–75, 272; on Next Generation Telescope Scientific Advisory Committee, 92; at Kitt Peak National Observatory, 106; on National New Technology Telescope project, 127
Burbidge, Margaret, 27, 47, 106–107
Bush, George H. W., 171

Bush, George W., 279
Bush, Vannevar, 20, 43
Butler, Paul, 262
Byrd, Robert C., 192

California Extremely Large Telescope, 302–303
Caltech: Greenstein appointment at, 19–21, 94–95; Greenstein AURA representation of, 95; University of California negotiations with, 149–150; Keck gift to, 149–151; Moore gift to, 302–303
Canada, in Gemini 8-Meter Telescopes Project, 188–189, 211
Canada-France-Hawaii Telescope, 189, 261–262
Carina Nebula, 237
Carleton, Nathaniel, 65–66
Carnegie-Caltech Observatory Committee, 24, 26
Carnegie Institution of Washington, 19, 20, 59, 77
Carnegie Southern Observatory, 77–79
Cassegrain focus, 71
Cassegrain telescope, 54
Cellular mirror, 51
CERN (European Organization for Nuclear Research), 72–73, 74–75, 104–106, 291
Cerro Tololo: 1.5-meter telescope of, 238–239; Wolff's visit to, 240–241
Cerro Tololo Inter-American Observatory, 77, 82, 238; funding of, 85
Chafee, Frederic H., 109
Challenger, 163
Charge-coupled device, 110–111
Chile: Gilliss expedition to, 237–238; astronomical conditions of, 238; European Southern Observatory in, 239, 245; in Gemini 8-Meter Telescopes Project, 241–242, 243, 244–246
Chrétien, Henri, 54
Code, Arthur B., 21
Cohen, Judith, 264
Columbia University, Angel appointment at, 121
Colwell, Rita, 3, 279, 281
Computers, 31–32, 70, 72–76, 248

Corbett, Ian, 194, 212–213, 226, 247
Corning Glass Company: Hale telescope mirror by, 23–24; mirror market and, 127, 153; Gemini bid of, 219–220, 221; Houck panel on, 231–232; Gemini mirrors by, 254
Côté, Stéphanie, 285–289
Crawford, David, 73

d'Amato, Alfonse, 229
Dark time, 26
Data analysis programs, 293–295
Davies, Roger L., 133–134, 169, 174, 190
Davis, Marc, 228
Decadal survey. *See* Panel on Astronomical Facilities
Dennison, Edwin W., 70–71
Dewars, 62
Disney, Michael J., 92–93
Djorgovski, George, 294–295
Domenici, Pete, 162
Dressler, Alan, 265, 304–305
DuBridge, Lee A., 23, 78

Edmunds, Michael G., 173–174, 177
Eidophor, 154–155
Eisenstein, Robert, 304
Electromagnetic radiation, 10–11
Ellis, Richard S., 177–178, 190, 191–192, 274
Eta Carinae, 237
European Organization for Nuclear Research (CERN), 72–73, 74–75, 104–106
European Southern Observatory: large telescopes meeting of, 72–74; optical telescopes meeting of, 104–106; New Technology Telescope of, 132, 180–181, 240; establishment of, 178–182; Chilean relationship with, 245

Faber, Sandra M., 114–116, 152, 165, 267, 298–299
Feinleib, Julius, 157
Fermilab, 79–80
Field, George B., 110, 112, 131–132, 163, 190
Flamsteed, John, 172
Focal ratio, of Hale telescope, 23–24

Ford Foundation, 77, 78–79
Fosdick, Raymond B., 22–23
Fowler, William A., 33, 47, 96
Fugate, Robert Q., 156–157, 198
Future Directions for NOAO Committee, 167–168

Gabor, George, 103
Galaxies, clusters of, 115
Gallagher, Jay, 224; Chile visit of, 240–241
Gamow, George, 18
Gardner, David P., 140, 150
Gehrz, Robert D., 134, 141–142, 235, 245–246
Geller, Margaret J., 165, 228
Gemini 8-Meter Telescopes Project, 2, 172; AURA plan for, 183–186; proposal for, 183–186; cost of, 184, 193–194, 279; mirror for, 184–185, 215–223, 226–227, 230–231, 254–255, 256; research proposals for, 185, 282–283; data collection from, 185–186, 283–289; funding for, 186–187, 192–194, 210–211, 227–228; international collaboration on, 187–191; Canadian participation in, 188–189, 211; United Kingdom collaboration on, 189, 191–192; Bahcall's decadal survey and, 202; management of, 206–215, 233–234; NOAO relationship with, 209–210, 214; infrared capabilities of, 211–212, 225–226, 251–253; Mountain appointment to, 221, 254; Corning mirror for, 221–225; NSF review of, 228–231; on NSF review, 231–232; corporate vs. entrepreneur conflict and, 233–234; image quality for, 235; modeling of, 235; design review of, 235–236; Chilean participation in, 241–242, 243, 244–246; Argentina in, 242; Brazil in, 242; Australian participation in, 242, 246–247; systems integration for, 248–249, 256–257; Nasmyth platform elimination from, 249–250; secondary mirror for, 251–253; imagers for, 257; instruments for, 257–264; instrument support structure for, 258; tertiary mirror for, 258; spectrographs

Gemini 8-Meter Telescopes Project *(continued)*
for, 258–260; adaptive optics for, 262–264, 276–277; observing time on, 270–271, 273–275, 282–283; instruments for, 271; dedication of, 275–281; ideal night at, 281–282; Operations Center of, 284; system support associates for, 284–285; real time monitoring on, 285; staff astronomers for, 285–288; OSCIR for, 287–288
Gemini Law, 245–246
Germanium bolometer, 62
Giacconi, Riccardo, 245, 304
Gillett, Frederick C., 87, 250–251, 254, 277
Gilliss, James M., 237–238
Goldberg, Leo: childhood of, 13–14; education of, 14; postgraduate fellowship for, 15; at McMath-Hulbert Observatory, 15, 16; University of Michigan appointment of, 17–18; Harvard appointment of, 19, 42–43, 84; on National Research Council, 37–38; at 1953 Lowell Observatory meeting, 39; on first decadal survey, 46; on ground-based astronomy, 46; at Association of Universities for Research in Astronomy, Inc., 49; on Astronomy Missions Board, 49; at Kitt Peak National Observatory, 49, 84–87, 95, 96, 98–101; on Carnegie Southern Observatory, 77–78; at International Astronomical Union, 95; compensation of, 98; on Hubble Space Telescope, 98; on Gilbert Lee, 98–99; retirement of, 99; on AURA, 99, 100, 101; resignation of, 101; on large telescope projects, 111–112; on National New Technology Telescope, 133; death of, 171
Goldberger, Marvin L., 150
Golden, William T., 97, 100
Goldwater, Barry, 54, 127
Gore, Albert, 205
Government Performance and Results Act, 273
Gramm-Rudman-Hollings Act, 160
Great Attractor, 165
Great Wall, 165

Greenstein, Jesse L.: childhood of, 13; education of, 14; postgraduate fellowship for, 14–15; at Yerkes Observatory, 14–15, 16; Caltech appointment of, 19–21; research by, 33–34, 57–58; on National Research Council, 38; on first decadal survey, 46; on Astronomy Missions Board, 49; at National Academy of Sciences, 49; on data collection methods, 74, 76; on Carnegie Southern Observatory, 79; on Kitt Peak telescope development, 80; decadal survey by, 80–83, 203; as Caltech AURA representative, 94–95; as AURA board chairman, 96–97; on Kitt Peak National Observatory, 97–98; on AURA, 100; on ground-based telescopes, 105–106; on Keck Telescope, 151, 162
Gunn, Jim, for Gemini review committee, 229

Hale, George Ellery, 23, 50, 296
Haleakala military telescope, 156, 157
Hale Observatories, 70. *See also* Mount Wilson Observatory; Palomar Observatory
Hale telescope, 23–25; research allocation on, 24, 26–27; spectrographs for, 27, 33; nightly experience with, 27–29; student training on, 29. *See also* Palomar Observatory
Hall, Donald N. B., 87, 89, 91, 93, 105, 212, 223–224
Harris, William C., 232
Harvard College Observatory, 42
Harvard University: Goldberg education at, 14; Greenstein education at, 14; Goldberg postgraduate fellowship at, 15; Goldberg appointment at, 42–43
Hayden, Carl, 41
Henry Norris Russell Lecture, 58–59
Herbig, George, 69, 73–74
Herzburg Institute of Astrophysics, 263
High-energy physics, vs. astronomy, 303
High Speed Photometer, 207
Hill, John M., 122
Hinman, Stephen F., 217
Hoffman, Marion O., 140

Hoffman, Thomas, 66–67
Honeycomb borosilicate mirrors, 119–125
Houck, James R., 228–231
Hough, James H., 177
Howard, William, 112
Hoyle, Fred, 21, 47
Hubble, Edwin, 19, 22, 74
Hubble Space Telescope, 98, 105, 111, 131, 202; vs. National New Technology Telescope, 135; challenges to, 160; Boland on, 162; mirror defect of, 204–206, 224; solar panels of, 207; data collection from, 268–269
Huchra, John P., 165, 228, 254, 267, 290–291
Hughes, Thomas P., 117
Hulbert, Henry S., 15, 17, 18, 34
Humason, Milton, 26, 57
Hyper-telescopes, 8, 248–257

Image, telescope location and, 146–147
Image Photon Counting System, 174
Infrared arrays, 199–200
Infrared astronomy, 62–63; Multiple Mirror Telescope for, 63–68; National New Technology Telescope for, 135–136; adaptive optics and, 155, 159; United Kingdom Infrared Telescope for, 175; detector development in, 199–200
Infrared radiation, 10–11, 62–63
Inouye, Daniel K., 212
International Astronomical Union: 1965 conference of, 59–61; Goldberg presidency of, 95
Inter-University Astronomical Observatory, 39
Isaac Newton Telescope, 173, 174
Isoplanatic patches, 155
Itek Optical Systems, 151, 155–156

JASON group, 157
Jefferies, John T., 138–139, 146, 162, 164, 177

Kapper, Francis B., 227
Keck, Howard, 149–151
Keck Telescope, 150–153, 217; mirror for, 151–152; management of, 152, 271
Kelly, Kevin, 227

Keyworth, George, 161
Kilgore, Harley M., 43
Kirshner, Robert P., 115, 134
Kitt Peak National Observatory, 9, 34, 39–42; site survey for, 40, 53–54; university consortium for, 41; Goldberg appointment at, 49, 84–87, 95, 96, 98–101; Meinel appointment at, 54–55; mirror for, 54–55; Mayall appointment at, 55; Baustian appointment at, 60; vs. university research applications, 80, 82; funding for, 85–86, 96; mission of, 85–87; 25-meter telescope project of, 88–92, 100; Next Generation Telescope program of, 92–94, 103, 106, 116; observing time at, 96, 97; Burbidge appointment at, 106–107; "Telescopes for the 1990s" conference of, 110–112, 114; in National New Technology Telescope project, 117; Wolff appointment at, 139
Klopsteg, Paul, 37
Kron, Gerald E., 31
Kuiper, Gerard P., 44, 53, 61
Kurz, Richard J., 253

Lagos, Ricardo, 277–278
La Palma: William Herschel Telescope on, 174–175; 8-meter telescope for, 190–191
"Large Astronomical Telescope at Low Cost, A," 65
Large Binocular Telescope, 145–146, 183
Large Telescope Panel, 177–178
Las Campanas, 239
Laser-beacon guide star, 156–158
Laser Interferometer Gravitational Wave Observatory, 192, 193, 211
La Silla, 178–179, 180, 239
Latour, Bruno, 137–138
Lawrence, Ernest O., 296–297
Lead-sulphide cells, 62
Lee, Gilbert L., 94, 98–99
Leighton, Robert B., 102
Lick Observatory, 45, 102; data collection at, 74–75
Lilly, Simon J., 189
Longair, Malcolm, 175, 176, 191, 212, 226; on Steward Observatory, 223

Low, Frank J., 62, 199, 250; at University of Arizona, 63, 64; for Future Directions for NOAO Committee, 167, 168; on Gemini, 225
Lowell Observatory, 1953 meeting at, 38–39
Lubliner, Jacob, 103, 107–108
Lunar and Planetary Laboratory, 61
Lynds, Beverly T., 95

M87, 288
Magellanic Clouds, 237
Magnusson, Warren, 41
Malow, Richard N., 159–160, 280; on Very Large Baseline Array, 160–161; on NNTT, 167; on Gemini, 188, 193, 210, 227; on Chilean Gemini participation, 242–243; AURA appointment of, 243
Marcy, Geoff, 262
Mast, Terry S., 103, 138
Mauna Kea, 138–139; for NNTT, 146–148; Keck Telescope on, 150–153; for Gemini, 191, 211–212, 223–224
Mauna Kea Observatory, 108
Mayall, Nicholas, 55, 57, 80, 86
Mayor, Michael, 262
McMath, Robert T., 15, 16, 18, 38, 39–40, 53, 54
McMath-Hulbert Observatory, Goldberg at, 15, 16
Meinel, Aden B., 1, 2, 40, 51; X-inch telescope design of, 51–53; on Kitt Peak site survey, 53–54; as Kitt Peak's director, 54–55; at University of Arizona, 55, 61–62; Project COLT of, 63–64
Meniscus mirror, 104, 118, 180–181
Menzel, Donald H., 15, 19
MIAMI (Minimum Inertia and Mass Instrument), 54
Mikulski, Barbara, 204, 232
Military research, 153–159
Milky Way, 237
Minimum Inertia and Mass Instrument (MIAMI), 54
Minkowski, Rudolph, 26, 27, 28–29
Mirror, 11, 12, 50; for Hale telescope, 23–24; cellular, 51; for X-inch telescope, 51–53; for Kitt Peak National Observatory, 54–55; for Project COLT, 63–64; for Multiple Mirror Telescope, 63–69; for PALANTIR, 90–91; for Steerable Dish, 103; segmented design for, 103–104, 105, 107–108, 125–130; meniscus design for, 104, 118, 180–181; for University of California 10-meter telescope, 105, 107, 108; designs for, 117–120; rotating furnace casting of, 119–125; for Keck Telescope, 151–152; Eidophor, 154–155; rubber, 155; for Starfire Optical Range, 158; for United Kingdom Infrared Telescope, 175; for New Technology Telescope, 180; for Very Large Telescope, 180–181; for Gemini 8-Meter Telescopes Project, 184–185, 215–223, 226–227, 230–231, 254–255, 256; for Hubble Space Telescope, 204–206, 224; support for, 230; temperature of, 230; for Overwhelmingly Large Telescope, 302
Mirror Lab. See Steward Observatory Mirror Lab
Missile Defense Alarm System, 199
Mitchell, Sir William, 191
Moore, Gordon, 302
Morgan, W. W., 45
Mould, Jeremy, 134
Mountain, Charles Mattias, 175–176, 226, 230, 231, 280; Gemini appointment of, 221, 254; on corporate vs. entrepreneur conflict, 233–234; on discovery, 266; on CERN, 291; on California Extremely Large Telescope, 303
Mount Graham, for NNTT, 146–148
Mount Hopkins, 64–65, 66–67, 68, 109–110
Mount Wilson Observatory, 19; guest astronomers at, 26. See also Palomar Observatory
Multichannel spectrometer, 71–72
Multiple Mirror Telescope, 63–69, 109–110; funding for, 65, 67; enclosure of, 67
Münch, Guido, 21

Nasmyth platforms, 249–250
National Academy of Sciences, 45–46;

on NOAO, 299. *See also* Panel on Astronomical Facilities
National New Technology Telescope (NNTT), 117; University of California mirror design for, 118–119, 125–126, 127–130, 141–142; University of Arizona mirror design for, 119–125, 128, 129, 141–142, 143–144; institutional conflicts and, 126–127; Burbidge on, 127; mirror selection for, 127–130, 133–142; NSF on, 132–133, 145, 164; funding for, 132–133, 160–164, 167; Goldberg on, 133; Scientific Advisory Committee for, 133–138; vs. Hubble Space Telescope, 135; infrared imaging and, 135–136; research priorities for, 135–136; performance requirements for, 136; Santa Cruz meeting of, 140–142; cost of, 144; ownership of, 145–146; site for, 146–148; Malow on, 160, 161; advocacy for, 163–164; institutional conflicts and, 164; Future Directions for NOAO Committee report on, 168, 169; cancellation of, 169; diminished ambitions of, 170–171
National Optical Astronomy Observatories (NOAO): Jefferies appointment at, 138–139; establishment of, 138–140; vs. University of Arizona, 145–146; 1987 crisis of, 165–170; Wolff appointment at, 166–167; Future Directions Committee of, 167–169, 170; Gemini 8-Meter Telescopes Project and, 209–210, 214; Gemini separation from, 297–298
National Radio Astronomy Observatory, 40, 42, 192
National Research Council, 37–38
National Science Foundation (NSF), 37, 38, 44, 76; 1953 Lowell Observatory meeting of, 38–39; facilities management and, 39–40, 41–42; allocations control and, 44–45; Kitt Peak funding by, 85–86, 96; on funding strategy, 112, 116; on National New Technology Telescope, 132–133, 145, 164; Very Large Baseline Array funding by, 160–161; Bloch appointment at, 161–162; Gemini funding by, 186–187, 192–194, 210–211, 227–228; Gemini 8-Meter Telescopes Project review by, 228–231; under Government Performance and Results Act, 273–274
National Virtual Observatory, 294–295
Nelson, Jerry E., 102–103, 107, 113, 125–126; NNTT mirror design of, 118–119, 136–137; on Keck Telescope project, 206
Neugebauer, Gerry, 102, 228
New Technology Telescope, 132, 180–181, 240
Next Big Machine, 296–297, 300–302; institutional factors in, 304–305
Next Generation Telescope program, 92–94, 103, 106, 116
Ney, Edward P., 250
NGC 628, 278
NNTT. *See* National New Technology Telescope
NOAO. *See* National Optical Astronomy Observatories
Noise, 63

Observation: classical, 265–266; queue, 266–270
Oertel, Goetz K.: AURA appointment of, 166–167; on Gemini, 172; on American-British partnership, 178; on international collaboration, 189–190; on Gemini, 229; on Chilean partnership, 241
Office of Naval Research, 36–37
Oke, J. Beverly, 31, 75
Optical Sciences Center, 61
Oschmann, Jacobus (Jim), 253
OSCIR, 287–288
Osterbrock, Donald E., 21, 29, 104
Overwhelmingly Large Telescope, 302

PALANTIR (Program for a Large Aperture Novel Thousand Inch Telescope), 89–92
Palomar Observatory, 19–20, 22–34; Bowen appointment at, 19–20; dedication of, 22–23; observation time at, 24, 26–27; photoelectric photometry at, 31; multichannel spectrometer for, 71–72. *See also* Hale telescope

Panel on Astronomical Facilities, 45;
  first decadal (1960s) survey of, 45–49;
  second decadal (1970s) survey of, 80–
  83; third decadal (1980s) survey of,
  131–132; fourth decadal (1990s) survey of, 195–202; fifth decadal (2000s)
  survey of, 295, 299–300, 302
Perkin-Elmer, 205
Perlman, Eric, 287–289
Pettit, Edison, 51, 54
Pettit, Marjorie, 51
Photoelectric photometry, 32
Photomultipliers, 30–31
Pinochet, Augusto, 240
Pipher, Judith L., 228
Porter, Russell W., 22
Primary mirror, 11, 12, 50. *See also* Mirror
Project COLT, 63–64
Project Mohole, 42
Publication, 293, 301
Puxley, Philip, 287

Quasars, 58, 109–110
Queloz, Didier, 262

Radio astronomy, 83
Randall, Larry K., 60, 89, 100, 112, 213–
  214, 231, 234, 253
Rayleigh scattering, 157
Reagan, Ronald, 157, 158–159, 160
Red shift, 115
Reference star, 156–157
REOSC, Gemini mirror polishing by, 255
Rice, Stuart A., 133
Ride, Sally, 250
Rieke, Marcia, 225
Ritchey, George W., 50–51, 54
Rocket-based astronomy, 18–19
Royal Greenwich Observatory, 172, 173,
  225, 300; Boksenberg appointment at,
  174
Royal Observatory, Edinburgh, 175, 176
Rubber mirrors, 155
Rubin, Vera, 27
Ruiz, Maria Teresa, 239–240
Rule, Bruce H., 45, 52, 60
Russell, Henry Norris, 14, 35

Sacramento Peak Observatory, 98, 162
Sagan, Carl, 250

Salpeter, Edwin, 18
Sandage, Allan R., 6, 21, 26, 29, 45, 46,
  57, 134
Sandia Optical Range, 156–157
Sargent, Wallace, 152
Saxon, David S., 104, 108
Schmidt, Bernhard, 107
Schmidt, Maarten, 58, 75, 161, 274
Schmidt telescope, 26
Schott Glasswerke, 107, 153, 181
Schwarzschild, Martin, 77, 240
Science and Engineering Research
  Council (SERC), 173, 177–178
Science–The Endless Frontier, 43
Scientific Advisory Committee for National New Technology Telescope,
  133–138
Seeger, Raymond J., 38
Seeing: telescope location and, 146–147;
  bad, 154–155
Serrurier, Mark, 24
Shane, C. Donald, 37
Shapley, Harlow, 18, 35
Singles Array, 92–93
Sky & Telescope, 22, 26–27
Smith, Gerald M., 152, 233–234
Smith, Harlan J., 116
Smith, William, 280
Smithsonian Astrophysical Observatory,
  64–69
Solar telescopes, 98, 162
Southern Observatory for Astrophysical
  Research, 183
Soviet Union, 6-meter telescope of, 59,
  108–109
Space-based astronomy, 18–19, 46; vs.
  optical astronomy, 19, 46. *See also*
  Hubble Space Telescope
Spain, United Kingdom collaboration
  with, 190–191
Spectra, 11–12
Spectrograph, 11; for rocket-based astronomy, 18; prime-focus, 27; coudé,
  33, 58; quasar observation with, 58;
  for Gemini 8-Meter Telescopes Project, 258–260
Spectrometer, multichannel, 71–72
Spitzer, Lyman, 19, 37–38
Sputnik I, 43
Stanford Linear Accelerator Center, 79

Starfire Optical Range, 157, 158
Stars, element formation in, 33
Star Wars, 157, 158–159
Steerable Dish, 103
Steward Observatory, 61, 119
Steward Observatory Mirror Lab: National New Technology Telescope mirror design by, 119–125, 128, 129, 141–142, 143–144; rotating furnace of, 124–125, 143–144; 1.8-meter mirror by, 144; mirror polishing by, 145, 158; 3.5-meter mirror by, 148–149, 158; expansion of, 153, 164; vs. commercial firms, 153; Jefferies on, 164; 6.5-meter mirror casting by, 216, 218; management of, 216–217; Gemini bid of, 219, 220–223; vs. Corning Glass, 221–225; operational style of, 223
Stone, Edward, 190
Strategic Defense Initiative, 157, 158–159
Stratton, Julius, 78
Strittmatter, Peter, 75, 119–120, 145–148, 183
Strom, Stephen E., 86, 167, 196, 228, 299
Struve, Otto, 14, 35, 38, 44
Subaru telescope, 217
Systems engineering, 8

Team research, 47–48
"Telescopes for the 1990s" conference, 110–112, 114
Thermal noise, 63
Thomas, Albert, 41
Thorndike, Alan, 293
Tinbergen, Jaap, 75
Tinsley Laboratories, 151
Trapezium, 158
Traxler, Robert, 210
Trimble, Virginia, 27

United Kingdom Infrared Telescope, 175
Universidad de Chile, 238, 244
University of Arizona: Meinel appointment at, 55, 61–62; Kuiper appointment at, 61; Low appointment at, 63; NNTT mirror design of, 119–120; Angel appointment at, 121; Hill's research at, 122

University of California: 10-meter telescope of, 102–105, 107–108; National New Technology Telescope mirror design by, 118–119, 125–126, 127–130, 141–142; Hoffman donation to, 140; Caltech negotiations with, 149–150
University of Hawaii, 108, 138–139
University of Michigan, Goldberg appointment at, 17–18
University of Texas, 113, 118

Van Citters, Wayne, 197–198, 200, 247
Van der Laan, Harry, 75
Vatican Observatory, 144
Vega, 250
Very Large Array, 83, 112
Very Large Telescope, 179–182, 240, 273, 304
Very Long Baseline Array, 132, 160–161

Walker, Arthur, 182
Wampler, E. Joseph, 104, 131
Watson, Daniel, 124–125
Weymann, Ray, 65, 109
Whipple, Fred L., 37–38, 64, 84
White dwarfs, 33
Whitford, Albert E., 6, 31, 38, 40, 45
William Herschel Telescope, 174–175
Williams, Robert E., 241
WIYN telescope, 182–183
Wolff, Sidney C.: at Kitt Peak National Observatory, 139; NOAO appointment of, 166–167; Chile visit of, 240–241
Woolf, Neville J., 112, 121–122
World War II: scientific research during, 15–16; federal science spending after, 16–17, 34–42; scientific research after, 16–22

X-inch telescope, 2; Meinel's design for, 51–53; cost of, 55, 57; first decadal survey on, 56; multiple-mirror approach to, 65–69

Yerkes Observatory: Greenstein at, 14–15, 16, 17; Meinel at, 51–53

Zerodur, 107, 181
Zwicky, Fritz, 20